IEE Energy Series 2
Series Editors: M. Barak and
Professor D. T. Swift-Hook.

Renewable energies:

sources, conversion and application

Previous volumes in this series

Volume 1	Electrochemical Power Sources M. Barak.

Renewable energies: sources, conversion and application

P.D.Dunn

Peter Peregrinus Ltd on behalf of the Institution of Electrical Engineers.

Published by: Peter Peregrinus Ltd., London, United Kingdom.

Dunn, P. (Peter)
Renewable energies: sources, conversion and application.—(IEE energy series; v. 2)
1. Renewable energy sources
I. Title II. Series
621.042 TJ163.2

ISBN 0 86341 039 1

Printed in England by Heffers Printers Limited, Cambridge.

Contents

Acknowledgements

This book has been written over a number of years, it has in fact taken an embarrassingly long time to complete. I have been greatly assisted by many colleagues and friends who have advised and commented on various chapters. In particular I wish to mention Monty Barak, Roger Bentley, John Burton, Joy Clancy, Bob Marshall, Peter Moore, Les Mullett, Alvaro Pinillo, Boonlong Piyawat and Graham Rice.

My thanks are due to Jan Stevens and Charles Preston for their help in preparation of diagrams and to Alvaro Pinillo for preparing the diagrams in Chapter 9. I also wish to thank the individuals and institutions who have generously allowed me to reproduce their photographs and diagrams. Credits are given where these appear in the book.

Since the book is based on lectures given over many years and covers a very broad range of material, it is not always easy to recall the original sources of this material. I have tried to acknowledge the work of others but should I inadvertently have overlooked or miscredited, I hope my apologies will be accepted.

The credit for the typing and preparation of the book goes to Mrs. J. Collins and Mrs. G. O'Farrell whose contribution is greatly appreciated.

P.D.D.

Foreword

By writing the following I want to show how much I am in sympathy with the aims of Professor Dunn in producing this book.

Man has used the renewable fuels, wind and water, for many centuries and the biofuels, wood and animal fodder, for several millenia before the Industrial Revolution. Coal has been mined for heat since Roman times, but it was not until coal was used for raising steam for Power that the tremendous expansion of factory production took place. The first steam engines used heat equivalent to some thirty times the power that was produced, and even now we, in the rich countries, throw away most of the heat that we obtain from fossil fuels. A larger part of the electricity and power produced is then used for purposes which are totally unnecessary for our lives, such as lighting an unoccupied area or overheating.

The industrial revolution was based on very cheap coal, and even now fossil fuels are so cheap in the rich countries that numerous known methods of saving fuel are regarded as too uneconomic or laborious to be installed. While the rich countries squander the earth's limited fossil fuel resources in this way, the Governments and rich people in the poor countries naturally strive to copy them, and the poor look on in desperation.

The ridiculous situation thus arises that poor countries want to install large central power stations, hundreds of megawatts in size, powered with water from very expensive dams, with nuclear energy or imported fossil fuels. To do this they have to borrow vast sums from abroad and find themselves saddled with national debts that require interest at rates some five times as large as those paid when Britain financed her Industrial Revolution. The cost of the machinery and fuel to build the stations has gone up in real terms by a comparable ratio; moreover if the electricity is to be supplied to the bulk of the population who still live in the rural areas, then it is necessary to install a huge distribution system which can cost up to 20 times as much as the power station itself.

The result is that not only is the country saddled with a national debt of totally unredeemable size, but every household has to use and pay for hundreds if not thousands of kilowatt-hours every year, if the capacity of the power station to produce electricity is to be used.

The local people neither want to change their habits to such an extent, nor can they ever conceivably afford to use electricity in this incredibly wasteful way. When one is considering the use of energy by the individual, Schumacher's 'Small is Beautiful' is an absolute truth. The only question is 'how small?'; what is the least energy which will just suffice to give a person in a poor village all the essential benefits of the industrial revolution, with as little disturbance to their traditional way of life as possible. These essential benefits include enough good food and clean water, hygiene, sanitation and medical care, the possibility of earning by their own reasonable amount of work a life style (including house and education) corresponding to that of the richer people in their area, and free energy for hobbies, and even some travel and communication.

An example, which can help us to assess this minimum energy is the need to have one room in each house illuminated for some hours in the evening. In the rich countries we have developed, as a convenience, strip light torches operated from dry cells which give enough light for several people to read by, which use less than 10 watts. From this example and others I have calculated that a 'village power station' providing some 5 w per person, ie some 40–50 kwh/year, would be sufficient for domestic electricity needs. Thus for a village of 1000 people a power station producing a steady output of 5 kw, and a peak of 10 kw would suffice.

A similar calculation for all the other needs of a village (such as a village workshop, water pumping for agriculture, heat for cooking with an efficient stove and a village bus/lorry) leads to figures of 75 kwh/year for Power and 0.18 tons of coal equivalent as heat. This result is somewhat astonishing, when one considers that, if one assumes that the total of 125 kwh of power is generated from heat with an efficiency of 10%, the total heat requirement comes only to 0.35 tons of coal equivalent. It is generally reckoned that the people of the poor countries use about 0.5 TCE/head-year, mainly in the form of wood, as heat for very inefficient cooking.

The gathering of firewood is becoming exceedingly laborious in many parts of the world and it is also one of the main factors causing the growth of deserts. Thus it is certainly no exaggeration to say that the provision of renewable energy in adequate quantities to all the poor people of the world is one of the three or four most important problems which must certainly be solved if our descendants are to live in a stable XXIst century. It is of comparable importance to the problems of world population growth and arms escalation; it is true to say that if we gave to all the people of the world these benefits of the industrial revolution, rapid population growth would be reduced by the resulting educational opportunities, and the risk of world war III would be reduced by the reduction in world tension as the lifestyle contrast was reduced.

This argument leads to the conclusion that this book by my friend Peter

Dunn, is a contribution to 'humane technology' which is far more vital to our future than all the literature on multi-megawatt power stations (whether fired with nuclear or fossil fuel or water-powered) or on computers or the space race. Professor Dunn has worked on, studied and directed research in this field for nearly two decades and has a world-wide reputation for his knowledge and skill in applying renewable energy.

I wish the book the success it richly deserves.

Professor M. W. Thring

Preface

The use of fire was a most important step in human development. With wood as the fuel, energy in the form of fire provided cooking facilities, heat for cold nights and a protection against marauding animals. Today, for a significant fraction of the world's population, wood for cooking and heating remains the major source of energy utilised. The world use of wood is generally under-estimated in published figures since these data are primarily concerned with commercial sources of energy, that is those types of energy which form part of the normal market economy for which statistics are readily available. Some authorities put the contributions of non-commercial energy to world energy use as high as 15% of the world total. The greater part of this is wood, although other biomass, such as vegetable waste and animal dung, make an appreciable contribution. Human muscle was early augmented by animal power, which is still used quite extensively. For example, even today the installed bullock power in India is greater than the installed capacity of the electricity generating supply. The earliest mechanical devices for providing mechanical power were the mills, which used either wind or water as their source of energy. The total contribution of these energy forms to the per capita use of energy remained small and it was not until the invention of the practical engine and the exploitation of coal reserves that human and animal muscle were supplemented to any great extent. The Industrial Revolution, with its dependence on the coal-fired steam engine, was the result. Towards the end of the nineteenth century the development of the internal combustion engine employing oil fuels enabled a further increase in energy use to be made. Although coal has retained a major position in commercial energy supply during the twentieth century, the increase in energy use over the past three decades has been largely met by oil and natural gas. In 1900 the annual per capita energy averaged for the world was 0·5 t.c.e. (tonnes of coal equivalent). By 1950 this had doubled to a value of 1 t.c.e and the rate of increase rose so that a further doubling occurred in the next 15 years. These figures are world averages and disguise the fact that the increased energy use has been primarily in the developed countries with little increased commercial energy available to the mass of the population in the developing countries. Currently the average annual per capita commercial energy use in the developed countries is around 6 t.c.e. This should be compared with a figure of less than a tenth of this value for the average developing country. The rapid increase in the consumption of oil and natural gas after 1950 to some extent reflects the very low cost of these energy sources. This situation changed dramatically in 1973 at the time of OPEC oil embargo. The price of Saudi Arabian oil increased \$2·70 per barrel to \$9·80 per barrel and similar oil price rises were made elsewhere.* For some years after 1973 the price remained fairly steady but a further sharp increase occurred following the Iran–Iraq war. In 1982 the price of oil was around \$30 per barrel.

Much more serious than the price rises is the finite nature of the reserves of oil and gas, which, at present rates of growth in consumption, can only maintain supplies for a few decades. It might be argued that the somewhat artificial oil crisis has served to focus attention on the problem of oil supplies and the need to find alternative energy sources which can replace them in the longer term. Coal reserves are larger than those of oil and gas and in the medium term it is expected that coal will fill the gap between demand and supply. However, coal itself is a finite resource and longer term alternatives are required. Since the Second World War nuclear fission has received a great deal of attention and considerable resources have been devoted to its development. Progress has been striking; nevertheless, nuclear energy today supplies only 3·5% of the world energy total and this small penetration is an indication of the problem of changing an industry of such size and complexity as the energy industry. Fusion power offers an attractive prospect but less effort has been given to its development and a sustained and controlled reaction has not yet been achieved. With the exception of biomass, which is a major energy source, and to a lesser degree hydro power, little progress has been made in exploiting the solar-derived renewable energy sources. These include the direct use of solar energy, wind power and wave power. Geothermal power, and to a less extent tidal power, might also make a useful contribution. There is no single solution to the problem of energy supply and all these forms of energy may be expected to play a part in satisfying the world's future energy needs. Unfortunately, the energy industry has considerable inertia, its size making rapid changes difficult, and hence there is an urgent need to pursue development programmes in all the various energy resource fields.

Energy conversion is an amazingly broad subject, which is one of its many fascinations. In this book I have tried to present energy and its conversion as a coherent subject. The book explains the basic physical principles underlying the various conversion methods and shows how they are applied in actual devices. Although the presentation is elementary, some of the concepts, such

* This data is also given in section 1.5.1 but is expressed in terms of 1974 \$

as availability, are not easy to grasp at first and require thought. Mathematical treatments have been avoided but considerable emphasis is placed on numerical magnitudes and rough estimates. It is important to remember a few typical magnitudes and to do this in a consistent set of units. Examples are given wherever possible in the text. Because of the breadth of topics covered in energy, new workers in the subject will find gaps in their basic knowledge. For this reason, relevant introductory matter has been included where it seemed appropriate. I hope that the reader with a chemistry background will be tolerant of the short excursions into elementary chemistry, that the mechanical engineer will not be bored by engineering theory such as heat transfer, that the physicist will nod approvingly at references to the Second Law of Thermodynamics, Planck and Wien, and that other specialists will forgive the rash sorties into their subjects. All this simple basic material is necessary, important and should be understood – and used.

By their nature, renewable energy devices tend to be small scale (though not always) and hence it is this aspect which is emphasised in the book.

Renewable energies can be expected to play a very significant role in developing countries in the future but they also have a potential in the developed countries too. The subject of energy is important and exciting and covers many other areas in addition to technology – for example, economics, politics (both domestic and international) and even morality – should we have nuclear power? All serious workers in the energy field will have views and many of us have pretty strong opinions on these topics. These matters, although interesting, are not, however, appropriate to the objectives of this book and therefore are not discussed here.

The material presented in the book has been chosen to provide a comprehensive account of renewable energy sources and conversion methods. However, since explanatory background material has been introduced where it is required, the chapters can be read separately. The book is intended for engineers and scientists who wish to be introduced to the subject, and it is hoped that it will also be of interest to the more general reader. For several years the material has been used to provide introductory lectures for a one year taught Masters course in 'Alternative Energy for Developing Countries'.

Chapter 1 describes energy use and introduces common units and magnitudes. Current consumption patterns are discussed for both developed and developing countries, together with world trends in energy consumption. Energy costs of materials and the less desirable side effects of energy use, such as air pollution, are referred to.

Chapter 2 summarises the nature and origin of both renewable and non-renewable energy resources. It includes solar radiation and its characteristics, photosynthesis and biomass production, fossil fuels, wind, waves, tides, geothermal energy, fission and fusion.

Chapter 3 summarises some of the more important conversion principles, the First and Second Law of Thermodynamics and their application to heat engines. Availability is introduced because of its importance in conversion devices. Some practical considerations, such as matching, irreversibilities and safety, are described.

Chapter 4 contains a discussion of heat engines using either vapour or gas as the working fluid. The Stirling engine is described in some detail. The heat pipe and fluidised bed are included. Refrigeration methods are explained.

Chapter 5 is concerned with internal combustion engines. The characteristics of spark ignition and compression ignition engines, together with their fuels (both conventional and unconventional), are explained. The Humphrey pump is described partly because of its inherent interest but also because this information is not readily available.

Chapter 6 summarises energy conversion methods. The relationship between Gibbs free energy and enthalpy is given in Appendix 2.

Chapter 7 surveys the applications of direct solar energy. The flat plate collector is discussed in some detail and the associated heat transfer methods explained. Applications to water heating and crop drying are given. The solar pond is introduced and methods of solar desalination described. Solar thermal engines and solar refrigeration are discussed, and finally, photovoltaic conversion is treated.

Chapter 8 gives a brief summary of the relevant chemistry of bioconversion methods and then discusses in turn: muscle power; combustion and cooking stove design; vegetable oils; pyrolysis; gasification; anaerobic fermentation, and alcohol production.

Chapter 9, on wind power, introduces the performance limits set by linear and angular momentum considerations, also the concepts of lift and drag and goes on to describe the performance of the different configurations together with their applications.

Chapter 10 describes the main types of turbine and their characteristics and applications. It also includes other hydro power devices such as the run of stream turbine and the hydraulic ram.

Chapter 1

Energy and its use

1.1 Introduction

In the Preface the way in which energy is used was outlined and attention was drawn to the large differences in energy consumption between the developed and the developing countries. This chapter provides a more quantitative background on energy use. It considers in turn the total world energy use, both the commercial and the non-commercial energy sources, how this energy is supplied, and national differences; historical trends and likely future projections are given and finally the use of energy in the developed and developing countries is described.

Although not always recognised as such, energy is a basic human need. It is required to enable us to perform all our everyday activities. Food, of course, may be regarded as the most essential of our energy requirements. Energy can be defined as the ability to do work, for example, to raise a weight.

1.2 Energy, fuel and power

1.2.1 Units

Available sources of energy are sometimes referred to as fuel. The term originally came into use to describe energy sources that could be burnt, e.g. coal, wood, and later oil and gas. It has now been extended to cover uranium and other materials used to produce nuclear energy. In this book energy will be used as the preferred description. As has been said in the previous section, energy is the ability to perform work; it is the equivalent, in terms of money, to capital in the bank. When money is spent the rate of expenditure is given by the amount of money divided by the time period over which it is spent; energy is used in the same way, and the rate of use of energy is called power.

In order to discuss energy quantitatively it is necessary to adopt a unit of measurement. The precisely defined, and recommended, scientific unit for energy is the joule, which has the symbol J. The joule is named after the 19th century Manchester brewer turned scientist, James Prescott Joule. Joule carried out fundamental experiments which demonstrated the equivalence of heat and work and his results are now incorporated in the First Law of Thermodynamics. Unfortunately, the joule is far too small a unit to be convenient in describing world energy supplies and resources. This objection can be overcome by the use of prefixes such as mega (10^6), giga (10^9) and tera (10^{12}).

1.2.2 Magnitudes

It is useful to remember the approximate values of the energy content for a few common fuels. Typical values* are:

Diesel and petrol approximately 45 MJ/kg
Coal, charcoal approximately 30 MJ/kg
Wood (dry) approximately 20 MJ/kg

Note that 1 kg of oil will release 1·5 times the energy of 1 kg of coal.

Natural gas is of similar chemical composition to oil and hence will have a similar heat content (approximately 50MJ/kg). Of course the density of a gas is much smaller (by about 1000) than that of a liquid, thus the energy of unit volume is correspondingly less. Oil has a typical density of 0·8 relative to water, or 800 kg/m^3, whereas natural gas has a density of around 0·6 of that of air (1·2 kg/m^3 at STP) or about 0·7 kg/m^3.

1.2.3 Practical Energy Units and Unit Conversion

The small value of the joule has been referred to earlier and is an inconvenience which can be overcome by the use of prefixes. A second and more serious objection to the joule as a measurement unit for large scale energy is that practical people measure the output of their oil well, or coal mine, in such obvious units as barrels of oil or tons of coal. To obtain an understanding of energy matters it is essential to be able to convert such units rapidly to a common equivalent. The tonne* of coal equivalent, or t.c.e., is such a unit and has been adopted by UN bodies. Since the chemical composition of coal is not a fixed quantity it is necessary to assume a representative figure. A commonly used value for the t.c.e. is $2{\cdot}9 \times 10^{10}$ J or 29 GJ. It is useful to remember a few approximate conversion factors to enable the rapid conversion of numbers expressed in other units to the t.c.e.

* Fuels such as oil are mixtures of many different chemical compounds and the relative proportions differ for oils from different sources. Hence there will be a range of values for heat content. Heat content itself is a somewhat vague and unscientific term; it is useful to enable rough calculations and comparisons to be made. For more accurate work scientific definitions should be used. The heat released from a reaction will depend on how the reaction is carried out, for example at constant volume or at constant pressure. Also, for fuels such as oil, the exhaust products include a considerable amount of water. If this water is released as a liquid then the latent heat will be included in the useful heat; if, however, the water is present in the exhaust as a vapour this energy is lost. The two calorific values are referred to in the literature as gross and net calorific value and can differ by as much as 10%. The calorific values for solids and liquids for constant volume and constant pressure reactions are not significantly different.

Since the metric ton or tonne is equal to 1000 kg*, the values of MJ/kg can readily be converted from kg to tonnes by changing MJ/kg to GJ/tonne and to t.c.e. by dividing by 29 GJ. 1 tonne of oil is approximately 1·5 t.c.e. The oil industry frequently expresses oil output in terms of barrels of oil a day (bbl/d). One barrel of oil per day is approximately equal to 50 tonnes of oil per year.

Electricity is normally quoted in terms of kilowatt hours or 'units'. Care must be taken in the use of a conversion factor to convert electrical energy to other energy equivalents since there are two different conventions used in the literature. The rough equivalent in energy terms is 8000 kWh per t.c.e. This factor gives the energy equivalent of electrical energy in terms of coal; however, where the electricity is generated by nuclear or hydro plant it is often the practice to regard the equivalence as 2700 kWh/t.c.e. The reason for this is as follows: had the electricity been generated by a coal-fired station whose average efficiency was 33%, then three times the coal would have been required to generate each unit of electricity. Some authors use oil replacement value in order to convert alternative sources to the equivalent oil value. In the case of hydro or nuclear power this is a factor of about 3, whereas for biomass it might be only 1/5. However, this interchangeability factor will depend very much on the particular application and should be used with caution.

Table 1.1 shows the approximate equivalents for some common units and approximate equivalents for common fuels. In Europe the average energy use is about 6 t.c.e. per head per year. The rate at which energy is used, or power, is usually expressed in kilowatts (kW) or horsepower (hp). A horsepower is about ¾ kW (0·746 kW). It is helpful to know that a tonne of coal per year is approximately equivalent to an average power of 1 kW or

$$1 \text{ t.c.e. per year} \simeq 1 \text{ kW}$$

Hence, the people of Europe, on average, are using energy at a rate equivalent to a 6-bar electric fire operated continuously. The average energy required by man is, of course, his food intake, which is equivalent in energy terms to 1/6 t.c.e. per year, or 1 lb of coal a day.

1.3 Energy conversion

In the previous section it was stated that the average energy requirement for an adult is equal to about 1/2 kg or 1 lb of coal a day. However, not even the best chef could make coal palatable, and in practice it frequently happens that the energy source in its original form is unsuitable for the application and it is necessary to convert to a more acceptable form. The major forms of energy are given in Table 1.2. They may readily be converted from one form to another; for example, oil can be used to drive an internal combustion engine, which in

* The tonne is the metric ton and is equal to 1000 kg or 2205 lb. The Imperial ton is equal to 2240 lb.

Table 1.1 *Unit conversion: some general unit conversion factors*

Length	1 ft = 0·305 m
Area	1 ha = 10^4m^2 = 2·471 acres
Volume	1 barrel = 35 Imperial gallons = 42 US gallons = 0·159 m^3
	1 litre = 3·533 × 10^{-2} ft^3 = 0·220 Imperial gallons = 0·26 US gallons
Mass	1 kg = 10^{-3} tonne = 2·205 lb
	1 metric tonne = 2205 lb
	1 short ton = 2000 lb
	1 Imperial ton = 2240 lb
Pressure	1 bar = 10^5 N/m^2 = 10^5Pa
	1 atm = 1·013 bar
Energy	1 BTU = 1055 J
	1 Calorie = 4·19 J
	1 Therm = 10^5 BTU
	1 Quad = 10^{15} BTU
	1 e.v. = 1·6 × 10^{-19}J
Approximate values for fuel equivalent	
	1 barrel oil = 6·1 × 10^9J = 5·8 × 10^6 BTU
	1 tonne oil (t.o.e.) = 4·47 × 10^{10}J
	1 m^3 natural gas = 3·7 × 10^7J
	1 Cu. ft natural gas = 1·05 × 10^6J
	1 tonne coal (t.c.e.) = 2·9 × 10^{10}J

turn drives a dynamo in order to convert the chemical stored energy of the oil into electrical energy. With the important exception of heat energy, all other forms of energy can, at least in theory, be freely converted from one from to another without loss. This principle, the Law of Conservative of Energy, or the First Law of Thermodynamics, is one of the most important physical laws. Of course, in practice losses will occur, since perfection can never be achieved in practical equipment; it is, however, important to know that there is no theoretical upper limit to the conversion efficiency. Heat is the exception to this conversion rule and the restriction on conversion efficiency from heat to other forms of energy is contained in the Second Law of Thermodynamics. There is no restriction in converting other energy forms to heat; for example, an electric fire can, in principle, and very nearly in practice, have a 100% efficiency. The two laws of thermodynamics have been summarised (in a rather pessimistic manner) as:

First Law: You can't win
Second Law: With heat you can't break even

Table 1.2 *Principal energy forms*

Heat
Mechanical
Electrical
Electromagnetic waves (light, radiowaves)
Chemical
Nuclear

The two laws are dealt with in more detail in Chapter 3, together with the general principles of energy conversion.

1.4 How much energy do we use? Renewable and non-renewable energy

Currently the total annual world energy use is around 10^{10} t.c.e. The different contributions to this total by the various energy sources are shown in Table 1.3.

Table 1.3 shows that about three-fifths of the world's energy demand is provided by the fossil hydrocarbons* oil and natural gas. Fossil fuel, including coal, amounts to 90% of the current world use. The nuclear contribution is small (3·3%) but in some countries is a major supplier of electrical energy. For example, over 50% of the electricity in Sweden is from nuclear power.

Most of the published energy statistics deal only with what is known as commercial energy sources; these consist primarily of the fossil fuels (coal, oil and natural gas) together with the much smaller contribution from hydro power, nuclear and geothermal sources. It is these fuels which dominate the energy market economy, and statistics on their use are readily available.[1,2] Such published figures do not usually include the non-commercial fuels, such as firewood, animal dung and vegetable wastes. Because this energy is not part of the normal commercial market it is difficult to estimate accurately the amounts involved. Some authors, for example Hall,[3] put the figure as high as

Table 1.3 *World primary commercial energy demand – 1982*

	%
Oil	41·7
Gas	18·7
Coal	29·3
Hydro	6·8
Nuclear	3·3

* Hydrocarbons are compounds containing only the elements hydrogen and carbon. They exist as solids, liquids and gases.

15% of the world total. Such a figure is comparable with natural gas consumption, two-and-a-half times hydro and almost five times the nuclear contribution. Biomass is clearly a very important source of energy in world terms. It is even more so when we consider its contribution in the developing countries. The classification of commercial and non-commercial is somewhat arbitrary since some biomass is included in the commercial sector. A better classification of energy sources is renewable and non-renewable.

1.5 World energy consumption: the past, the present and the future

1.5.1 Historical trends in energy consumption to the present

Fig. 1.1 shows how commercial energy use has increased over the past three decades and brings out clearly how the hydrocarbons, oil and gas, have met the

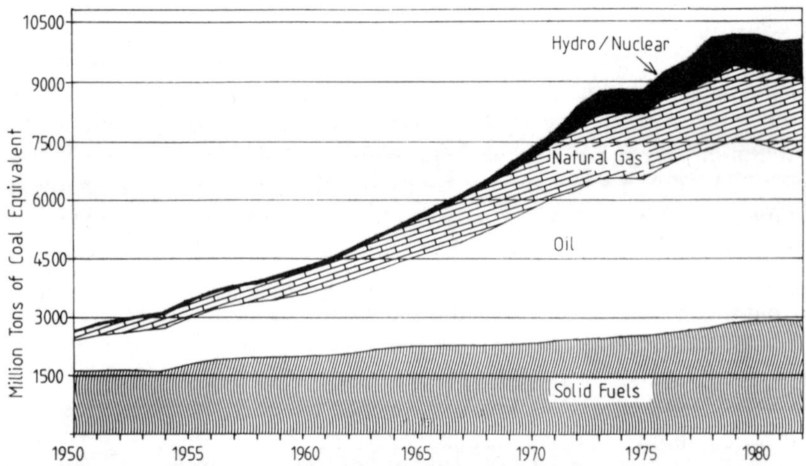

Fig. 1.1 *Growth of world commercial energy demand (Ref. 2)*

increased energy demand, with coal use increasing only slowly. This is because of the convenience in use, and, until recently, very low cost of oil and gas. On average, oil use has increased by 7%/year over the period 1945–1973 corresponding to a doubling time* of 10 years. This rate of increased use changed dramatically after 1973, following the OPEC price increase. The average increase in oil use from 1973–1980 has been reduced to 1·25%/year. Until 1973, oil prices, in real terms, had kept remarkably low and had even

* Increases in demand are often expressed in terms of percentage increase over the year. Another way of expressing this is in terms of doubling time, that is, the time for the demand to double if the percentage rate remains constant. The relation between doubling time τ and annual percentage increase α is given by: $\tau = 69{\cdot}3/\alpha$ years.

fallen from \$4/bbl in 1950 to \$1·6/bbl in 1970 (in terms of 1974 dollars). The price then rose to \$3/bbl in 1973 and then suddenly to \$9/bbl. A further increase occurred in 1978, following the Iran–Iraq war, to \$18/bbl. This represents, in real terms, an order of magnitude rise in a decade.

Coal still supplies about 29% of the total, though most of the coal is consumed in the country in which it is mined and only 10% is exported internationally. This contrasts sharply with the international trade in oil. The share of the market provided by coal has increased from 26·5% in 1973 to 29·3% in 1982.

Fig. 1.2 gives examples of per capita energy use in both developed and developing countries. In Fig. 1.3 per capita energy use has been plotted against GNP per capita for both developed and developing countries for the year 1969. It is interesting to note the almost one-to-one relationship between the two quantities. This is not particularly surprising since in a technological civilisation it might be expected that there would be some correlation between wealth and energy use. This is because things require energy in their

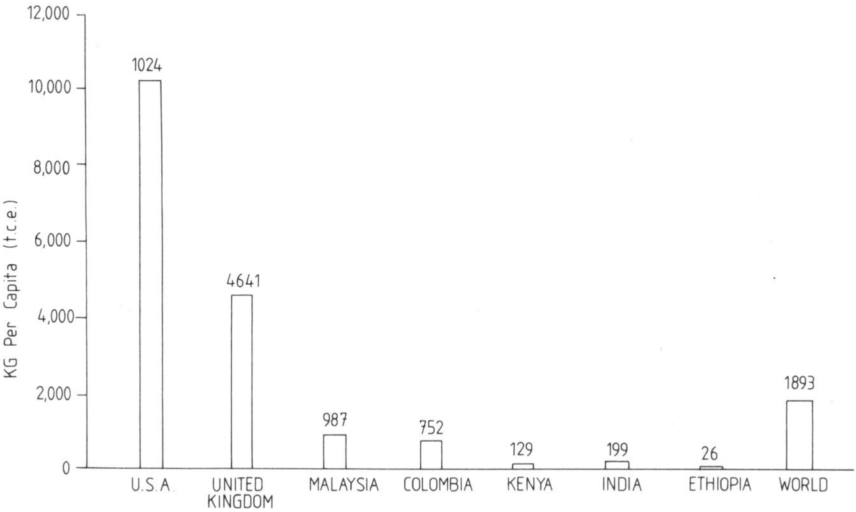

Fig. 1.2 *Per capita commercial energy use for selected countries (Ref. 1)*

manufacture. Although this seems to be generally true there are interesting differences between countries in the figure, for example, between the energy consumption of UK, Germany and France. It is sometimes said that this correlation between energy and GNP is misleading and simply reflects the effect of cheap energy over the last two or three decades. This view is not supported by Fig. 1.4, taken from Reference 4, which shows the energy per capita in the USA plotted against GNP per capita from 1890 to 1973, there appears to be a close correlation. Reference 5 gives a similar curve for the UK from 1700 to 1975 with again the same general trend, but the authors,

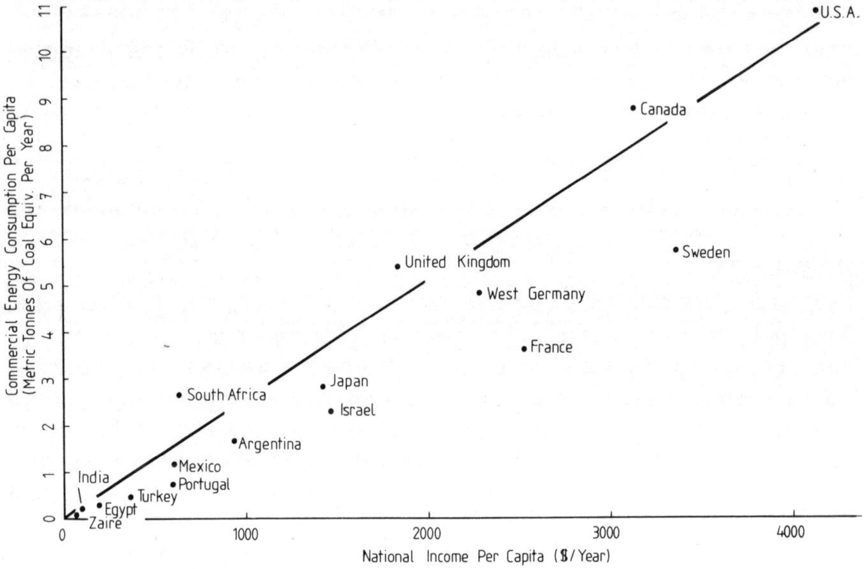

Fig. 1.3 *Per capita commercial energy consumption versus per capita national income for 1969 (Ref. 1)*

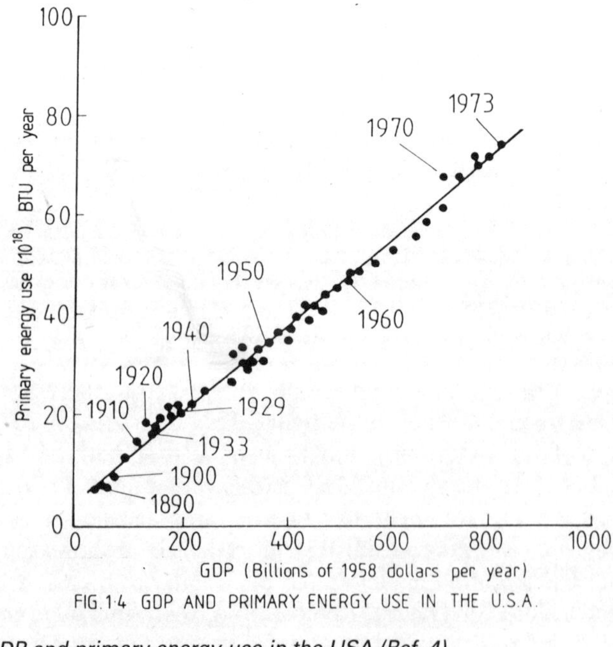

Fig. 1.4 *GDP and primary energy use in the USA (Ref. 4)*

Humphrey and Stanislaw, do point out the difficulty in obtaining reliable early data. On the other hand, Baguant[14] shows that the trend for Germany over the period 1950–1975 has been a sharp decrease in energy/GNP, whereas for developing countries the ratio is tending to increase as the degree of industrialisation is increased, see Fig. 1.5. More recently the effect of the increase in oil prices has caused the developed countries to look hard at means of conserving energy and some success has been achieved in reducing the energy/capita to GNP/capita ratio.

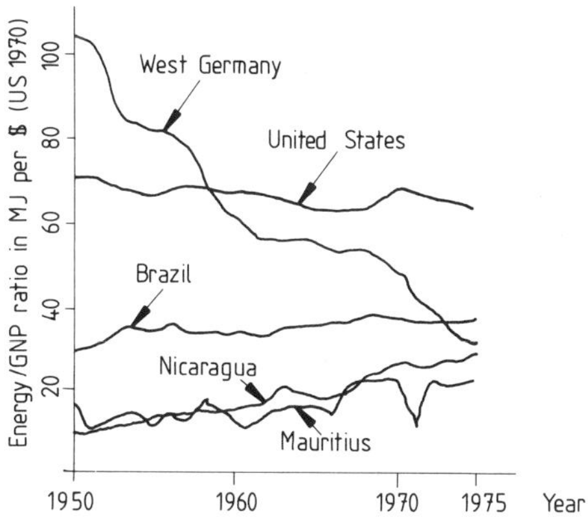

Fig. 1.5 *Energy/GNP ratio in MJ per $ (US) at 1970 value (Ref. 14)*

1.5.2 Pollution and environmental problems

Several problems have arisen from the increased use of energy, for example oil spillage resulting from extraction. On a global scale concern has been expressed over the release of carbon dioxide from the burning of fossil fuels since the level of this gas in the atmosphere could in the future exceed the natural level and may lead to climatic change.

On a more regional level air pollution in large cities can be a severe problem. The burning of coal in London led to the Victorian 'pea soup' fogs and caused major health problems until the effects of legislation for the use of smokeless fuel eliminated the trouble in recent years. Automotive exhaust has now replaced smoke from coal as an air pollutant and has become a serious health hazard in some areas, notably Los Angeles and Tokyo, where local geographic features accentuate the problem. Acid rain caused by coal and oil from electricity generating stations and automobile exhausts is a further regional problem.

The safe storage of nuclear waste has not yet been achieved to the satisfaction of many.

At a domestic level, smoke from wood burning stoves is the cause of widespread eye and chest complaints in many of the developing countries.

1.5.3 Future world energy consumption

Views on the future rate of increase of energy use differ. Reference 6 gives projections on energy use to the year 1990 and suggests an average annual increase of 3% from 1975 to 1990. Table 1.4 gives the corresponding per capita figures which include the effect of population increase. These projections suggest that the energy gap between the developing countries and the rest of the world will continue to increase and the per capita average in developing countries will remain at less than one-fifth of the world per capita average.

For the next two or more decades the fossil fuels will continue to provide the greater part of the world's commercial energy needs; during this period steps will have to be taken to develop the alternative renewable sources. Efforts will also be needed in conservation and in moving towards a less energy demanding way of life. Such a move would not necessarily lead to an impaired quality of life, but, by paying attention to environmental and ecological effects, could well result in a better life style. In the developing countries it is urgently necessary to increase the energy availability per capita, and we must not overlook the implications of population increase on the total world energy requirement.

1.6 How world energy is used

World and national energy consumption figures are so large that it is difficult to visualise their meaning. It is much more helpful to express these numbers in terms of energy per capita per year. In these terms the average use of commercial energy in the world is 2 t.c.e. per capita per year. The inequitable distribution of this commercial energy between the developed and developing countries is brought out in Fig. 1.2.

There are great differences in the type and amount of energy use within the developing countries themselves. There is usually a dual society, the more wealthy living in a manner similar to their counterparts in the developed countries and with similar energy requirements, whereas the poorer section of the community has a very low per capita commercial energy availability. The differences in energy use between the rural and the urban areas is also pronounced. Fig. 1.6 illustrates this in the case of commercial energy use in Thailand.

The importance of non-commercial fuels for very large numbers of the world population is not generally appreciated. With the progressive removal of trees and other vegetation there is an energy shortage in many developing countries, for example the sub-Sahelian countries of Mali and Upper Volta. For people in these regions the firewood crisis is much more urgent and real than is the oil crisis in the developed countries. This is not to say that oil prices have no effect on the less developed countries; consumer and manufactured goods in these countries are just as dependent on oil as are those in the developed countries;

Table 1.4 *1975 commercial energy use with projections for 1990 (from Reference 6, Tables 13.33, 13.34)*

	1975		1990		1975–1990	
Region	% of world total energy use	Per capita energy use as fraction of world average	% of world total energy use	Per capita energy use as fraction of world average	Average annual % increase Region	Per capita
United States	29	5·53	26	5·86	2·3	1·6
Other industrialised countries	27	2·27	33	3·25	4·2	3·6
Centrally planned countries	31	0·97	27	0·9	1·9	0·8
OPEC	2	—	4	—	6·1	—
Less developed countries	10	0·18	11	0·19	3·3	1·6
World	100	1	100	1	3·0	1·2

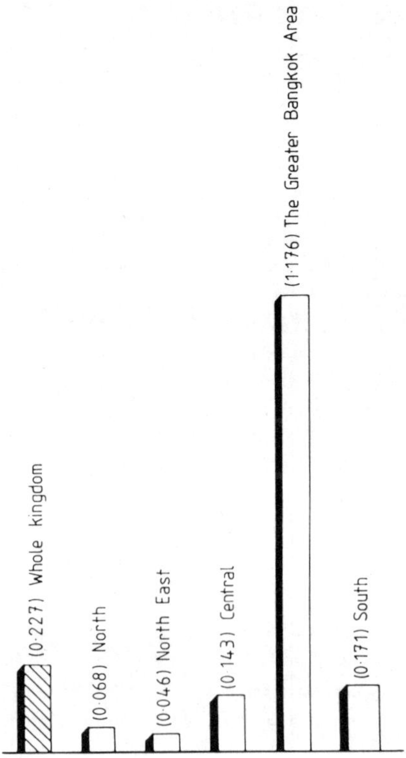

Fig. 1.6 *Commercial energy use in Thailand by region (t.c.e per capita) (Ref. 7)*

an increase in oil prices can have serious effects on the economy and can affect non-oil users by diverting resources which might otherwise go into rural development. Oil cost is now a large, and increasing, fraction of the import bill for developing countries. They earn their foreign currency by exporting agricultural products and raw materials whose prices are lagging behind the costs of oil. For example, in Thailand, a major rice exporter, the amount of rice necessary to export in order to buy a barrel of oil rose by a factor of five from 1968 to 1979. The energy mix for developed countries is very different from that of the developing countries. This is brought out in Table 1.5, which compares the energy mix of the Sudan with that of the USA. The dependence of the former on biomass and of the latter on fossil fuels is clear from these figures. It should be noted that these percentages disguise the relative absolute amounts of energy consumed per capita in the two countries. The USA figure is 12 t.c.e. per capita per year compared to a figure of 0·3 t.c.e. per capita per year in the Sudan. Table 1.6 gives similar energy breakdown for a number of other developing countries. It is seen that even India depends on biomass for almost half its energy.

Table 1.5 *Sudan/USA energy (from Reference 8)*

Sudan			USA	
		Energy Input		
oil		38%	oil	46%
firewood/charcoal		57%	natural gas	31%
bagasse		2·2%	coal	18%
hydro		1·8%	hydro/nuclear	5%
muscle		0·6%		
		Energy Use		
direct heat			commercial and domestic	19%
domestic	61%		industrial	26%
industrial	7%		transport	25%
total		68%	electricity generation	25%
transport		20%	non-energy use*	5%
agricultural tractors/pumping		7%		
electricity generation		5%		

*This heading refers to the petrochemical industry.

Table 1.6 *Energy use in developing countries (various sources)*

Country	Total energy use t.c.e./capita/annum	% Commercial	% Non-commercial
India (1970–71)	0·63	52	48
Mali (1973–74)	0·29	8	92
Pakistan (1975–76)	0·4	47·5	52·5
Bangladesh (1973–74)	0·12	25	75
Kenya (1980)	0·6	29	71
Thailand (1977)	0·7	50	50

1.7 Use of commercial energy in developed countries

1.7.1 Introduction

This section considers the way in which energy is used in the developed sectors of the world economy. The average per capita annual energy consumption in the developed countries is around 6 t.c.e. and in the USA almost double that figure. In the previous section, in order to compare the energy use by different countries, it was convenient to give the figures for total primary energy use. When discussing energy use within a country or sector, it is more helpful to present the data in the form of energy delivered to the end user, and in this case electricity is included as an energy source. The difference between the two totals, primary energy consumed by the country and energy delivered to the user, is made up of conversion and distribution losses. These are high in the case of electricity generation where as much as two-thirds of the primary energy input is lost. Further losses are incurred by the user, but, since electrical appliances such as electrical heaters are more efficient than those using fossil fuels, the low efficiency of large scale electrical power generation is to some extent offset by higher user efficiency.

The major users in a developed economy are industry (which includes the energy industries themselves), domestic and transport. The remaining users, such as street lighting, shops, and the non-energy users of fuel (tarmac, plastics industry, etc.), make up the fourth sector.

Fig. 1.7*a* shows how the total UK energy supplied to the user was made up in 1978. Fig. 1.7*b* indicates the manner in which this total was distributed to the user sector. Fig. 1.7*c* shows how total energy was shared out between the sectors. A striking feature in Fig. 1.7*c* is the dominant position of oil in transport.

Because of its convenience to the user, electricity is of considerable importance as a secondary source of energy. Electricity generation is treated separately in the next section since it consumes such a large fraction (24% of total world energy consumption in 1972) of the total primary energy input. In 1972, 74.5% of the world electricity was generated from fossil fuel, 23.7% from hydro power and geothermal and 2% by nuclear energy.

1.7.2. Electricity generation

The trend over recent years in electricity generation has been towards larger and larger generation plant and the siting of power stations has tended to be away from centres of population, partly for reasons of pollution and partly from the need for cooling water. The overall generation efficiency has progressively improved but it is still not much more than 30% when transmission and other losses are included. It is unlikely to be significantly increased in the foreseeable future. Thus, of the energy consumed, only 30% is supplied to the user in the form of electricity and the remainder appears as low grade heat in the cooling water and in the atmosphere. There are a number of ways in which this situation can be improved, for example the total energy schemes by which electricity generation plant is used both to provide electricity and also low grade heat for district heating or process heat in industry. The overall efficiency of such total energy schemes is typically 65% and can, in principle, be higher. Load matching can be a difficult problem. Total energy

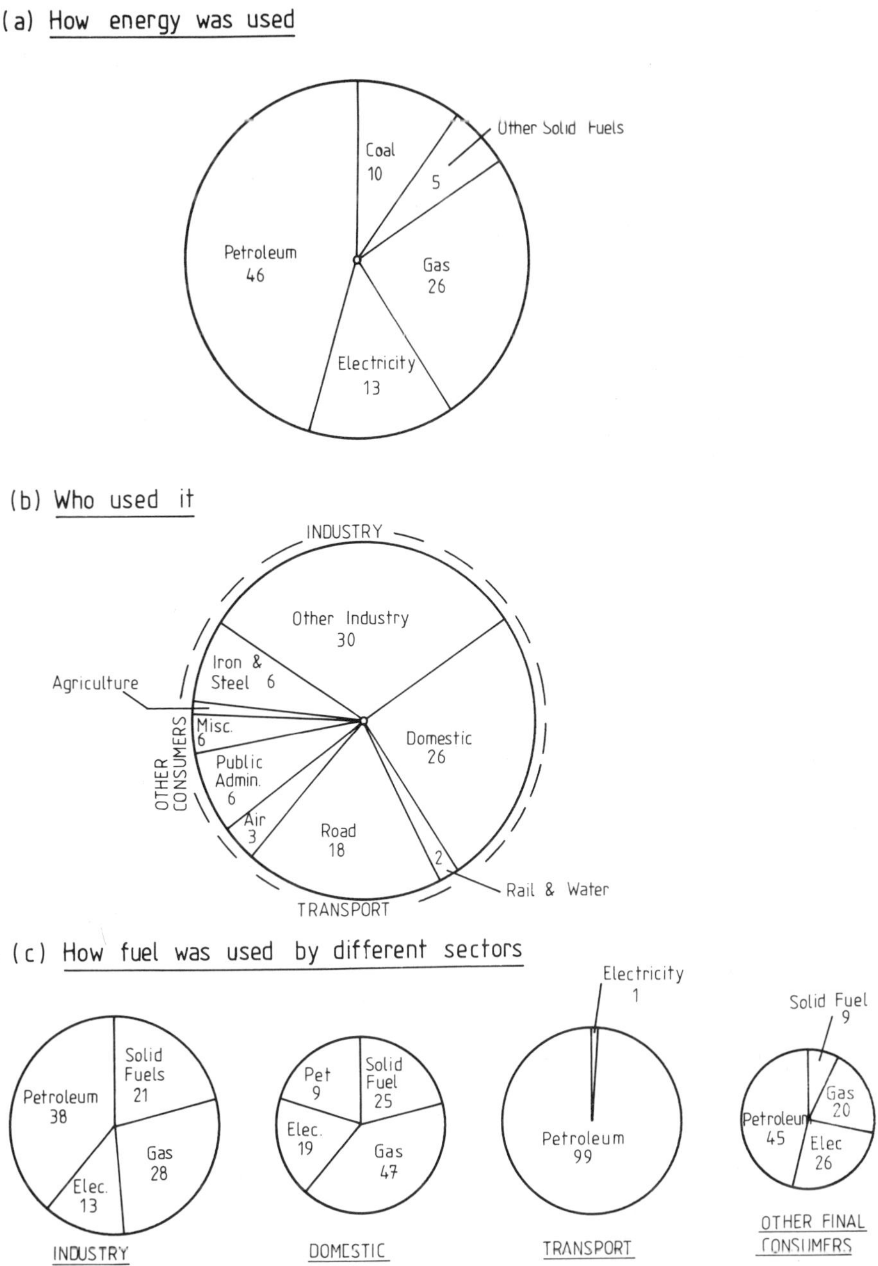

Fig. 1.7 *UK energy consumption of final users (heat supplied basis). Based on Reference 9*
a How energy was used
b Who used it
c How fuel was used by different sectors

schemes generally make use of smaller plant than is customary in central power stations since it is expensive to transmit heat over great distances. In particular, gas turbines of a few megawatts capacity are often used, and diesel engines and quite small gas engines of a few tens of kilowatts are also being developed for this purpose.

Losses in transmission and distribution depend on the population density and electrical energy use per capita. In developing countries the effective line length per consumer is higher than is the case in developed countries and the electrical losses correspondingly higher. For example, in 1975 the average loss in transmission and distribution in the European OECD countries was 7·5%, whereas in the South East Asia ESCAP region the average figure was 15·7%. The losses in particular countries can be considerably higher than this average would suggest, for example, 25% in Indonesia and 28% in Nepal.[10]

1.7.3 Transport

15% of the total energy consumed in the UK is used in the transport industry and the greater part of this in the private car. In spite of great efforts over the last 30 years the private car is still not efficient in terms of passenger miles/gallon and this situation is unlikely to change in the foreseeable future. It is interesting to compare the efficiency of different forms of transport in terms of passenger miles/gallon.

Table 1.7 illustrates the advantage in fuel economy of multiple passenger vehicles. One objection to such vehicles is the inconvenience of public transport for short journeys and for journeys from home to and from main termini. Such objections might be overcome by the use of small vehicles – either privately owned or as self-drive taxis – for these short journeys. While there is little to be gained from fitting smaller engines into large cars, considerable saving would result from the use of small cars with small engines.

When one takes into account efficiency of electricity generation, the electric car has no significant efficiency advantage over the internal combustion engine. Of course, if electricity was produced by a station incorporating total energy, then this might no longer be the case. If the electricity is generated from nuclear power or hydro power, efficiency is not necessarily the most suitable basis for comparison.

1.7.4 Industry

As might be expected, industry consumes a significant fraction of an industrial nation's energy supply, with the steel industry being the largest single consumer. As the price of energy has risen, industry has become increasingly aware of the need to avoid energy waste. In some cases, such as the manufacture of cement, energy is the main cost component. In other energy intensive processes, such as glass making, it amounts to only 10% of the manufacturing cost.

Table 1.7 *Energy consumption in passenger transport (based on Reference 11)*

Vehicle	Primary energy consumption				
	Fully loaded		Average load factor	With average load	
	MJ/seat km	Seat miles per gallon		MJ/ passenger km	Passenger miles per gallon
Passenger liner	7	16	100%	7	16
Concorde	5	22	90%	5·6	20
Jumbo jet	2·5	43	65%	3·9	28
Electric express train*	0·45	(243)	45%	1·0	(109)
Diesel express train	0·41	270	45%	0·9	122
Express coach	0·26	421	65%	0·4	274
Car – motorway (4 seats)	0·80	124	2 passengers	1·6	62
Car – rural (4 seats)	0·85	118	1·7 passengers	2·0	50
Car – urban (4 seats)	1·2	85	1·5 passengers	3·1	32
London Underground train*	0·22	(488)	14%	1·6	(68)
Urban bus (70 seat DD)	0·2	547	25%	0·8	137
Rural train (diesel multiple unit)	0·32	342	20%	1·6	68
Rural bus (45 seat SD)	0·21	521	15%	1·4	78

* Efficiency of electric generation = 24%.
Assumptions:
Petrol – 1 Imperial gallon 160 MJ, miles/gallon = 99.4/MJ/km.
Diesel – 1 Imperial gallon 176 MJ, miles/gallon = 109·4/MJ/km.

Energy is required in order to make things; it is needed to separate metals from their ores and subsequently to fabricate the metal into the required shape and form. The amount of energy required to produce a material is called the energy cost of that material. Energy costs are described in the next section.

1.7.4.1 Energy cost of materials. Light metals such as aluminium form very stable oxides which comprise the ores. Very large amounts of energy are needed in order to yield pure metal. Table 1.8 gives a few typical values; the figures are expressed both in GJ and also t.c.e. The latter is useful in enabling one to visualise the requirement in terms of tonnes of coal required to produce a tonne of the material. In the case of aluminium, around eight times its weight of coal is needed, whereas only around a tonne of coal is needed to produce a tonne of steel. These figures are averages and the precise amount of energy will depend on the process used and the age, size and efficiency of the plant and the degree of finishing given to the material. They should therefore be regarded as illustrative only. Having obtained the basic material, further energy may be needed in order to fabricate to the required form. Some common devices are highly energy intensive. For example, the colour television requires 1 tonne of coal for its material and manufacture. Considerable energy savings can be made by recycling energy intensive materials such as aluminium. Typically, recycling a metal requires only one-tenth of the energy needed for its extraction. A further significant saving can be made by designing devices so that they can be readily maintained and repaired in order to give long life. The practice of providing 'new' car models annually in which the differences are little more than a cosmetic redistribution of decoration or shape should be discouraged, and more attention paid to maintenance and life.

Table 1.8 *Approximate energy requirement of some common materials*

	GJ/tonne	t.c.e./tonne
Steel	25·1	0·87
Aluminium	230	8·0
Plastic polymers	90	3·1
Cement	13	0·45
Bricks	4	0·14
Glass	24	0·83
Pottery	10–250	0·34–8·6
Paper and board	62	2·1

Based on data in Reference 12.

For a fuel, such as coal or uranium, to be worth while the energy it will yield must be considerably greater than the energy required to mine and process into a form suitable for use. For example, natural uranium exists as a mixture of two isotopes, uranium 235 and uranium 238. thc formcr, uranium 235, is the isotope which is consumed in the conventional, thermal, nuclear reactor. This isotope is 135 times less abundant than U238 which makes up the major part of natural uranium. U238 can be processed in breeder reactors to produce a fuel, plutonium 239. By using breeder reactors, natural uranium will yield around 50 times the energy when only U235 is burnt. The breeder reactor makes it worthwhile to use ores would not be energetically worth while if only the U235 was utilised. This very greatly extends the availability of nuclear fuel, see Chapter 2.

In the case of renewable sources of energy, such as wind or solar energy, the 'energy pay back time', or time for the device to produce as much energy as was required to build it, can be significant. For example, for the single crystal silicon solar cell the energy pay back time may be several years.

1.7.5 Domestic

Figures suggest that energy in the home is used at an overall efficiency of around 60%. This is somewhat misleading since it merely means that most of the heat content of the fuel is released in the house. This heat is then subsequently lost by conduction through windows, walls, etc., and by air changes. There is still considerable scope for reduction of energy demand in the home by better thermal insulation, and in larger buildings by the use of heat recovery systems and possibly heat pumps. Most of the energy load in the home is in the form of low grade heat and – as mentioned previously – it is possible for this to be supplied from district heating associated with a total energy power station.

1.8 Developing countries

1.8.1 Energy needs in rural areas

The greater part of the world population lives at a low standard, and a quarter of the population, or about 1 billion, have an extremely low standard of life. Energy is one of a number of essential requirements if the quality of life of these people is to be raised from its present level of grinding poverty.

Fig. 1.2 indicates the large differences in commercial energy use between the developed and developing countries and Table 1.5 compares the way energy is used in a particular developed country, the USA, and in a developing country, the Sudan. In addition to the big difference in per capita energy use, the other feature is the high dependence of the latter on biomass, a characteristic of energy use in developing countries which is also brought out in Table 1.5. In order that the developing countries can raise their living standard, the per capita energy availability must be increased, particularly for those living in the rural areas.

What then are their needs and what energy resources might be employed to satisfy them? These basic energy requirements are summarised in Table 1.9. In terms of energy forms they can be classified as:

1 heat or cold
2 mechanical shaft power
3 light
4 sound (for communication, radio).

In addition to the primary energy requirements the energy cost of materials such as cement and fertiliser should be included in the overall energy needs. Table 1.10 gives approximate energy cost of fertiliser and pesticide.

Table 1.9 *Energy requirements in rural areas*

Domestic
Heating/cooling
Cooking
Lighting
Agricultural processes
Land preparation
Sowing
Weeding
Harvesting
Crop processing and storage
Drying
Grinding
Refrigeration
Small industry
Power machinery
Process heat
Water pumping
Domestic use
Cattle
Irrigation
Transport
Schools/clinics
Communication. Radio

The energy sources which might be used are discussed in Chapter 2. They are listed in Table 1.11. A number have been omitted from this list. Nuclear power is unlikely to make any contribution since inevitably it will require large central installations rather than the diffuse low cost sources required in rural areas. There might eventually be a case for the use of nuclear-powered rural

Table 1.10 *Fertiliser and pesticides (Reference 13)*

	Energy cost MJ/kg
Nitrogen	80
Phosphate	14
Potassium	9
Pesticide/herbicide	100

The amount of fertiliser used varies widely from country to country, for example, rice farmers in the USA use 135 kg/ha of nitrogen, whereas in the Philippines the amount used is only 31·5 kg/ha; the yield/ha is considerably increased with the higher rate of fertiliser use.

Table 1.11 *Energy sources for rural areas*

Fossil fuels
Coal
Oil
Natural gas
Peat
Direct solar energy
Indirect solar energy
Biomass
Wind
Hydro power

electrification networks but this possibility is unlikely to be economic in the foreseeable future. Geothermal power, except in those few areas where wet steam arises, is likely to prove too expensive for small scale plant. Wave energy also suffers from the difficulty of scaling down, though this may prove worth while in some situations.

The present position is that for most people in the rural areas, the sun, in association with traditional building practices, provides environmental comfort conditions supplemented when necessary by wood fires. Cooking is largely by wood or its derivative, charcoal, with cattle dung being used to a lesser extent. Mechanical power is provided by animals and diesel or gasolene engines. Some cooking and lighting is by kerosene. It should be recognised that this situation is unlikely to be changed for the next two decades; however, because of the need to increase energy availability and also to find alternatives to the rapidly decreasing wood supplies in many areas, it is necessary that a vigorous programme researching into alternative renewable energies should be set up immediately. There should be much more realism in the formulation of such programmes, for example, it is no use providing a solar powered pump at a price competitive with a diesel for someone who cannot ever afford a diesel engine. There seems to be a general feeling that the developing countries should concentrate on the renewable fuels. This is undoubtedly right in the long term, but not necessarily so for the short term. There may well be an argument for the provision of kerosene at a subsidised price for lighting and cooking purposes; while such a policy would increase world consumption of scarce oil the amount is quite small compared with that used in the developed countries and could well be supplied by reasonable economies by the latter. The urgent problem is to increase the energy available per capita to the very poor since this is necessary if they are to rise above their present level of extreme poverty. In the short term the West should be moving more into alternative energy since they have the resources and the capital to support such a programme. Another area in which energy availability could be helped in the developing countries, where firewood has become scarce, is by improvements to traditional cookers and ovens to raise the efficiency of fuel burning, and by planting fast growing trees to provide a constant fuel supply. Although unspectacular this could well prove to be an important contribution to the problem of rural energy and might be instituted quickly at little cost.

1.8.2 Alternative renewable energy sources for rice production in Thailand
Returning to the problem of replacing fossil fuel by renewable energies, it is necessary first to identify the energy needs in the rural areas. Reference 15 describes such a study carried out in Thailand. Fig. 1.8 shows the use of energy in Thai rice production, the staple food and major export of the country. Fertiliser is seen to be the single largest energy cost item. In the long term, land area availability will set an upper limit to total rice production. Thailand, with its high solar radiation and good water supply, is well placed to exploit biomass production and the question is 'how can this be done without increased demands on imported fossil fuel? There are three possible solutions to the problem.

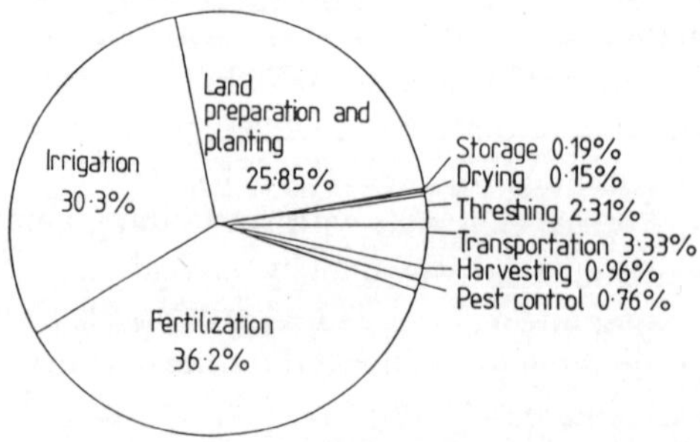

Fig. 1.8 *Proportion of current use of energy to produce rice (on farm, wet and dry seasons) (Ref. 15)*
Transitional system. Energy input 14551 MJ/ha.
Rice output 5560 kg/ha.

1 Produce with greater energy efficiency.
2 Reduce losses.
3 Substitute renewable for non-renewable energy sources.

It is not obvious where energy savings can be made, although there may be such possibilities

Reductions of fossil energy required for land preparation may be possible in the following ways.

(i) The use of power on the farm can be reduced by minimum cultivation. This can be applied to the minimum ploughing practices and techniques for land preparation on the rice field.

(ii) Economies are also possible in terms of better use of machinery by regular maintenance and skill in operation. Most of the fuel is used by tractors and a 10–15% reduction in fuel consumption can be achieved by correct adjustment and maintenance of equipment.

(iii) Use of vegetable oil as an alternative to diesel oil and gasohol as an alternative to petrol.

Losses and wastage occur at all stages of rice production both on-farm and off-farm. An estimate of annual total post-harvested losses in Thailand is about 35% of the total rice production. 10% of this occurs in drying (spoilage by birds, insects, dust). This could, for example, be reduced by using simple solar dryers.

As far as substitution of non-renewable by renewable energy sources is concerned, since fertiliser incurs the major energy subsidy, nitrogen fixation by Azolla could seem to offer an immediate and important contribution.

Oil for irrigation is another major energy subsidy. Wind pumps look very suitable for irrigation, both technically and economically. They are not used sufficiently widely at present and further work is required to produce optimised designs suitable for local manufacture. A further difficulty which is quite common in the encouragement of greater use of renewable energy sources is the high capital cost compared with the diesel engine: low interest loan arrangements could help appreciably here. Spark ignition and compression ignition engines are themselves cheap and readily available and the provision of fuel from biomass (ethanol, vegetable oils and biogas) could have an important role to play.

Because of the high percentage of diffuse radiation, concentrating collectors are not suitable for use in Thailand, but photovoltaic convertors with electric motors are possible.

Milling by diesel engine (small-scale rice mill) required only 0·13 MJ/kg of rice; nevertheless, diesel oil required in this process would be saved if the rice husk was used as the prime energy source for a small steam engine. The rice husk is an available byproduct of milling and free for the rice mill. However, the economics of energy replacement in this way need to be further investigated.

Perhaps the most important energy problem identified in this study is the use of charcoal for cooking. In quantity it represents 50% of the total Thai energy use and is causing a severe and growing deforestation problem. An experimental study showed that the traditional design of the Thai stove performs its functions very well and is insensitive to dimensional change. The technical improvements considered, although possible, do not justify the additional complexity or cost. Economic operation of the stove is now understood and practised within Thailand. However, the conversion process of wood to charcoal is generally inefficient. For these reasons it is recommended that the direction for improvement should probably be, in the first instance, improvement in the efficiency of the wood to charcoal conversion process. As a longer term possibility the use of biogas generation for cooking using waste vegetation and livestock dung should be investigated.

The study made the following recommendations for a future programme:

1 Assessment of the biomass-to-fuel potential (ethanol, vegetable oils, biogas) particularly for tapioca, sugar cane, peanuts, sunflower, vegetable wastes and livestock dung.
2 Investigation of the use of these fuels in internal combustion engines.
3 Development and use of nitrogen fixation.
4 Development of solar dryers.
5 Development of wind-driven pumping systems.
6 Consideration of funding/loan arrangements for renewable energy convertors.
7 The development of wood–charcoal convertors of higher efficiency than the traditional convertor.
8 Development of biogas generators/cookers at community level. This problem is more sociological than technical.
9 There is a need for further work on the effect of combined fertiliser and irrigation on yields to enable a better optimisation of energy requirement to be established.

Although the details will be different for different areas, this study describes a situation which is typical of rural areas in developing countries.

The scientific and technical basis of these and other conversion systems are described and explained in the subsequent chapters of this book, together with an indication of the current stage of development and performance.

References

1 'United Nations Statistical Year Book' (1981)
2 'BP Statistical Review of World Energy, 1981' (British Petroleum Co., London, 1982)
3 HALL, D. O.: 'The present status of solar energy for fuel', UNESCO 2nd. International Forum on Fundamental World Energy Problems.
4 STARR, C. and FIELD, S.: 'Economic growth, employment and energy', *Energy Policy*, 1979, **7**, 1.

5 HUMPHREY, W. S. and STANISLAW, J.: 'Economic growth and energy consumption in the UK 1700–1975', *ibid.*

6 'The global 200 report to the President', (Penguin, 1982)

7 'The energy situation in Thailand' (National Energy Authority, Bangkok, Thailand, 1976)

8 OBAYAZID, O. M.: 'Prospects of fuel and energy in the Sudan', Council for Scientific and Technological Research, C.S.T.R./12/D/3/8, Khartoum, 1975.

9 'Digest of UK energy statistics' (Department of Energy, HMSO, 1979)

10 'Electrical power in Asia and the Pacific 1975/1976' (ESCAP UN, New York, 1978)

11 Road Research Laboratory Report. Crowmarsh.

12 'Making the most of materials', Science Research Council Report, SRC, Swindon, 1979

13 'Energy for world agriculture' (FAO, Rome, 1979)

14 BAGUANT, J.: Energy and economic development strategies system studies for LDC's, Dissertation for PhD, University of Tennessee, Knoxville, June 1980

15 DUNN, P. D., SAMOOTSAKORN, Poonsab, TUNTAWIROON, Nart, and SPEDDING, C. R. W.: 'Alternative renewable energy resources for rice production in Thailand', *Agricultural Systems*, 1983, **11**, pp. 195–209

Chapter 2

Energy resources

2.1 Introduction

The earth has a continuous power input of $1{\cdot}73 \times 10^{14}$ kW from the sun – an annual energy 'income' of 1.5×10^{18} kWh ($1{\cdot}9 \times 10^{14}$ t.c.e.). This is a figure so much larger than the magnitudes to which we are accustomed that it is difficult to appreciate its significance. It is in fact about ten thousand times larger than the current annual world energy consumption. Our first reaction to solar energy is, therefore, that there is a lot of it. We note also that it is free. Unfortunately, it is spread over a large area and the peak power density is only 1 kW/m^2 at the earth's surface at noon in the tropics. Power density also varies considerably with latitude, season and, of course, time of day. Solar power is of particular interest to the developing countries, since most of these lie in latitudes which receive the most solar energy. In fact 80% of the world's population lives between the latitudes 35°N and 35°S with 3 000 to 4 000 hours of sunlight per year. In energy terms this is around 2 000 kWh/m^2 year (0·25 t.c.e./m^2 year). Power density is relatively high and in these low latitudes there is little seasonal variation.

Nature's way of harnessing solar energy is by means of photosynthesis in plants, and it has been pointed out in the previous chapter how important biomass is as a world fuel, and particularly so in the developing countries. Only about 0·1% of the solar radiation arriving at the earth's land surface is used in photosynthesis, but this is nevertheless a very large amount of energy and is roughly ten times present world energy use. However, by far the greater fraction of biomass returns its energy to the atmosphere, partly by respiration of the living plant and by oxidation of the carbon when the plant dies and rots. A very small fraction of the carbon fixed by photosynthesis is used to form fossil sediments which eventually transform to the fossil fuels, coal, oil and natural gas. This fraction is less than one part in a million of the carbon fixed by photosynthesis per year. Although the amount of fossil fuel formed annually is small it has built up reserves over very long periods of time, and these reserves are now very substantial. It is these fossil fuel reserves which satisfy almost all our present energy needs, particularly in the developed countries.* When we use fossil fuel we are in effect drawing on our energy capital, and although the reserves are large they are not inexhaustible, and in particular oil reserves are not able to sustain the present rate of growth in demand for more than a few decades. (The extent of the fossil fuel reserves are discussed later in this chapter.)

In addition to biomass there are other indirect forms of solar energy which are of interest. These include wind energy, waves, which themselves are due to the interaction between the wind and the surface of the oceans, and hydro power. There are also other forms of indirect solar energy, such as ocean temperature gradients, which might prove useful in the long term. Wind, wave and hydro power potential are discussed in later sections of this chapter.

So far energy income in the form of indirect and direct solar power, and also the fossil fuels, or energy capital, which has arisen from stored biomass has been considered. There are several other important energy sources which should not be overlooked. These are nuclear fuels, geothermal power and tidal power.

There are two principal sources of nuclear energy, one which arises from the splitting of the nuclei of heavy atoms, that is nuclear fission, and a second from the joining together or fusion of the nuclei of light atoms, nuclear fusion. The current generation of nuclear power stations are based on nuclear fission and use the lighter isotope of uranium, U235, as their fuel; these reactors are known as nuclear burners. The later generation of reactors will convert the heavy isotope of uranium, U238, to a fissile fuel and are known as breeders. Fission power is now available and many countries have considerable programmes for building power stations. Fusion in usable form has not yet been demonstrated although one should not forget that the sun itself is a fusion reactor.

Geothermal energy is principally due to the decay of radioactive elements in the earth's core. So far exploitation has been confined to a few areas of the world where the geothermal energy appears in the form of hot water or steam. However, very large amounts of energy exist and a number of programmes are now in hand to assess its more general use.

Finally there is tidal energy, which arises from the potential and kinetic energy of the earth moon system. Tidal energy is only of interest in a few special geographic situations but is of local importance in these cases.

Since the surface temperature of the earth is constant it must be losing as much energy as it receives. The only way in which heat can be lost in the vacuum of space is by thermal radiation. Fig. 2.1 shows the overall energy balance.

* The energy stored in biomas (over 90% in trees) is roughly equal to the proven fossil fuel reserves.[1]

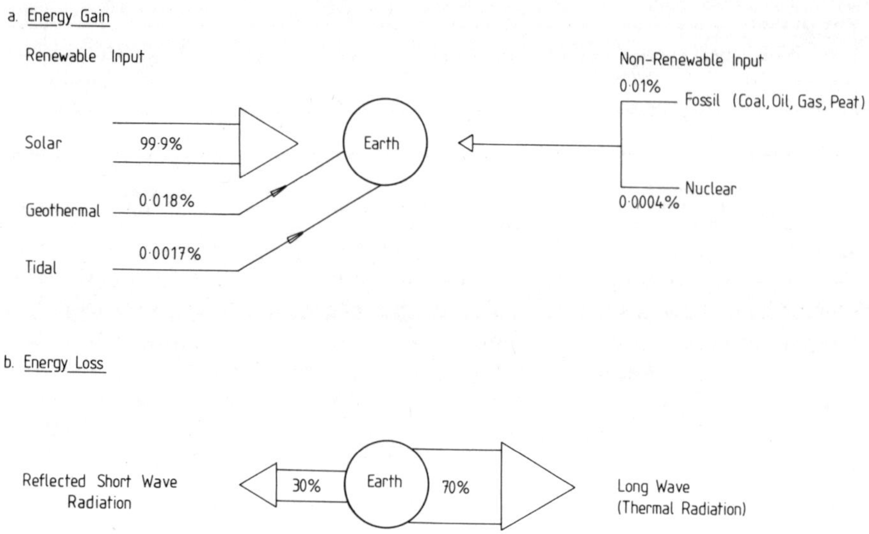

Fig. 2.1 *Overall energy balance of the earth*
a Energy gain
b Energy loss

2.2 Direct solar energy

2.2.1 Introduction

In considering the use of solar energy[2] it is necessary to know the energy falling on unit area of the earth's surface per year, and also the seasonal and other variations must be taken into account. We will now look at some of the more important factors which affect the distribution of solar energy over the earth's surface. The surface temperature of the sun is around 5 800 K and the radiation corresponding to this temperature can be roughly divided into high frequency, visible and low frequency, see Table 1.1.

Energy is continuously released from the sun by a fusion reaction which produces $3{\cdot}94 \times 10^{23}$ kW of power. Radiation from the sun takes about 9⅓ min to cover the 93 million miles to the earth. The earth receives only a small fraction of the total energy emitted by the sun, an amount $1{\cdot}73 \times 10^{14}$ kW or 340 W/m^2 averaged over the whole earth's surface.

The average intensity or power/unit area falling on a surface of unit area placed at the earth's outer atmosphere and at right angles to the incident radiation is called the solar constant. In SI units* the solar constant is 1·36 kW/m^2.

* The SI unit has now displaced the earlier unit for the solar constant which was Langleys/minute. The Langley is a unit of energy density defined as 1 g calorie per square centimetre. In these units the solar constant is 2 Ly/min.

Table 1.1

Frequency	Radiation type	Wavelength* μm	Fraction of energy
High frequency	Ultra-violet	0·2–0·4	9%
Visible frequency	Light	0·4–0·7	41%
Low frequency	Infra-red	0·7–3	50%

* Wavelengths of light are often expressed in micrometres (or microns) (μm). The micrometre is one-millionth part of a metre, 10^{-6} m.

2.2.2 Isolation at the outer atmosphere

If the earth is represented as a sphere situated as in Fig. 2.2*a* then at the equator a horizontal surface at a point immediately under the sun would receive 1·36 kW/m^2 continuously. Horizontal surfaces on the same longitude but different latitudes would receive correspondingly less, as shown in Fig. 2.2*b*. If the earth were to rotate about an axis vertical to the earth's sun plane then all points at the same latitude would receive the same average radiation. This would be a maximum for points at the equator and zero for points at the pole, see Fig. 2.3*a*.

a.

b.

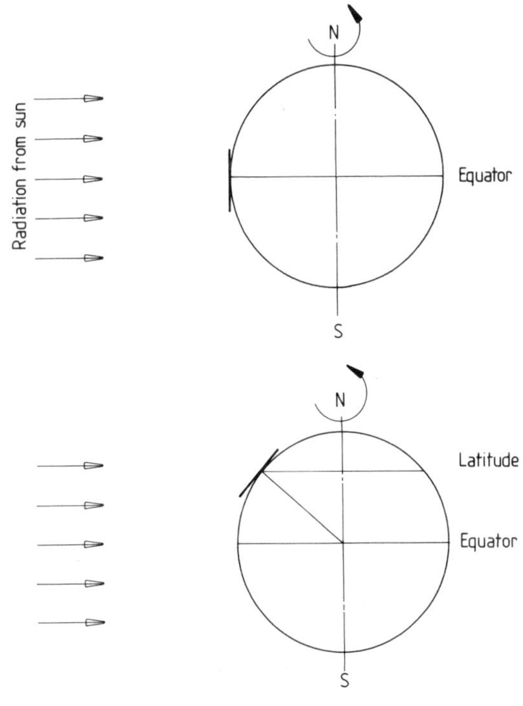

Fig. 2.2 *Effect of latitude on solar radiation intensity*

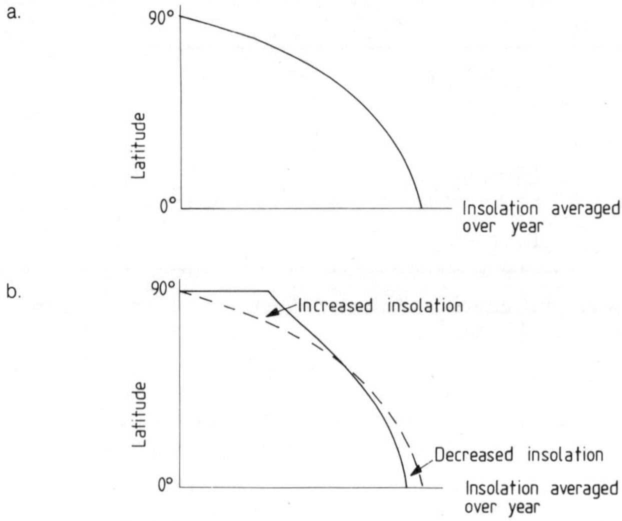

Fig. 2.3 *Variation of insolation with latitude*
a Insolation at different latitudes assuming earth's axis of rotation normal to orbital plane around the sun
b Effect of earth's declination

In fact the earth does not rotate about an axis which is perpendicular to the plane of the orbit around the sun, but instead rotates about an axis which is inclined at an angle of 23½° to it, see Figs 2.4*a* and *b*. This modifies the solar radiation as shown in Fig. 2.3*b*. The effect of this modification is to give rise to a seasonal variation of solar radiation as shown in Fig. 2.5. Surprisingly, more radiation falls on the polar regions in the summer than at the equator. An

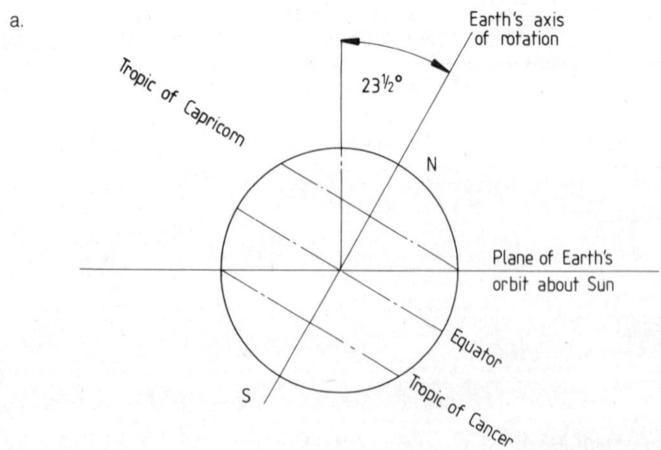

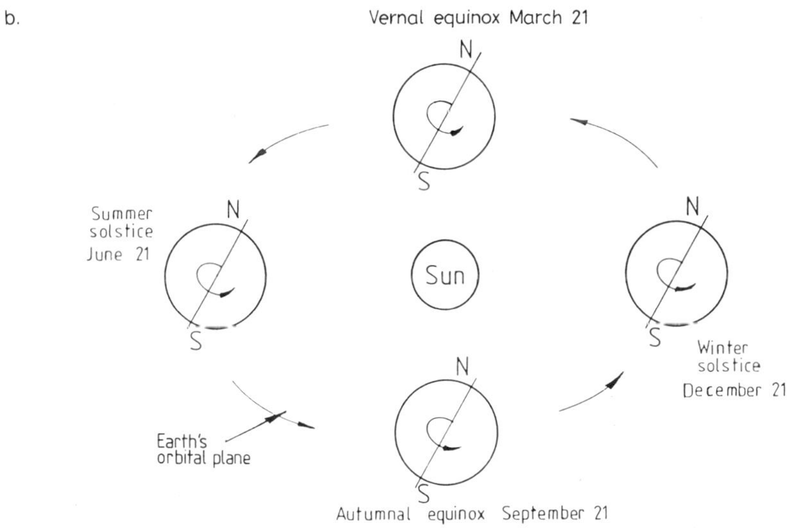

Fig. 2.4 *Rotation of the earth about the sun*
The sun can never be directly overhead for latitudes greater than 23½°N (Tropic of Capricorn) and greater than 23½°S (Tropic of Cancer)
Twice in a year the earth's axis of rotation will be at right angles to the sun's rays. The length of daylight will equal the length of night (12 hour each). These times are called equinoxes

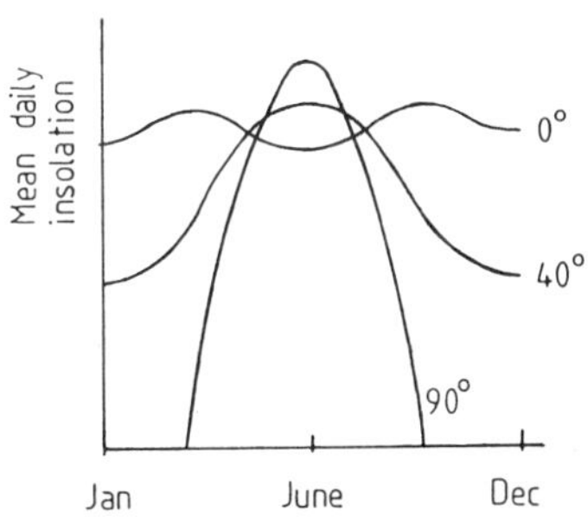

Fig. 2.5 *Variation of insolation with season for various latitudes in the northern hemisphere*

important feature is the absence of seasons at the tropics and the extremes of six month summer and six month winter at the poles. Another feature of significance is the 12 hour day, 12 hour night in the tropics compared to the short night/long day summer cycle in temperate areas and the reverse in winter, see Fig. 2.6.

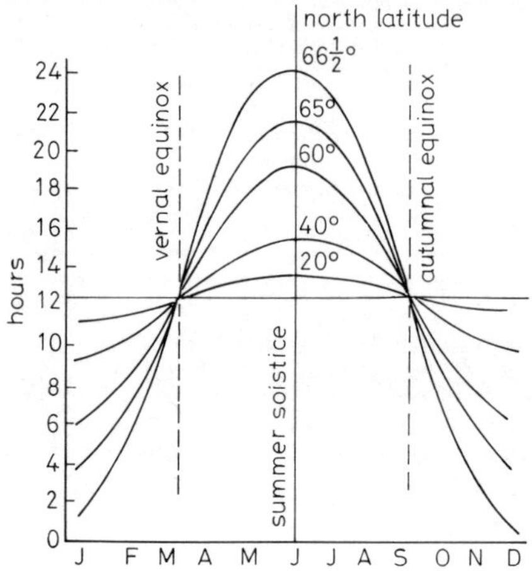

Fig. 2.6 *Duration of the day at various latitudes throughout the year. Vertical scale gives number of hours the sun is above the horizon*

2.2.3 Effect of the earth's atmosphere

As might be expected the earth's atmosphere considerably modifies the simple model given in the previous section. The length of path through the atmosphere will depend on the angle that the sun makes with the vertical at a particular position on the earth's surface, see Fig. 2.7.

Some of the incident radiation is directly reflected by the earth's atmosphere, the remainder is absorbed either by the atmosphere or the earth's surface.

Above 150 km the incident radiation is not affected by the atmosphere: by 88 km the X-rays and some of the ultra-violet have been absorbed. Between 18 and 15 km the greater part of the ultra-violet has been removed. As the radiation approaches the earth's surface, gas molecules in the air cause scattering, particularly at the shorter wavelength end of the spectrum. This effect gives rise to the blue sky, and the red appearance of the sun. The latter is particularly pronounced at sunset and sunrise owing to the increased path length through the air. Clouds and dust also cause scattering and absorption. The overall energy balance is shown in Fig. 2.8.

It is seen that about 30% of the incident solar radiation is reflected by the atmosphere, a further 20% is absorbed on passing through the atmosphere and the remaining 50% arrives at the earth's surface, where 2% is reflected and the remainder absorbed. An amount equal to 23% of the original solar energy incident on the outer atmosphere is used in evaporation and the remainder is lost by long wave radiation. Since the atmosphere radiates long wave radiation to the earth's surface the net long wave loss from the earth's surface will be the difference between these two heat fluxes. About 0·2% is absorbed by the

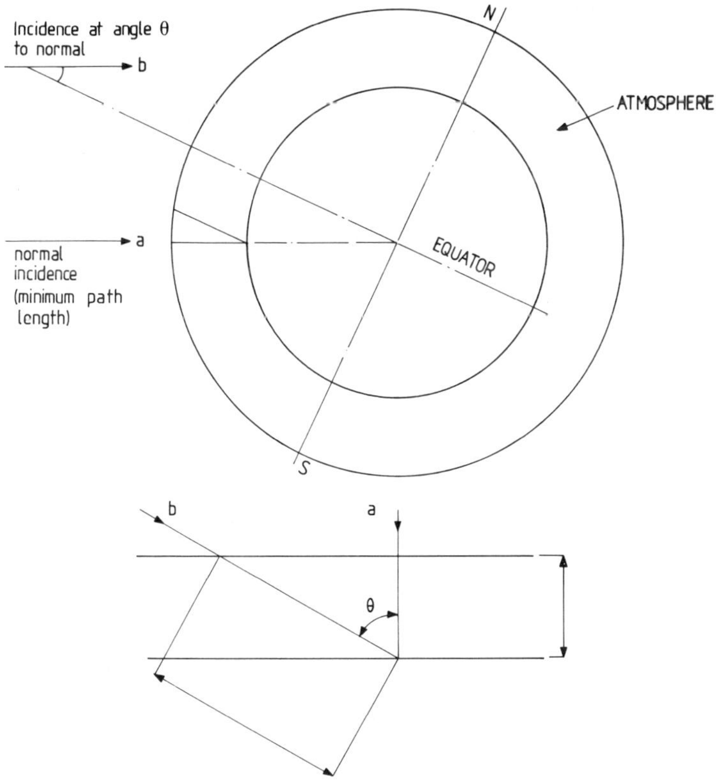

Fig. 2.7 *Variation of path length through earth's atmosphere*

atmosphere in the form of winds and ocean currents. Photosynthesis accounts for only 0.05%.[3]

Fig. 2.9 shows the annual mean irradiance on a horizontal plane at the earth's surface averaged over 24 hours. For the most favoured regions the average flux density is as high as 300 W/m^2 and the average for the tropics is about 250 W/m^2. In more temperate regions, such as most of Europe, the figure is around half this value, but with the marked seasonal variation referred to in a previous paragraph.

2.2.4 Characteristics of solar radiation

Solar radiation is made up from electromagnetic waves which travel from the sun to the earth with the speed of light c. The wavelength λ of the wave is related to the frequency ν by the equation:

$$c = \lambda \nu$$

The velocity of electromagnetic waves when transmitted through materials is less than the value c for vacuum (3×10^8 m/s). The electromagnetic spectrum is

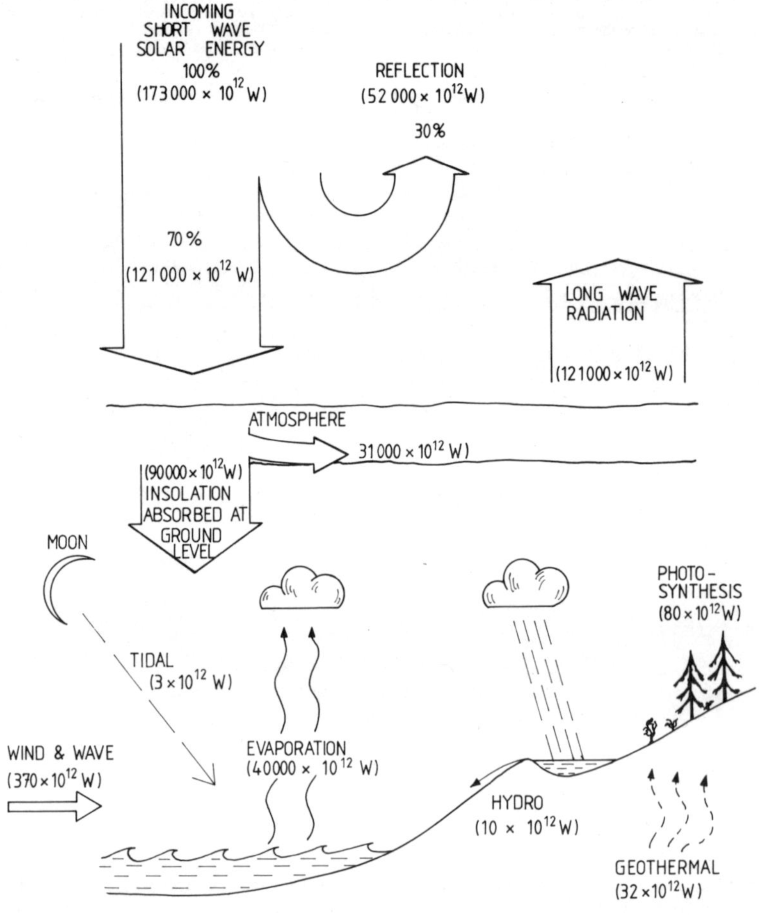

Fig. 2.8 *Detailed global energy balance*

shown in Fig. 2.10. It was Newton who first demonstrated that sunlight or 'white light' can be split up into its different component wavelengths, or colours, known as the solar spectrum, see Fig. 2.11. Newton's experiment made use of the fact that different wavelengths have slightly different velocities in glass and hence are refracted, or bent, by a prism by slightly different amounts, thus separating out the colours.

Returning to the solar spectrum, it contains wavelengths which are too long to be seen by the naked eye (the infra-red) and also wavelengths which are too short to be visible (the ultra-violet).

Table 1.1 gives a rough indication of the variation of intensity with wavelength of solar radiation. The actual variation is shown in Fig. 2.12. The vertical axis is given in units of watts per square meter per micrometre of

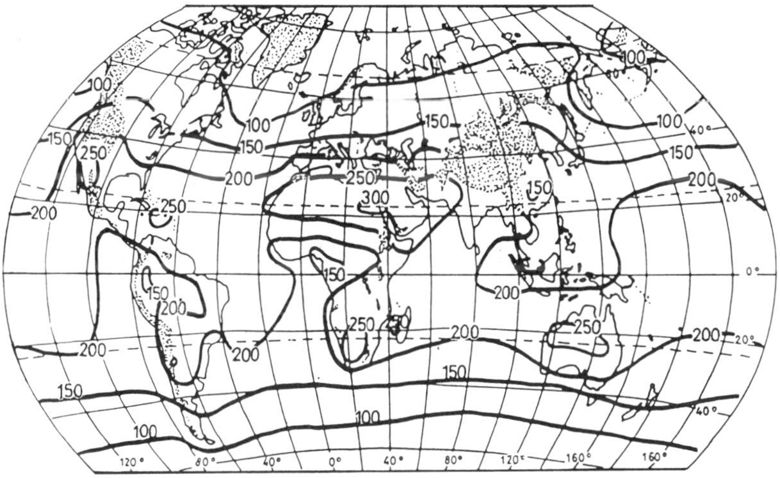

Fig. 2.9 *Annual mean global irradiance on a horizontal plane at the surface of the earth (W/m^2 averaged over 24 h) (Ref. 7. Courtesy U.K. I.S.E.S)*

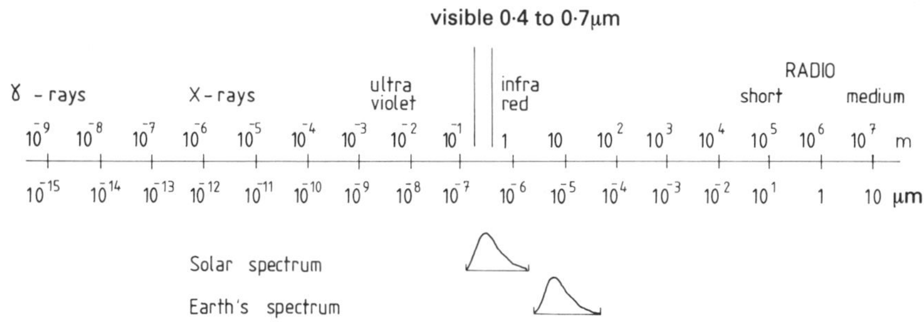

Fig. 2.10 *Electromagnetic spectrum*

wavelength; that is, it gives the power per unit area between the wavelength range of λ and $\lambda + 1$, where λ is measured in micrometers. With these co-ordinates the area under the curve gives the total power per square metre radiated by a surface at the specified temperature.

The solar spectrum is roughly what would be given by a perfect radiator (black body) at a temperature of 5 800 K. Water vapour, dust, and absorption by various molcculcs in thc air absorb certain frequencies strongly so that the spectrum received by the earth's surface is modified as shown in Fig. 2.13.

The earth itself also radiates as a black body, but at much lower temperature than the sun, around 300 K. The relation between the sun's spectrum and that of the earth is shown in Fig. 2.14. It is seen that the earth receives its radiation at short wavelength around a peak of 0.5 μm whereas it radiates to space at a

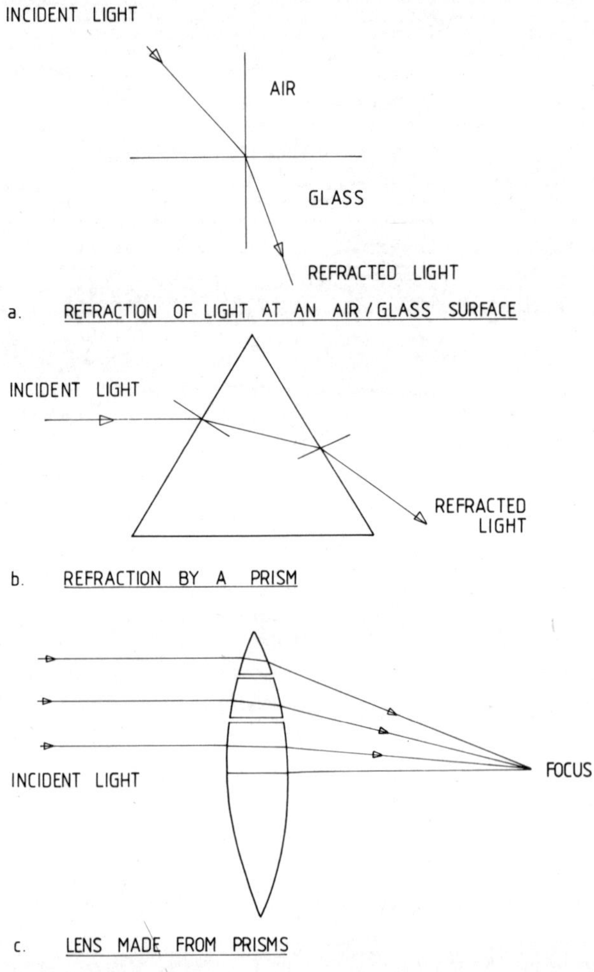

Fig. 2.11 *Refraction of light*
a Refraction of light at an air/glass surface
b Refraction by a prism
c Lens made from prisms

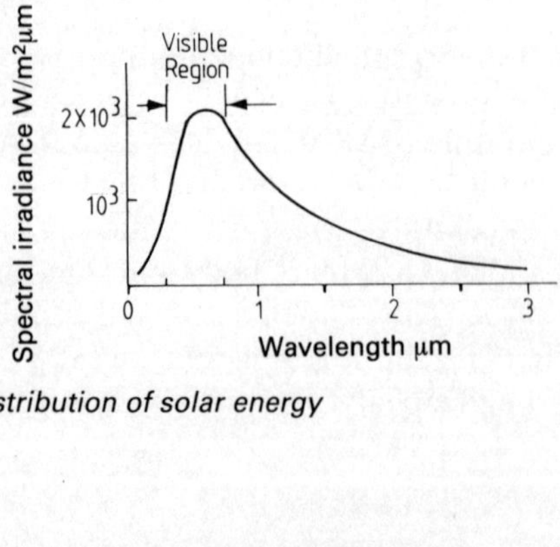

Fig. 2.12 *Spectral distribution of solar energy*

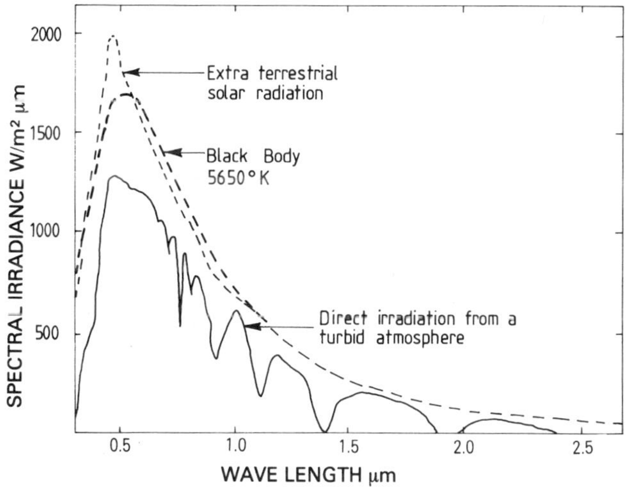

Fig. 2.13 *Solar spectrum (Ref. 7. Courtesy U.K. I.S.E.S)*

much longer wavelength around a peak value of 10 μm, which is well into the infra-red. The relationship between the wavelength λ_{max} at which the power radiated is a maximum and the temperature of the body T is given by Wien's Law, which states

$$\lambda_{max} \times T \simeq 3 \times 10^{-3} \text{ mK}$$

2.2.5 Quantum effects

In the previous paragraph's the sun's radiation was described as electromagnetic waves obeying the relationship:

$$c = \lambda\nu$$

This is quite true, but it is also found that electromagnetic waves show properties that might be expected from particles. In particular they behave as

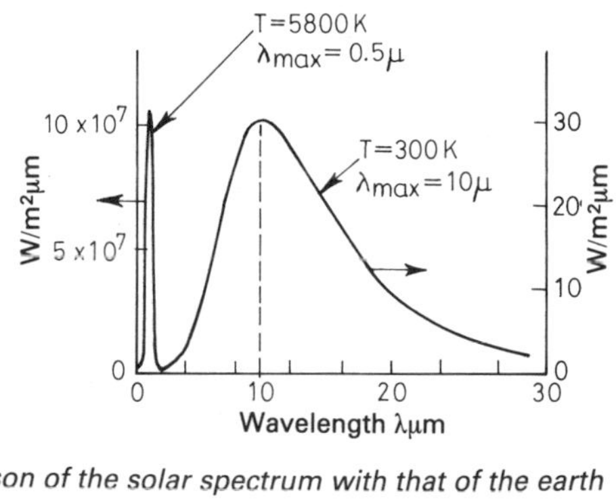

Fig. 2.14 *Comparison of the solar spectrum with that of the earth*

if they were made up of packets of energy, having an energy E which is related to frequency ν by the equation:

$$E = h\nu$$

where $h = 6{\cdot}626 \times 10^{-34}$ Js. These packets of wave energy are called photons and the equation which relates the photon energy to the frequency is called Planck's equation, after the physicist who first developed the theory.

Biomass is produced by photosynthesis and is discussed in the next section. The photosynthetic process depends on the quantum characteristics of solar radiation. The photocell also uses this mechanism.

Quantum conversion mechanisms rely on a minimum energy to remove an electron from its parent atom. The energy required to move an electron from one level of energy to another may be expressed in a diagram of the form shown in Fig. 2.15. This diagram shows the energy gap E_g that the photon must

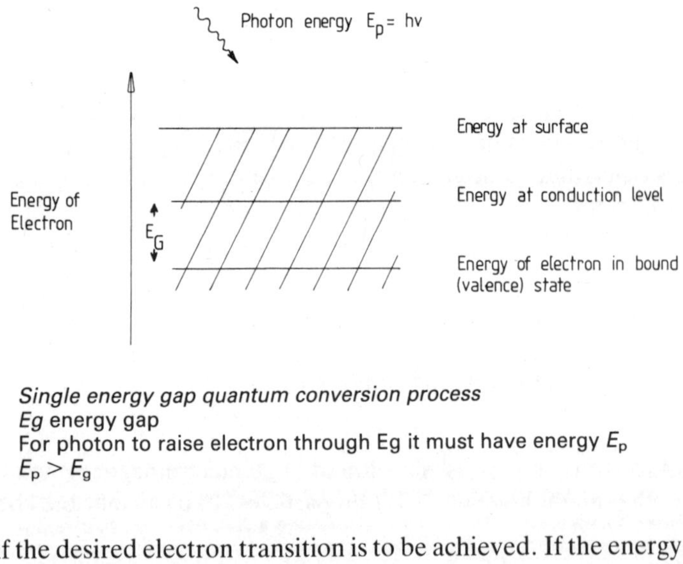

Fig. 2.15 *Single energy gap quantum conversion process*
Eg energy gap
For photon to raise electron through Eg it must have energy E_p
$E_p > E_g$

supply if the desired electron transition is to be achieved. If the energy E of the photon is less than the energy gap E_g then the effect will not occur. If the energy of the photon is exactly equal to the gap then the energy required will be available without waste. However, if the photon has a higher energy than is required then the excess energy will be wasted and appear as heat. The choice of energy gap to achieve maximum power conversion is therefore a critical factor in the choice of materials for photovoltaic convertors. If the energy gap is small then a large number of photons will have sufficient energy to raise an electron through the gap, but since the total energy converted is given by the number of photons multiplied by the energy gap the total energy will be smaller. Similarly, if the energy gap is large only a few photons will have sufficient energy to cause the transition so the total energy converted will again

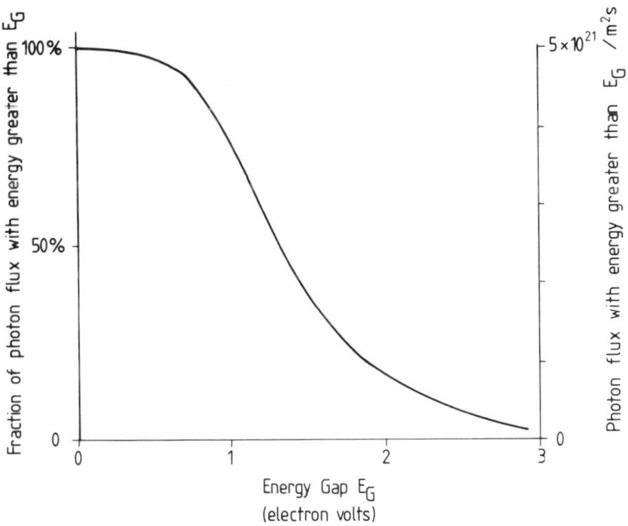

Fig. 2.16 *Fraction of total photon flux with energy greater than E_g*

be small. Fig. 2.16 shows how the number of photons in solar radiation, having energy greater than an amount E_g, varies as E_g increases. There will be an optimum value of E_g to give maximum conversion efficiency and this is shown in Fig. 2.17. Fig. 2.17 shows how the theoretical efficiency, by which solar energy can be converted, varies with energy gap.[4,5] The maximum achievable efficiency is seen to be about 50%. The photovoltaic cell operates in this way, however, other losses occur and the theoretical maximum efficiency of a photocell is around half that for the ideal convertor of Fig. 2.17. In Fig. 2.18

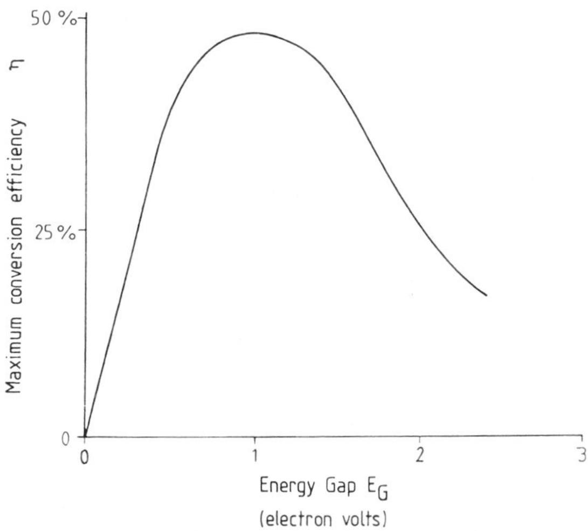

Fig. 2.17 *Maximum conversion efficiency as a function of energy gap*

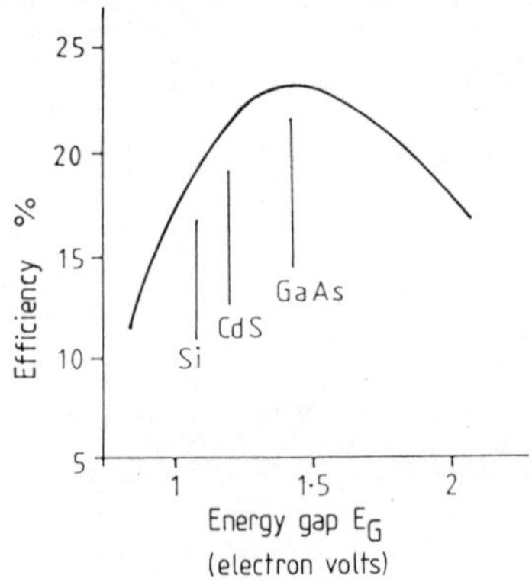

Fig. 2.18 *Theoretical maximum efficiency of photovoltaic conversion*
Si: silicon, Eg = 1·12 eV
CdS: cadmium sulphide, Eg = 1.2 eV
GaAs: gallium arsenide, Eg = 1·35 eV

the energy gaps of a number of semiconducting materials are indicated on the efficiency diagram. The energy gap of silicon is somewhat less than is required for optimum conversion efficiency but there are practical reasons for its widespread use in photo cells. With the energy gap for silicon the maximum efficiency to be expected theoretically is about 20% and in practice the figure for commercially produced cells is around 12%. While energy gaps can be expressed in the SI unit the joule, in practical materials the gap is very small and it is more usual to use the electron volt as the unit. The electron volt* is the energy acquired by an electron in falling through a potential difference of 1 V; in these units energy gaps are of the order of 1 or 2 electron volts. Photosynthesis also involves energy gap transitions but the actual mechanism is more complicated than the simple single transition discussed here and involves several transitions.

2.3 Biomass energy

2.3.1 Introduction

The conversion of solar energy to stored energy in the form of biomass is essential to human life. All of our food is produced in this way, almost all our fuel (either in the form of stored biomass as fossil fuel or as vegetation), and much of our material for industry and building (more wood than steel is used even in the developed countries). We have seen in Chapter 1 that vegetation is an important fuel even in world terms, and, if one considers numbers of people rather than total energy use, undoubtedly a greater number of people depend

* 1 eV = $1{\cdot}6 \times 10^{-19}$J.

more on vegetation for their energy needs than on fossil fuels. An important characteristic of the conversion of solar energy to biomass is that the energy is converted to a form in which it can readily be stored.

In order to put the biomass production into perspective it should be noted that about 0·1% of the energy incident on the surface of the earth is converted and stored in the form of biomass. This is roughly about ten times the current world energy use.

2.3.2. Photosynthesis

In the photosynthetic process solar energy is trapped by chlorophyll in the plant and the energy used to convert carbon dioxide and water into organic material:

$$\text{carbon dioxide} + \text{water} = \text{organic material} + \text{oxygen}$$

or in chemical symbols:

$$6CO_2 + 5H_2O = C_6H_{10}O_5 + 6O_2^* \qquad (2.1)$$

The efficiency of the photosynthetic process in converting solar energy into stored energy in the resulting biomass is defined as:

$$\text{efficiency} = \frac{\text{heat of combustion of dry mass}}{\text{incident solar energy}}$$

An average figure for the heat of combustion of air dried biomass is 15×10^6 J/kg (6 000 BTU/lb). Air dried biomass has a moisture content of around 10–15%; if the material is oven dried then the heat content is increased to $17{\cdot}7 \times 10^6$ J/kg (7 500 BTU/lb). The maximum efficiency of photosynthetic conversion is theoretically about 10%. Values as high as 7·5% have been reported for sunflowers during the rapid growth period. A more usual average figure is 5·5%. Fig. 2.19 shows the losses in the photosynthetic process which

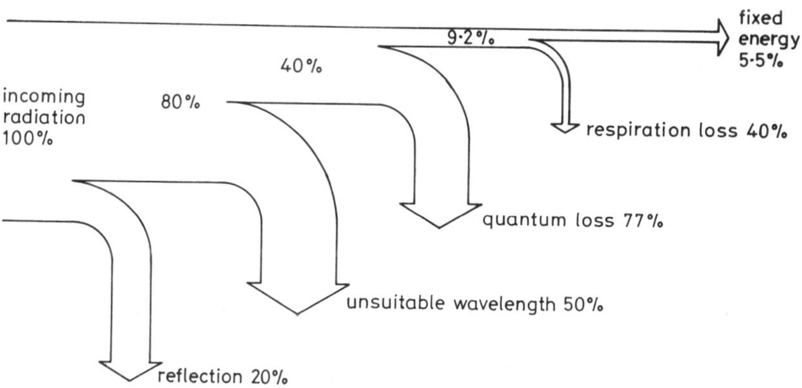

Fig. 2.19 *Photosynthetic conversion*

* $C_6H_{10}O_5$ is a molecule of cellulose, a typical carbohydrate. All carbohydrates have the chemical form $(CH_2O)n$.

lead to the figure of 5·5%. First, only light having wavelengths which lie between the limits of 0·4 μm and 0·7 μm actually takes part in the reaction. This reduces the energy availability to half that of the incident energy. 20% of this light is reflected from the plant, and of the remainder which is absorbed 77% is lost by the losses in the internal photosynthetic conversion. These losses result in around only 9·2% of the incident radiation becoming converted in the form of fixed carbon. The plant itself will use up 40% of this fixed carbon, mainly by respiration,* leaving a net energy storage of 5·5%. In practice there are other sources of loss which further reduce the overall efficiency of a field crop. These will include:

Partial cover by the crop (particularly at the seedling stage)
Inability to convert high light intensities
Shortage of water or nutrients
Damage by animals or disease

Hall[6] states that, in practice, under optimum field conditions 3 to 5% should be achieved during the growing season. For temperate areas the average photosynthetic efficiency is between 0·5 and 1·3% and sub-tropical areas 0·5 to 2·5%.

2.3.3 *Plant types*

There are several ways in which the initial process of fixing atmospheric carbon, by photosynthesis, in plants is achieved. Plants are classified as follows:

1 C_3 plants in which the initial process involves organic molecules containing three atoms of carbon. They require the following conditions:

High carbon dioxide concentration
Low light intensity (higher intensities are not converted)
Low temperature range 10 to 15 °C

C_3 plants are also characterised by high photorespiration, that is they require considerable quantities of water. Plants of this type are grown in temperate climates, wheat is a typical example. In the past not enough attention has been given to the suitability of such plants for more tropical regions and they have been introduced in areas where they are not particularly suited.

2 C_4 plants. In these plants four carbon molecules take part in the initial process. Maize is a typical example. Their characteristics include:

Low carbon dioxide requirement
Lower water requirement than C_3 plants
Ability to use high light intensity
High temperature range 20 to 25 °C
Will withstand seasonal droughts

* On respiration organic material is combined with oxygen (burnt) in order to release energy. The process is the reverse of equation 2.1.

3 CAM plants. The crassulacean acid metabolism or CAM plants are suited for arid conditions. They include the cacti. Their characteristics include:

Reduced daytime carbon dioxide requirement
Low water demand
Low growth rate

Considerable efforts are now being made to develop, by genetic engineering and other methods, plants having higher yields for use in particular areas.

The major characteristics of C_3, C_4 and CAM plants are summarised in Table 2.2.

Table 2.2 *Comparison of C_3, C_4, CAM plants*

	C_3	C_4	CAM
Typical examples	wheat, potato	sugar, maize	cacti, stonecrop
Respiration in day	yes	no	sometimes
Rate of photosynthesis	moderate	high	low
Preferred climate	temperate	hot semi-arid	arid
Water need	high	moderate	low
Photosaturation	1/5 full sunlight	no	sometimes

2.3.4 Crop yields

Table 2.3 shows the short term, dry weight yields for a number of crops and Table 2.4 gives the 'average to good' annual yields of dry matter production. These figures suggest that it should be possible to produce between 20 and 80 tonnes/hectare year which is equivalent to 3×10^{11} J and $1{\cdot}2 \times 10^{12}$ J per hectare year or approximately 10 to 40 tonnes coal equivalent per hectare year. The average solar energy falling on a square metre of area each year varies from around 100–150 W/m^2 in temperate regions to 250–300 W/m^2 in tropical regions. Since there are $31{\cdot}5 \times 10^6$ seconds in a year this means that $31{\cdot}5 \times 10^8$ J of energy falls annually on a square metre of surface for which the average insolation is 100 W/m^2. A hectare (10^4 m^2) is a more useful unit of area when considering biomass yields, hence the corresponding annual energy/hectare is $31{\cdot}5 \times 10^{12}$ J. The energy of sun-dried biomass is around 15×10^6 J/kg. If the conversion was 100% efficient the yield/hectare would be

Table 2.3 *Dry weight yields and short term photosynthetic efficiency (Reference 6)*

Crop	Country	Typical g/m² day	Photosynthesis efficiency % of total radiation
Temperate			
Tall fescue	UK	43	3·5
Rye grass	UK	28	3·5
Cock's foot	UK	40	3·3
Sugar beet	UK	31	4·3
Kale	UK	21	2·2
Barley	UK	23	1·8
Maize	UK	24	3·4
Wheat	Netherlands	18	1·7
Peas	Netherlands	20	1·9
Red clover	New Zealand	23	1·9
Maize	New Zealand	39	2·7
Maize	USA, Kentucky	40	3·4
Sub tropical			
Alfalfa	USA, California	23	1·4
Potato	USA, California	37	2·3
Pine	Australia	41	2·7
Cotton	USA, Georgia	27	2·1
Rice	South Australia	23	1·4
Sugar cane	USA, Texas	31	2·8
Sudan grass	USA, California	51	3·0
Maize	USA, California	52	2·9
Algae	USA, California	24	1·5
Tropical			
Cassava	Malaysia	18	2·0
Rice	Tanzania	17	1·7
Rice	Philippines	27	2·9
Palm oil	Malaysia (whole year)	11	1·4
Napier grass	El Salvador	39	4·2
Bullrush millet	Australia, NT	54	4·3
Sugar cane	Hawaii	37	3·8
Maize	Thailand	31	2·7

Table 2.4 *Average to good annual yields of dry matter (Reference 6)*

	Tonnes/ hectare year	g/m² day	Photosynthetic efficiency % of total radiation
Tropical			
Napier grass	88	24	1·6
Sugar cane	66	18	1·2
Reed swamp	59	16	1·1
Annual crops	30		
Perennial crops	75–80		
Rain forest	35–50		
Temperate (Europe)			
Perennial crops	29	8	1
Annual crops	22	6	0·8
Grassland	22	6	0·8
Evergreen forest	22	6	0·8
Deciduous forest	15	4	0·6
Savanna	11	3	
Desert	1	0·3	0·02

around 2 000 tonnes or 20 tonnes at a conversion efficiency of 1%. The general relation is:

$$\text{Annual yield/hectare} = \frac{20 \times (\text{percentage efficiency}) \times (\text{average annual insolation in w/m}^2)}{100} \text{ tonnes/ha year}$$

Aquatic plants should not be overlooked: for example, the water hyacinth, which grows prolifically in many of the developing countries, can yield 15 tonnes of dry matter per hectare year. Algae too offer the possibility of high yields.

2.3.5 Carbohydrates and hydrocarbons: direct production of hydrocarbons by photosynthesis

So far only the production of carbohydrates by means of photosynthesis has been discussed. Plants are mainly made up from carbohydrates in the form of cellulose $C_6H_{10}O_5$. They are formed by the combination and partial reduction of carbon dioxide and water. Reduction is the name given to the process of removing oxygen from molecules. In a carbohydrate the carbon dioxide could be regarded as completely reduced to carbon, leaving the water molecules unaffected. This is an over-simplification of the chemical reduction process but makes the point that carbohydrates are the result of only a partial reduction. If the oxygen were to be completely removed the result would be a hydrocarbon, for example (CH_2). Carbohydrates such as wood, starch and sugar are solids with a calorific value per unit mass of around 18 MJ/kg. They can be converted to a liquid, ethanol, having a calorific value per unit mass of about 30 MJ/kg, or a gas, biogas of about 20MJ/kg. Fuel oils such as diesel fuel have a calorific value per unit mass of around 45 MJ/kg or twice that of wood and with the added advantage of liquid form. Fuel oils are hydrocarbons and there are obvious advantages of growing plants which can complete the reduction process to give liquid hydrocarbons of similar specific energy to conventional liquid fuels. Vegetable oils, obtained from seeds, nuts, etc, are of this type and their use is discussed further in Chapter 8.

Rubber trees, such as the *Havea brasielieris*, a species of the genus* *Havea*, produce a hydrocarbon in the form of sap. Yields of 2½ tonnes/ha/year are possible. It is obviously of interest to consider the rubber tree and other hydrocarbon producing plants as a source of liquid fuel. The *Havea* is a genus of the Euphorbiaecae family. Another genus of the same family is *Eurphorbia* which contains more than one thousand species, all of which produce latex (a 30% solution of rubber in water). The rubber molecule has a molecular weight around one thousand times that of fuel oil and there would be considerable value in either discovering or developing a high yield plant which produces hydrocarbons of lower molecular weight than the rubber tree. Plants which produce a suitable product for internal combustion engine use exist although work on them is still at an early stage.

2.3.6 Energy inputs

The energy requirements in the form of fertiliser and other inputs should not be forgotten when considering the use of biomass as an energy source. Traditional agriculture is extremely economic in energy, for example, the ratio of the energy value of rice grown by traditional methods to the energy required for its cultivation is between 20 and 40.† However, the yield per hectare for such agriculture tends to be low and developments in modern agriculture over the past few decades have been concentrated on raising yield with striking success. This success has been at the expense of the energy input and western food now typically requires three to five times its energy content for its production. This aspect is of particular relevance when considering biomass

* Botanists classify plants in families which are divided into genus. The genus in turn is subdivided into species.

† The energy ratio is defined:

$$ER = \frac{\text{energy in food}}{\text{energy input}}$$

In calculating energy input, solar energy is not included, neither is the total food bill for the farmer and his family. If the latter is included the ER is considerably decreased.

for energy use. One important contribution which can be made to reduce the energy input is the development of nitrogen fixing varieties which will not require the addition of a nitrogen fertiliser. Either plants having this property can be used or they can be grown in association with other plants having this property. An interesting and important example of this is the use of a small plant Azolla in association with rice in wet or paddy rice growing. The Azolla has a symbiotic relationship with a blue-green algae which fixes nitrogen.

2.4 Stored biomass: the fossil fuels

2.4.1 Characteristics of fossil fuels

2.4.1.1 Solid fuels: peat, lignite and coal: The solid fossil fuels have been formed by vegetation from the primeval forests. This decaying vegetation has, over many years, built up very large masses of material. The initial material, mainly wood, has been changed in composition initially by fermentation and later by pressure and temperature increases of geological origin. The formation of peat is the first stage in the process, and very large peat deposits exist in many parts of the world. Further stages in the conversion process lead, in turn, to lignite, bituminous coal and finally anthracite. The sequence is primarily the reduction in the oxygen content of the original woody material.

It is, however, a smooth transition and the commonly used names such as bituminous coal are simply a convenient method of describing a stage in the process rather than having a precise meaning, see Table 2.5. Coals are often referred to by 'rank'. Rank increases as the oxygen content is decreased.

Table 2.5 *Composition of wood and wood derived fossil fuels**

Material	Carbon C	Hydrogen H	Oxygen O	GJ/tonne
	(% by weight on dry ash free basis)			
Wood	50	6	42·5	
Peat	57·2	6	35	
Lignite	68	5	23	15
Bituminous coal	82·3	5·5	9·2	29
Anthracite	94·6	3·3	1·3	33

*The percentages refer to the analysis of typical materials from each group.

2.4.1.2 Liquid and gaseous fossil fuels: oil and natural gas: Unlike the solid fossil fuels whose origin is wood from the early forests, oil and natural gas are thought to be derived from the remains of small marine creatures which have accumulated in porous rocks. Oil and natural gas often occur together, natural gas consists very largely of methane, CH_4. The composition of crude oil depends on its source and oils are usually classified under three main headings:

(i) Paraffin type base. The crude oil consists mainly of the saturated hydrocarbons ranging from methane to the waxes. The distillation residue is solid paraffin wax.

(ii) Asphaltic type base. The principal components of this type are not paraffinic but contain either considerable aromatic hydrocarbons or napthenes. The oils also contain some paraffins. The distillation residue is asphalt or bitumen.

(iii) Mixed type base. This category contains both (i) and (ii) and may be regarded as intermediate between them. Wax appears in the distillates but the residues are asphalt.

Crude oil is mainly a mixture of hydrocarbons but there are also smaller quantities of elements of which sulphur is the most important. When crude oil is first extracted it is accompanied by water and sand which must be removed before the oil can be used. The oil itself is a mixture of many hydrocarbons and it is necessary to separate it into a range of components, each of which has particular commercial applications. This separation is carried out by fractional distillation. In fractional distillation the oil is heated to successively higher temperatures and the components of different volatility separated out. The major classes are shown in Table 2.6.

Table 2.6 *Typical composition of a barrel of crude oil*

	Distillation Temperature range, °C	%
LPG	—	5
Gasolene	40–180	23
Kerosene	140–250	20
Diesel	200–320	
Light fuel oil	250–340	52
Heavy fuel oil	>320	

The typical percentage content of a barrel of crude oil is shown in the third column of Table 2.6. This 'straight run' distillation, as it is called, may not give the proportions required by the market. The percentage of lighter oils which are required for transport is usually higher than that obtained by simple distillation. The percentage of light oil in the barrel can be increased by breaking down the heavier molecules to increase the lighter fraction, a process known as 'cracking'.

In the first stage of the distillation process hydrocarbons such as propane, C_3H_8, and butane, C_4H_{10}, are released. These materials are gases at normal temperatures and pressures but can readily be liquified by storing under pressure. Liquified petroleum gas, or LPG as it is known, is an important source of commercial energy. It is sold for the domestic market as commercial butane or Calor gas, which is largely butane, and for the industrial market primarily in the form of propane.

Natural gas consists mainly of methane but also contains other gaseous hydrocarbons such as butane and propane. The composition varies with the source, some natural gas contains significant (14%) amounts of nitrogen. Unlike LPG, natural gas cannot be liquified at normal temperatures and if a liquid is required it must be cooled to a very low temperature (pure methane has a boiling point of $-161{\cdot}5\,°C$ at atmospheric pressure). Natural gas is converted to liquid form, LNG, in this way for transport in tankers. The alternative transport method is in gasesous form in pipe lines.

2.4.2. *Fossil fuel reserves*

The fossil fuel reserves are derived from residues from vegetation and marine creatures which have been stored for many millions of years. The process is still taking place today but the rate of energy storage is small. It is estimated that only 0·0001% of the annual biomass production is stored in this way, and this represents only around 0·001% of our current world energy use. Therefore fossil fuel has to be regarded as a non-renewable resource. Since fossil fuel provides most of the world's commercial energy supply it is important to know how long this may be expected to continue. The conventional wisdom on the subject seems to suggest that oil/natural gas will only last for a few decades* whereas coal might be expected to continue as a major energy source for several hundred years. We should look into the underlying assumptions in a little more detail. The time that a resource will last depends on two factors. First, how much there is of the resource that can be economically extracted, and secondly, how fast it is being used up. Unfortunately there is considerable uncertainty in estimates for both of these quantitites. One such estimate for the ultimately recoverable fossil fuel energy is given in Table 2.7.[9] These figures include both the proven recoverable reserves, that is those which can be economically worked using current practice, and estimates for future discoveries.

The second factor in the equation is the rate of demand. At present the average commercial energy use per capita is around 2 t.c.e. per annum, but as has been seen this is made up from a highly inequitable distribution. Estimates

* Estimates for oil and gas reserves are frequently given in terms of proved reserves. These reserves represent known reserves which can be extracted economically at prevailing cost and price levels. These figures are considerably less than the estimates for ultimately recoverable reserves given in Table 2.7. It is interesting to note that at current rates of use at the time of estimate proved oil and gas reserves are usually adequate to maintain supplies for 20–40 years. For example, in 1950 the supply on this basis was 22 years, 1960 was 37 years and 1970 was 34 years.

Table 2.7 *Ultimately recoverable world non-renewable energy resources (based on data in Reference 9)*

	$J \times 10^{22}$	t.c.e. $\times 10^{11}$	%
Oil	1·0	3·5	6
Natural gas	0·9	3·1	5.4
coal	12·7	44	75
Shale oil	2·1	7·3	12·5
Uranium*	0·2	0·7	1·2
Total	17·0	58·7	100

*Based on light water reactor technology.
World energy consumption 1976 = $2{\cdot}64 \times 10^{20}$J ($9 \times 10^9$ t.c.e.).

for fossil fuel use will vary as a function of percentage rise in consumption per year. If world energy use were to remain at 1976 levels (9×10^9 t.c.e.) then the 60×11^{11} t.c.e. would be expected to last about 600 years. At 2% annual increase this falls to 133 years and at 5% the time is reduced to 70 years. This rate of rise will be affected by many factors. There is a limit to the amount of energy that will be required by any one person. Currently in the USA the figure is 12 t.c.e. per annum and hopefully it can be kept to this value by conservation measures and in the longer term even be reduced somewhat without lowering of the quality of life. At the other end of the scale, where the use of commercial energy is as low as 0·1 t.c.e. per annum, clearly this must be raised. An average world figure to aim for might be around 4 t.c.e. per capita per year. This figure, taken with the rise in world population, suggests that the world energy demand should rise by a factor of five. Hopefully the very poor will be able to make increasing use of the renewable energy resources and this will reduce the rise in demand for the non-renewables. It should however be stated quite firmly that in the short term the best way to raise the energy per capita use in the poorer areas of the world is by increasing the availability of fossil fuel. It is the developed countries who, by means of conservation programmes and investment in renewable energy, should make more of the fossil fuel available for this purpose.

2.5 Wind energy

There are two principal reasons for the movement and direction of motion of the earth's atmosphere; the unequal amounts of solar energy received at different latitudes and the rotation of the earth. Superimposed on the global atmospheric movement or wind systems due to these two factors are modifications arising from local disturbances, such as the presence of large continental land masses. Fig. 2.3*b* indicates how the solar energy falling on the outer atmosphere varies with latitude. The resulting energy unbalance between radiation loss and the input energy is given in Fig. 2.20. The band lying roughly 35°N and 35°S receives more solar energy than it radiates away to space, whereas the remaining area radiates more energy than it received from the sun. The balance is restored partly by a flow of heated air from the equator to the poles and partly by a similar flow of the ocean currents. The wind, which includes the latent heat of the evaporated water, accounts for about 87% of the total heat transport.

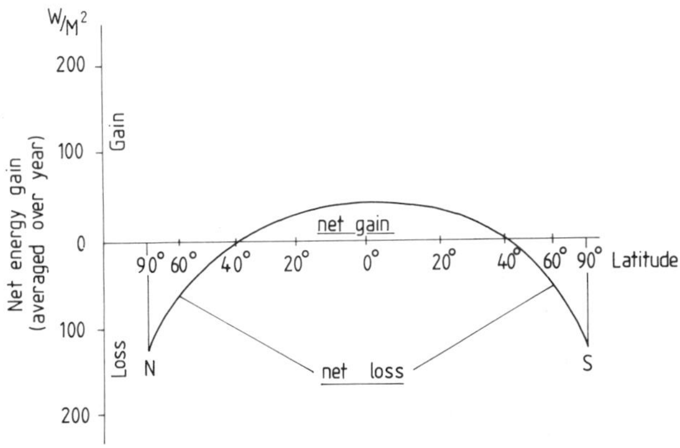

Fig. 2.20 *Variation of net annual energy exchange between earth and space with latitude.* The horizontal scale is proportional to the earth's surface area between given latitudes

If the earth rotated very slowly (so that the solar radiation would not fall only on one side) then the global wind system might be expected to be as shown in Fig. 2.21. That is, hot air arising from the equator would move at high level to the poles and return at the earth's surface as a north wind in the Northern hemisphere and a south wind in the Southern hemisphere*. In this simple picture the wind system forms two cells, known as the Hadley cells after the eighteenth-century scientist who first put forward this model. One might expect that the earth's rotation would add an easterly component to this north–south flow pattern, and this is roughly what happens in the tropics between latitudes 30°N and 30°S. The actual situation is more complicated than this simple picture would suggest. Fig. 2.22 shows the result of averaging sea level pressures and winds over the earth's surface and is useful in identifying the most important global wind characteristics.

The major wind patterns were first observed by seamen, to whom they were of overriding importance. Starting at the equator there is a region of little and unpredictable wind known as the doldrums. Moving north we enter the region

* It is customary to refer to wind direction in terms of its source.

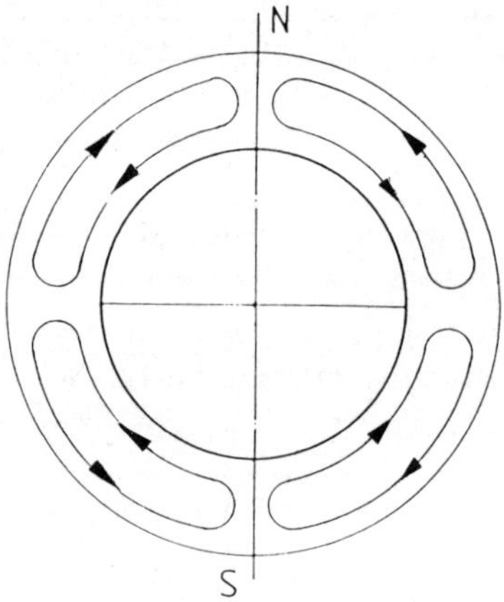

Fig. 2.21 *Non-rotating earth with Hadley cells*

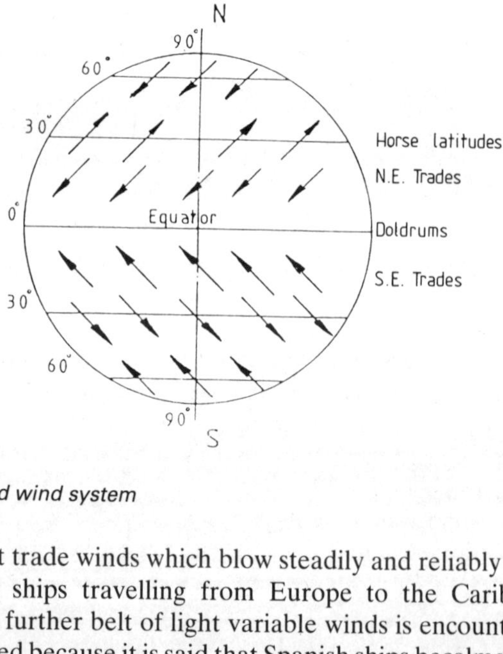

Fig. 2.22 *Idealised wind system*

of the north east trade winds which blow steadily and reliably and were much used by sailing ships travelling from Europe to the Caribbean. Around latitudes 30°N a further belt of light variable winds is encountered, the horse latitudes, so called because it is said that Spanish ships becalmed here disposed of their horses in order to conserve water for the crew. Moving further north there are the south-west trade winds and further north still the polar easterlies. Continental land masses alter this global pattern since they heat up more rapidly than the surrounding ocean in the summer and cool down more rapidly in the winter, giving rise to the seasonal monsoon-type wind systems which dominate the wind pattern of South East Asia. Hence in the northern hemisphere the general pattern is significantly modified by the presence of the large continental land masses. This is not the case to the same extent in the southern hemisphere where very strong westerly winds are encountered (known to sailors in terms of latitude as the roaring forties and screaming fifties). At a more local level coastal areas experience sea breezes which flow from the land to the sea at night and in the reverse direction in the heat of the day. These sea breezes can be of considerable importance for wind power.

This outline is intended to provide a background for the understanding of wind power. The generalisations should be treated with caution since local variations in wind are often of major importance in deciding on its economic and technical feasibility. There are areas of the world where, from general considerations, one would conclude that wind power is unsuitable but in fact one finds it to be exploited widely and with success.

2.6 Hydro energy

Nearly one-quarter of the solar radiation received at the earth's surface is used to evaporate water, mainly from the oceans. The water vapour rises, condenses to form droplets in clouds, and later is returned as rain and snow. The water falling on the higher ground is returned to the sea in rivers and streams and the initial potential energy partially converted into kinetic energy of flow and partly lost in friction. Ultimately the remaining kinetic energy is dissipated as heat when the river enters the sea. The total power corresponding to this hydrologic run off is approximately 9×10^{12} W. Estimates of the amount of energy which can be extracted as useful power are given in Table 2.8.

While much smaller than the total wind power which is available, it is

Table 2.8 *Regional distribution of hydro resources (Reference 10)*

	Potential 1000 MW	% World total	Developed 1000 MW
North America	313	11	59
South America	577	20	5
West Europe	158	6	47
Africa	780	27	2
Middle East	21	1	—
South East Asia	455	16	2
Far East	42	1	19
Australia	45	2	2
USSR, China	466	16	16
	2857	100	152

nevertheless a very significant energy source. Hydro power is a more concentrated source of energy than wind power and therefore can generally be extracted more economically if a suitable site can be found.

2.7 Wave energy

Waves are caused by friction between the wind and water surfaces, such as the oceans.[11–15] Where prevailing winds blow across long unimpeded surfaces, such as the Atlantic Ocean, then a steady wave system will build up. The waves will persist for periods of twelve hours or so over distances of thousands of miles. They thus store the energy from the wind and tend to smooth out its fluctuations in power. A typical wave amplitude is 2 to 3 m and the corresponding energy is 10 kJ/m^2 of surface area. The average velocity is about 10 m/s so the power per metre at right angles to the wave direction is around 50 kW/m. A 1 500 m length of convertor off the NW Scottish coast would receive 100 GW of power or twice the UK current demand of electricity. This potential has long attracted the inventors, but only recently have serious experimental and theoretical studies been set up to assess the potential value of wave energy. Major problems include the design of cheap long lived devices to extract the power and the cost of the transmission of power to the shore. Wave power may be of importance in the future but the ocean schemes will inevitably be large scale and capital intensive.

There may be some situations, e.g. on the African lakes and in coastal waters, where small scale wave devices have application.

2.8 Geothermal energy

Heat is continually produced in the earth by the decay of radioactive elements.[16–19] In Fig. 2.8 it is indicated that the total rate of heat production from this source is about 32×10^9 kW. Dividing this figure by the area of the earth's surface, $5{\cdot}1 \times 10^8$ km^2, gives an average flux of 62 kW/km^2 or 0·062 W/m^2, a very low value which is doubled in certain more favourable regions.

Most of the geothermal energy reaches the earth's surface by conduction through the rock and the temperature increase with increasing depth is between 8 and 20 °C/km; a small fraction (about 1%) is convected in the form of hot springs. The energy stored in the hot rocks forming the earth's core works out to an enormous figure – 10^{35} J. The problem, of course, is how much of this can be extracted at an economic cost.

The surface of the earth appears to be made up of a dozen or so solid 'plates' of rock of the order 100 km thick and several thousand kilometres across. Fig. 2.23 shows the principal plates. These 'tectonic plates' float on a molten rock

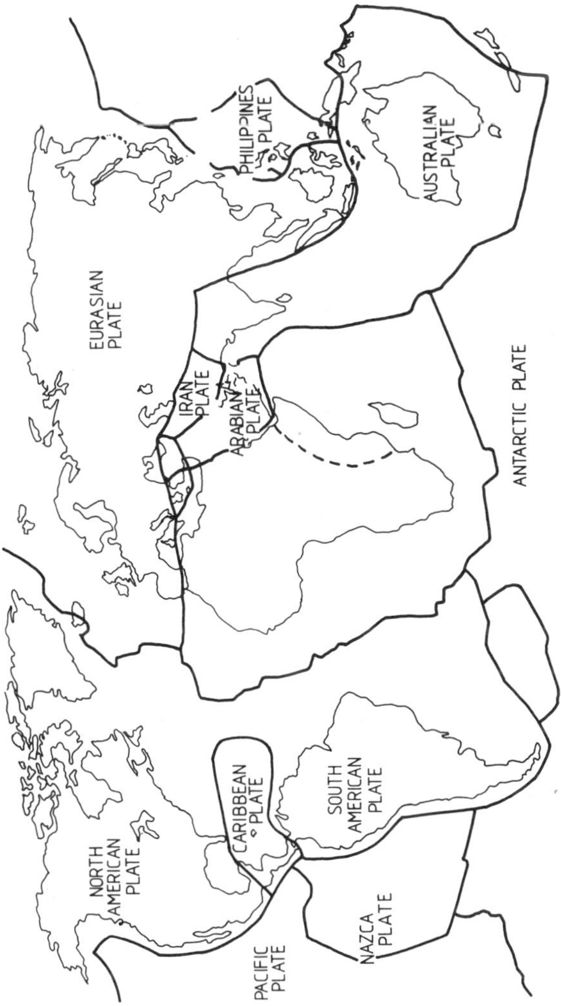

Fig. 2.23 *Main tectonic plates on the earth's surface*

core. The maximum geothermal effects generally occur at the edges between the plates. Molten rock or magma rises in the interfaces and can reach the surface, to give volcanic effects or solidify below the surface to give intrusions. Ground water percolates downwards to the heated intrusion, becomes heated in the porous layer, where it is either trapped or rises to the surface through a further fault. Depending on the conditions the water may emerge as superheated steam, saturated steam or simply hot water. Natural geysers like these have been known for many years. More recently walls have been drilled into the reservoirs of superheated water to give steam which has been used to power turbines for electricity generation. It has been estimated by White (Reference 19) that of the 1% energy in hydrothermal springs the usable energy is about 4×10^{20} J and if this is converted at an efficiency of 25% over a period of 50 years around 60 000 MW of electricity would be generated.[18] The total power produced in this way today is around 2500 MW. One of the earliest examples of the use of geothermal steam to generator electricity was at Lardello, Tuscany, Italy, in 1904. The present installed capacity is around 400 MW. Table 2.9 lists a number of geothermal electricity generating plant.

Table 2.9 *Geothermal electricity generation capacity*

Country	MW
USA (West Coast)	1005
Philippines	446
Italy	398
New Zealand	140
Mexico	180
Japan	164
El Salvador	65
Kenya	15
Iceland	8
World	2422

Comparison with Fig. 2.23 shows these sites to lie on the edges of the tectonic plates. The water from geothermal sources is rich in minerals which can present an environmental problem.

A new approach to exploration of thermal power is by drilling down to the subterranean rock at depths of the order 5 km. In these dry holes water is pumped down from the surface and returned as steam through a second hole. It is too early to comment on the economic feasibility of the method which might well be of considerable importance in the future.

2.9 Tidal energy

The ebb and flow of the tides have tempted coastal dwellers to try to harness the energy to perform useful tasks.[20–22] For example the principle was used in the mills along the Breton coast which were built as early as the twelfth century.

Unlike solar, wind or hydro power, tidal power is predictable. The pattern repeats every 18 years and the variation from month to month is small (about 5%).

Since tidal power is, by nature, intermittent, storage will be required if power is to be guaranteed at any time, and correspondingly large volume storage facilities will be necessary.

Along the Atlantic coast of both America and Europe tides rise and fall twice in approximately every 24 hours. The times of high and low tide vary in a predictable manner, and so also does the amplitude or distance between high and low tide. Though always predictable the behaviour is not necessarily the same in all areas, for example, in Vietnam the high tide occurs only once in 24 hours and in Tahiti the time of high tide does not alter.

The amplitude varies widely from place to place and can be, for example, as high as 17 m (50 ft) in the Bay of Fundy on the Atlantic coast of Canada. Other locations where large tidal amplitudes are experienced are Alaska 11 m (35 ft), Colorado river estuary 10 m (30 ft), River Severn estuary 15 m (45 ft), and on sites on the costs of Argentina, India, Korea, Australia and the USSR. A site of particular interest is the Rance estuary, 13 m (40 ft) in Brittany, France, where an electricity generation plant has been installed.

The formation of tides is largely due to the gravitational pull exerted by the moon, see Fig. 2.24. That there should also be a similar high tide on the

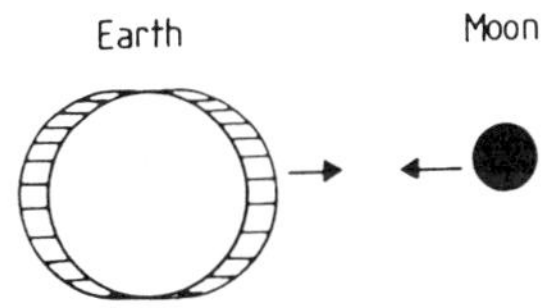

Fig. 2.24 *Origin of tides*

equator at a point in the far side of the earth, with respect to the moon, is at first sight unexpected. The result is of course to give two tides in 24 hours. The reason is that the earth/moon system rotates about the common centre of mass. Owing to the large mass ratio between the earth and moon this point occurs about 3000 miles from the centre of the earth. The centripetal force due to this rotation results in a water movement on the far side of the earth, with respect to the moon. Since the period of the earth's rotation is 23 hours 56 minutes 4.08 seconds, the time peak of the tide will change in a cyclic manner.

We have so far neglected the effect of the sun. The gravitational attraction exerted by the sun is about half that of the moon. Twice a month the moon and sun will be in line and the effects will reinforce, giving rise to the spring tides at full and new moon. At first and third quarter the sun's effects will tend to offset that of the moon and the reduced amplitude 'neap tides' will result.

The effect of the above is to generate tidal waves which move out from the equator. In addition local configurations may significantly modify the amplitude at a particular situation. An obvious effect is the funnelling in estuaries, e.g. the Severn, resulting in very high local amplitudes. It is such locations which are of interest for tidal power generation sites.

It is of interest to ask what is the source of the tidal energy? The gravitational pulls themselves do not provide energy, they in fact cause a frictional drag on the rotation of the earth which is the source of tidal energy. The earth's rotational energy is very large ($2{\cdot}6 \times 10^{29}$ J), and the slowing down (10^{20} J/year or 3×10^{19} W) negligible. Are we increasing the rate of slowing down by tidal power schemes? This is not the case since the energy has already been put into the tidal motion and would in any case be dissipated by friction if not used to produce power.

The simplest method of generating tidal power is to operate the turbine at high tide by filling a basin through a turbine. The basin is then emptied by opening a sluice gate at low tide and then the whole process being repeated. Alternatively, of course, one can make it work in the reverse direction by operating the turbine to allow the water to flow out during the low tide period and refilling it at high tide.

Since tidal generators are normally connected to the power network, if the turbines can be reversed and used as motors there is some advantage in pumping additional water into the container basin at high tide. As has already been pointed out, the time for high tide varies and it may well not be possible to operate the tidal generator when the main demand is required. This is a serious disadvantage. It can, to some extent, be overcome by using two adjacent basins, one of which operates on out-flow and one on in-flow.

The dependence on tide times can largely be avoided by a tidal basin which interconnects. It is not usual, however, to find such situations naturally and to construct them tends to be rather expensive.

The Rance river estuary tidal power station is the most advanced so far built. The plant site is located on the estuary of this 60 mile long Breton river, about 2 miles upstream of the medieval port town of St Malo. The tides occur twice daily reaching an amplitude range of about 14 m maximum. The Rance dam contains a water volume of 184×10^6 m^3 forming a basin of area 22 km^2. The 24 10 MW turbo-alternator units give an installed capacity of 240 MW and generate 500 GWh/annum. The rated head is 5·65 m.

2.10 Nuclear energy

2.10.1 Nuclear fission

Nuclear fission is based on the splitting, or fission, of the nuclei of certain heavy elements of which uranium is the only naturally occurring one of commercial interest. The fission reaction releases a great deal of energy. Only 1 tonne of fuel* is burnt in a year in a 1000 MW power station; a coal burning station of similar output would require 3 million tonnes of coal. Natural uranium contains two different isotopes: uranium 238 and uranium 235. The isotopes of an atom have the same chemical properties but are of different mass as a result of having different numbers of neutrons in their nuclei. In the case of the two isotopes of uranium the difference in mass is very small, it is in fact the difference between the two isotopic numbers 238 and 235. This difference is nevertheless of great practical importance since only the U235 isotope will allow fission. Unfortunately, only about one part of natural uranium in 135 is U235. The remainder is the non-fissile isotope U238. Reactors is the name given to assemblies which fission 'burn' the fissile U235 to produce heat. This heat is used to raise steam in a boiler in the usual way and the resulting steam operates a turbine which in turn drives an alternator to generate electricity. These reactors are known as burner reactors and have been in use in the UK for over 20 years for the generation of electricity. In 1982, 14% of the electricity generation in the UK was by burner reactors; the contribution to electricity generation from nuclear power is higher in certain other countries, notably France, Belgium, Sweden and Germany. The total energy content of workable uranium deposits is about the same as the proven world oil reserves if the uranium is to be used in burner reactors. Thus fission using burner reactors is only a relatively minor source of world energy. However, it is possible to convert the non-fissile uranium isotope U238 into a fissile material, an isotope of plutonium, Pu 239. This can be achieved in a type of reactor called a breeder reactor, which can be used both to produce power and to breed a fuel from the U238. Not only does this hold out the possibility of increasing the energy which may be obtained from a given amount of natural uranium by a factor of perhaps 50, but it also makes it worth while to mine relatively poor uranium ores which would otherwise be economically and energetically unsuitable. If breeder reactors are used they will make available very much more energy than is stored in the fossil fuel reserves. This is the argument of the protagonists of nuclear power. There is, however, a strong anti-nuclear body of opinion made up of those who claim that hazards associated with the nuclear fission programme are so great as to more than offset its advantages. In this book we are primarily concerned with small scale

* This figure refers to the fissile isotope U235. In terms of natural uranium, 1 tonne is equivalent to 20 000 tonnes of coal.

power generation. Nuclear fission is unlikely ever to lend itself to small scale application and therefore will not be discussed further.

2.10.2 *Nuclear fusion*

When the nuclei of light atoms are caused to collide they fuse to form nuclei of heavier atoms and the process is accompanied by the release of energy. In principle, the simplest way of achieving such collisions is to take a gas made up of hydrogen and to raise it to a very high temperature and pressure. This essentially is what takes place in the sun though the detailed reactions are complex. Thermonuclear fusion, as such a reaction is called, has been achieved in the form of the hydrogen bomb but so far it has not proved possible to obtain a controlled reaction such as would be required in a power station. The raw materials required for the thermonuclear reaction are virtually limitless and the reaction itself is safer and does not produce the long lived radioactive isotopes which occur in a fission reaction. In the long term thermonuclear power would appear to be a very important source of energy. It is unlikely to become commercially feasible in quantity for two and possibly three decades and may not be economic until fossil coal reserves are much more scarce. Since thermonuclear power, like nuclear fission, is not likely ever to be a small scale power source we will not discuss it further.

References

1 WHITTAKER, R. H. and LICKENS, C. E.: 'Primary productivity of the biosphere' *in* ed. LIETH, H. and WHITTAKER, R. H. (Springer-Verlag, 1975)

2 SELLERS, W. D.: 'Physical climatology' (University of Chicago Press, 1965)

3 HUBBERT, King. M.: 'Energy resources of the earth', *Scientific American* Sept. 1971

4 LOFERSKI, J. J.: 'Theoretical considerations governing the choice of the optimum semi conductor for photovoltaic solar energy conversion', Conference on Solar Energy, Tuscan, Arizona, Nov. 1955

5 WOLF, M.: 'Photovoltaic solar converters', *Proc. IRE* July 1960, **48**

6 HALL, D. O.: 'The present status of solar energy for fuel', UNESCO 2nd International Forum on Fundamental World Energy Problems

7 'Solar energy – AUK assessment' (UK Section of International Solar Energy Society, London 1976)

8 PIMENTAL, D. and PIMENTAL, M.: 'Food, energy and society' (Arnold, 1979)

9 'The global 2000 report to the President', report prepared by the Council on Environment and the Department of State (Penguin, London, 1982)

10 HUBBERT, King M.: 'Energy resources', Resources and Man, National Academy of Sciences. N.R.C. Freeman San Francisco. 1969

11 COUNT, B. M.: 'The absorption of energy from ocean waves', Conference on Future Energy Concepts, IEE, London Jan/Feb 1979

12 GLENDENNING, I.: 'Energy from waves' (Oceanology International, 1979)

13 MOLLISON, D.: 'Wave power availability in the North East Atlantic', *Nature*, 1976 **263**

14 FARLEY, F. J. M., PARKS. P. C. and ALTMANN, H.: 'A wave power machine using free floating vertical plates', Conference on Future Energy Concepts, IEE, London Jan/Feb 1978

15 SALTER, S. H.: 'Wave power', *Nature* June 1974, **249**

16 DUNHAM, K.: 'Geothermal energy and heat pumps', Conference on Future Energy Concepts, IEE, London Jan/Feb 1979
17 LEARDINI, T.: *Phil. Trans. Royal Society A*, 1974, **276**
18 McMULLEN, J. T. and MURRAY, R. B.: 'Energy resources' (Arnold, 1979)
19 WHITE, D. E.: 'Geothermal energy' (US Geological Survey Circ. 1965)
20 'Fourth Report from the Select Committee on Science and Technology. the Exploitation of Tidal Power in the Severn Estuary' (HMSO, 1977)
21 Bay of Fundy Tidal Power Review Board and Management Committee. Reassessment of Fundy Tidal Power, Ottawa. Nov. 1977
22 COBTLEN, J.: 'La Rance tidal power station, review and comments', Proceedings of Colston Symposium, University of Bristol UK', 1978

Chapter 3

Energy and its conversion

3.1 Some general considerations

A simple example of energy is the effort required to raise a mass from a low to a high level. In order to do this, it is necessary to apply a force slightly greater than the downward force exerted by the weight of the mass. By lifting the weight work is done; the amount of work is defined as the product of the force and the distance moved in the direction of the force.

The work done is equal to the gain in potential energy of the mass and this work may be recovered by allowing the mass to fall back, giving up the potential energy and doing work on the way down. Another form of potential energy is the energy stored in a spring. These are forms of mechanical energy. Kinetic energy, the energy due to motion, is yet another form of mechanical energy. Fig. 3.1 shows some common examples of energy forms.

Energy can be defined as the ability to perform work

3.2 *Conservation of energy – the First Law of Thermodynamics*

A number of different forms of energy have already been discussed. One of the most fundamental discoveries in physical science was the understanding that:

*the total sum of all the forms of energy in a closed system is constant**

This principle is called the First Law of Thermodynamics or the Law of Conservation of Energy. Devices which purport to produce more energy, continuously, then they absorb are known as perpetual motion machines of the first kind. Fig. 3.2 shows several examples of such machines; all contain a basic fallacy. The First Law of Thermodynamics asserts that the sum of the various energies contained in a closed system remains constant. Hence, if in a closed system a process occurs which increases one energy form, the other energy

* This includes the mass equivalent of energy $E = mc^2$

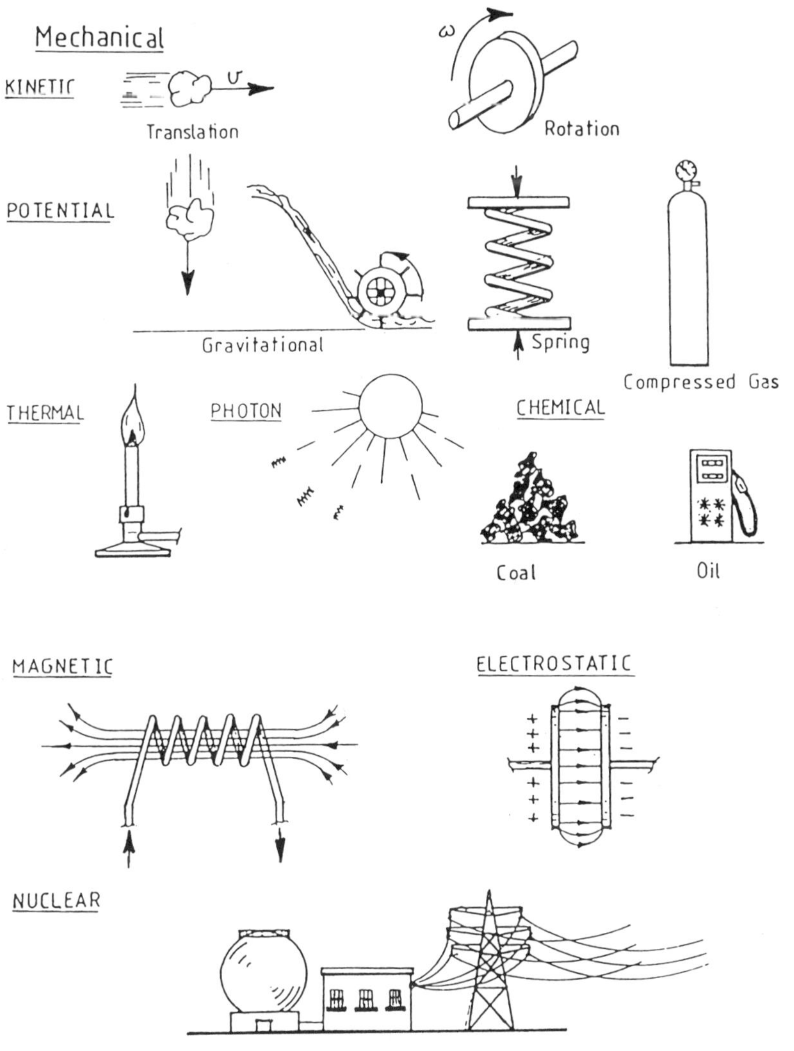

Fig. 3.1 *Some energy forms*

forms in the system must be decreased by the same amount. this exchange of energy may take place freely in either direction. For example, in the case of the pendulum the potential energy of the bob when stationary in its highest position is converted to kinetic energy at the lowest part of the swing. In the electrical analogy of the pendulum, the *LC* resonant circuit, magnetic energy in the inductor is changed to electrical energy in the capacitor and back again (Fig. 3.3).

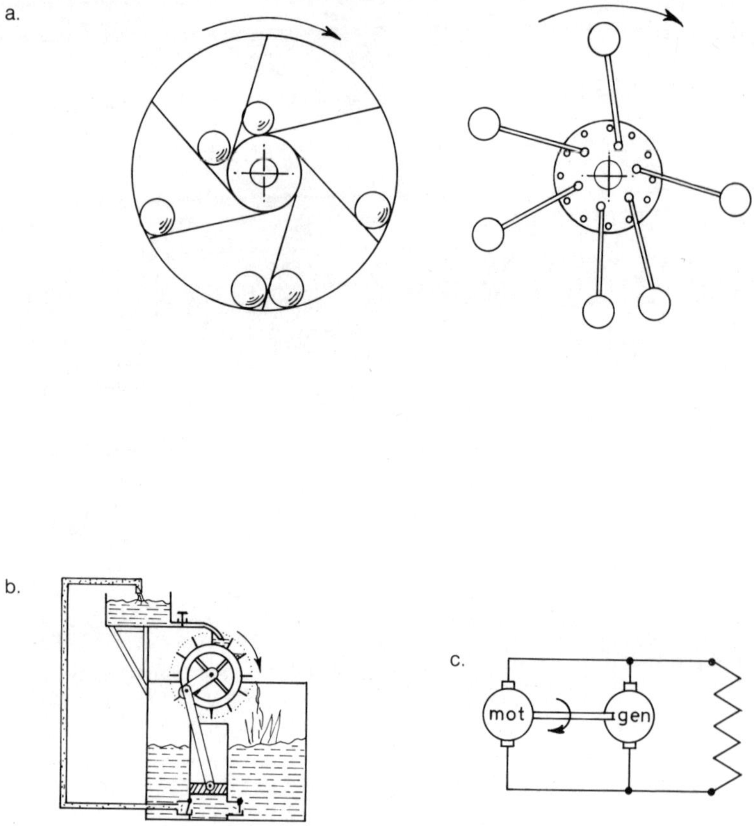

Fig. 3.2 *Perpetual motion of the first kind*
a Unbalanced wheels. A variant by Sir William Congreve employed sponges which were soaked on one side and squeezed dry on the other
b A water wheel operates a pump which returns the water
c Another variant employing an electric motor and a generator

In the previous paragraph we considered the simple pendulum and asserted that if no external energy was added to the pendulum then by the First Law the sum of the kinetic energy and potential energy would be constant and the form of energy would change from potential to kinetic and back again to potential energy as the pendulum swung from one side to the other. It is, however, common experience that friction will exist and eventually the pendulum will come to rest. How is this reconciled with the First Law? What will happen is that the mechanical energy of the pendulum will slowly be converted to heat in the air. The amount of heat energy so produced will exactly equal the original mechanical energy (potential plus kinetic) of the pendulum, thus no discrepancy exists. In studying the pendulum motion, frictional loss can usually be neglected, and this term can be ignored in applying the First Law to the problem.

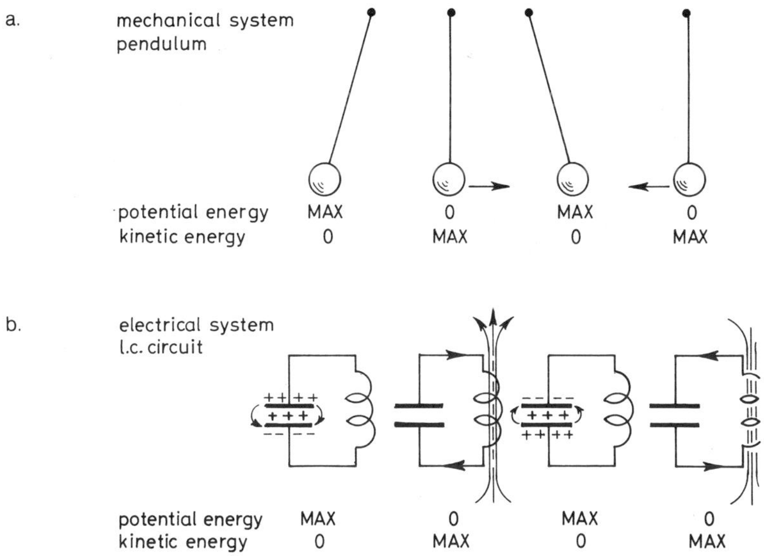

Fig. 3.3 *Resonant systems*
a Mechanical system: pendulum
b Electrical system: *LC* circuit

3.3 Second Law of Thermodynamics

There is, however, one form of energy which is not freely interchangeable with other energy forms – thermal or heat energy. A gas can be pictured as a large number of moleules in random motion. The molecules possess kinetic energy and the mean kinetic energy is used to define the temperature of the gas.

If a volume V of this gas is isolated by a container, Fig. 3.4*a*, then the gas molecules will strike the walls of the container, and since there will be many such collisions/second the effect will be of a steady pressure acting on the inner walls of the container. Because the container is immersed in gas, a similar pressure will be exerted on the outside of the walls which will just balance the pressure on the inside. Thus there will be no resultant force on the container walls. Now if the temperature of the gas inside the container is raised the average velocity of the gas molecules will be increased, Fig. 3.4*b*. They will make more collisions with the walls and the force of collision will be greater. Hence the inside pressure on the container walls will exceed that on the outside. By using a piston arrangement it is therefore possible to extract some of the excess energy. Whenever there are bodies at different temperatures it is possible to extract some of this excess thermal energy in the form of useful work. If we think about this we will see that this implies that we cannot extract useful work from a body by trying to cool it below the temperature of its

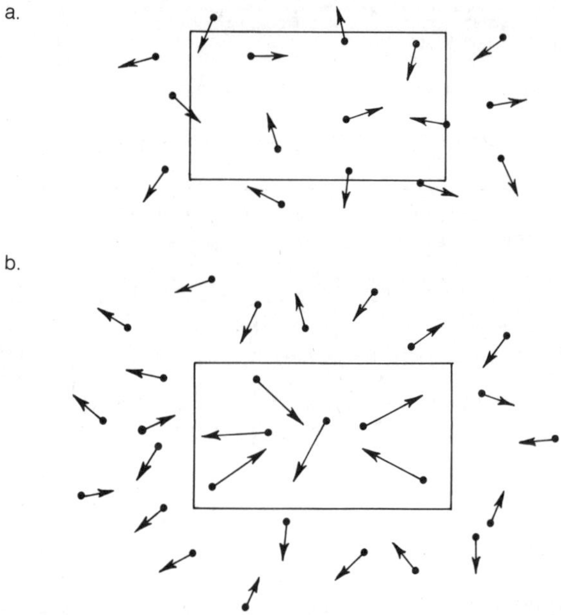

Fig. 3.4 *Molecular motion in a gas*
a Random motion of molecules in gas (arrow lengths represent average velocity)
b Effect of temperature of enclosure raised above that of the surroundings

surroundings. It turns out that not all the excess thermal energy can be converted into useful work. If, for example, the gas temperature in the container were to be raised from 300 K to 600 K only about 0·3 of the thermal energy required to cause this rise in temperature can be converted to useful work. The remaining 0·7 of the heat must be rejected at a temperature of 300 K.

This essentially is what the Second Law is all about. It has been expressed in various ways, e.g. heat will not of its own accord flow from a cold to a hot body. An important conseqence of the Second Law is the restriction it places on the amount of thermal energy which can be converted to other forms of energy, that is, it sets an upper limit on the conversion efficiency. This topic, the study of heat engines, will not be discussed more fully.

3.4 Heat engine – the Carnot efficiency

3.4.1 Heat engines

An energy convertor which is operated in a cyclic manner, and takes in energy from a high temperature source, converts some of this energy to useful work and rejects the remainder to a cooler sink, is called a *heat engine*, see Fig. 3.5.

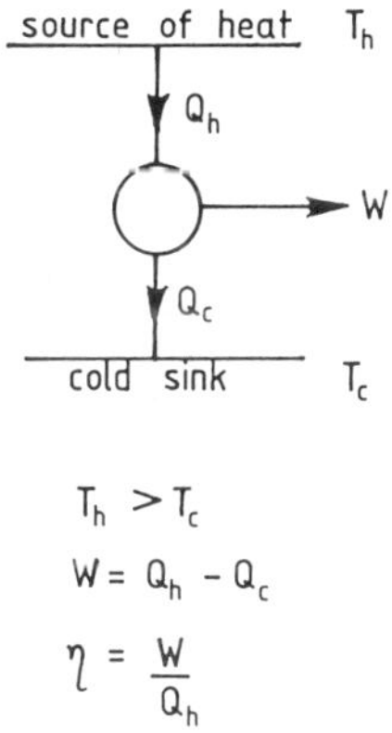

Fig. 3.5 *Heat engine*

Kelvin stated the Second Law of Thermodynamics in terms of the heat engine as follows:

> It is impossible to devise an engine, which, working in a cycle, shall produce no effect other than the extraction of heat from a reservoir and the performance of an equal amount of mechanical work.

A corollary of this definition, due to Clausius, and sometimes used as an alternative statement of the Second Law is:

> It is impossible to devise an engine which, working in a cycle, shall produce no effect other than the transfer of heat from a colder to a hotter body.

The efficiency of any heat engine is the fraction of the thermal energy taken from the high temperature source which is converted into work.

Efficiency can be expressed by the relation:

$$\eta = \frac{Q_h - Q_c}{Q_h} \tag{3.1}$$

where Q_h is the heat taken from the heat source and Q_c is the heat rejected to the cold sink.

An ideal or perfect heat engine is one in which friction is absent and heat is transferred, without temperature drop, at the hot and cold reservoirs. This means that the work and heat quantities crossing the boundaries can be reversed in sign without any change in magnitude. Hence an ideal engine is perfectly reversible since all the processes taking part in the cycle can be reversed by an infinitesimal change in conditions.

A reversible engine is one which may also be driven and cause heat to flow from the cold sink to the hot source – a refrigerator.

Carnot considered the limitations on heat engines and showed that it is impossible to make an engine that is more efficient than a perfectly reversible engine. The proof, which is given in Appendix 1, is quite straightforward and simply shows that if a perfectly reversible engine is driven backwards by an irreversible engine of a higher efficiency, then the net effect will be for heat to flow from a cold to a hot body, which is contrary to the Second law of thermodynamics. Hence all perfectly reversible engines operated between the same temperatures must have the same efficiency.

3.4.2 Mechanically reversible engines and perfectly reversible engines

An ideal, or perfectly reversible, heat engine will take in heat Q_h from the high temperature source and convert some of this heat to mechanical work W. Practical heat engines will contain friction and other irreversible processes and the work output W_1 from such an engine, for the same heat input Q_h, will be less than W.

The perfectly reversible engine can be reversed and if driven by work W will return heat Q_h to the hot reservoir. Many practical engines can also be reversed, that is they are mechanically reversible, but the work W_2 required to drive the engine to return heat Q_h will be greater than W the work required by the perfectly reversible engine. Hence

$$W_1 < W < W_2$$

Not all practical heat engines and refrigerators are mechanically reversible. For example, the domestic compressor type refrigerator (see Section 4.5.2) cannot be operated as a heat engine.

3.4.3 Carnot efficiency

So far we have talked about hot and cold in a non-quantitative way. We must now define these conditions by introducing *temperature*. Kelvin pointed out that the efficiency of an ideal heat engine depends only on the temperatures of the two thermal reservoirs and is independent of the nature of the working fluid in the engine. The heat Q exchanged with each reservoir can be used as a measure of temperature T of that reservoir.
That is:

$$\frac{Q_c}{Q_h} = \frac{T_c}{T_h}$$

The maximum possible efficiency of an ideal engine will occur when the cold sink is so cold that the molecular motions are zero. At this point – called absolute zero – Q_c and T_c both have diminished to nothing. Having determined the zero of the temperature scale it is then necessary to fix the units in which temperature is to be measured.

The practice is to select the difference in temperature between the melting point of ice and the boiling point of water at atmospheric pressure as 100 degrees interval. Temperatures in the kelvin scale may readily be converted to the centrigrade scale by subtracting 273. With this unit the melting point of ice becomes 273 K and the boiling point (at atmospheric pressure) of water 373 K. Hence, since

$$\frac{Q_c}{Q_h} = \frac{T_c}{T_h}$$

the efficiency of the ideal engine η_c can be expressed in terms of the source and sink temperatures T_h and T_c measured on the kelvin scale*:

$$\eta_c = \frac{T_h - T_c}{T_h} \tag{3.2}$$

Eqn. 3.2 is called the Carnot efficiency.

The Carnot efficiency as given by eqn. 3.2 is plotted in Fig. 3.6*a* against the maximum temperature T_h for a fixed value of T_c = 300 K (27°C) which corresponds roughly to ambient temperature.

It is seen that the Carnot efficiency increases rapidly at first as T_h is increased. As T_h is raised from 300 K to 600 K the Carnot efficiency rises from a value of zero to 50%. A further increase in T_h of 300 K increases the maximum conversion efficiency from 50% to 66%, a rise of only 16%, and an additional rise in temperature from 900 K to 1200 K gives an efficiency improvement of only 8%. Increasing T_h gives a diminishing return. On the other hand, the effect of changing the sink temperature T_c can be very significant, see Fig. 3.6*b*.

The Carnot efficiency is a most important relationship since it fixes the maximum efficiency that a heat engine may achieve. Engines which purport to convert thermal energy to other energy forms with an efficiency greater than the Carnot efficiency, or do not reject heat to a cooler sink, are called perpetual motion machines of the second kind. They still crop up from time to time, and, while one can immediately declare the proposal unsound on general thermodynamic grounds, it is often difficult to see the particular step in reasoning which is at fault. It is said that the Patent Office will not knowingly accept patent applications for such engines, see Fig. 3.7.

It should be noted that the Second Law restriction on efficiency applies to the continuous conversion of heat energy to other forms of energy. Single processes such as the conversion of heat energy to work by expanding a gas at constant temperature are in principle both 100% efficient and completely reversible. However, it is not possible to make a continuously acting engine which only uses this process, since a second compression process at a lower temperature is required, in order to give a net work output, and once more we are back to the Carnot limit on efficiency of the overall conversion process.

A heat engine can use solids, liquids or gases as the working fluid. Fig. 3.8 shows the cycle for a Carnot engine using a perfect gas as the working fluid. The gas is expanded at constant temperature T_h absorbing heat from the hot

* The temperatures must be expressed on an absolute temperature scale, that is, a scale where zero corresponds to absolute zero. The kelvin scale is such an absolute scale.

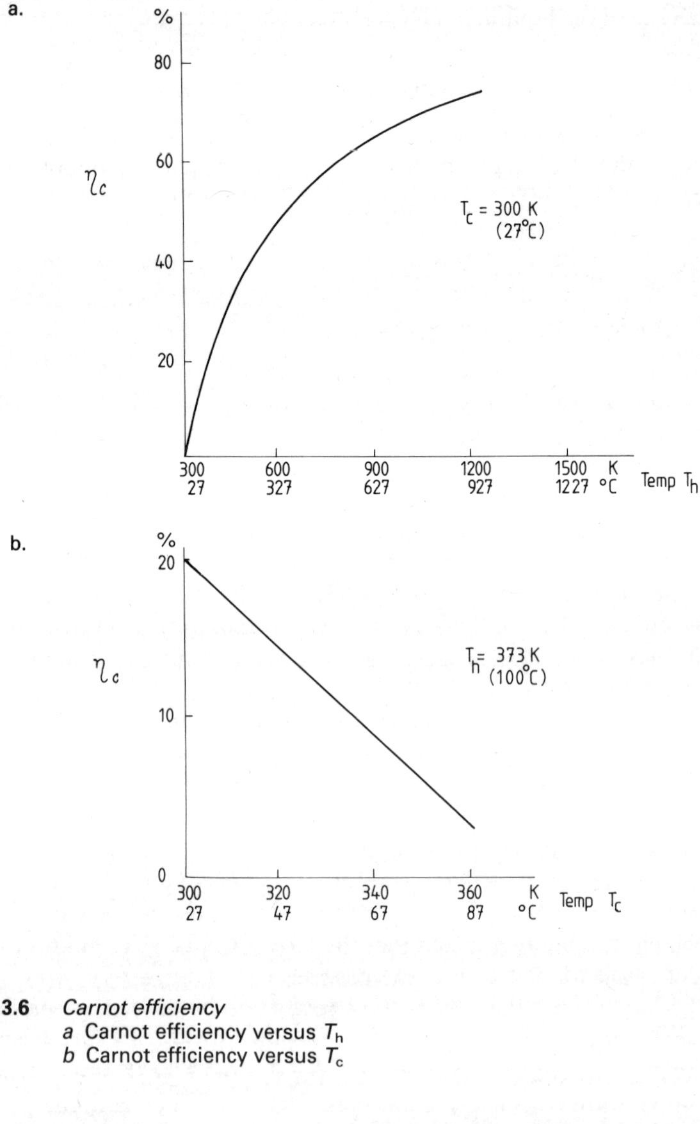

Fig. 3.6 *Carnot efficiency*
a Carnot efficiency versus T_h
b Carnot efficiency versus T_c

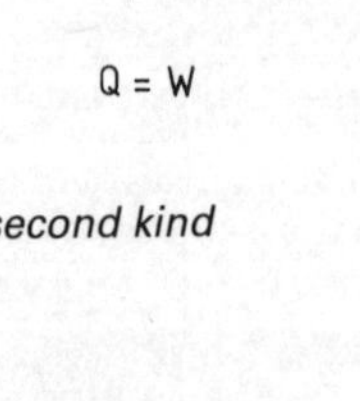

Fig. 3.7 *Perpetual motion of the second kind*

reservoir 3–4. It is then insulated from the surroundings and further expanded until its temperature falls to T_c, the temperature of the cold sink 4–1. The cold gas is compressed at constant temperature 1–2, giving up heat to the sink and finally restored to temperature T_h by compressing while insulated from the surroundings 2–3. The work done per cycle W is the difference between the two heat quantities $Q_h - Q_c$ and is equal to the closed area.

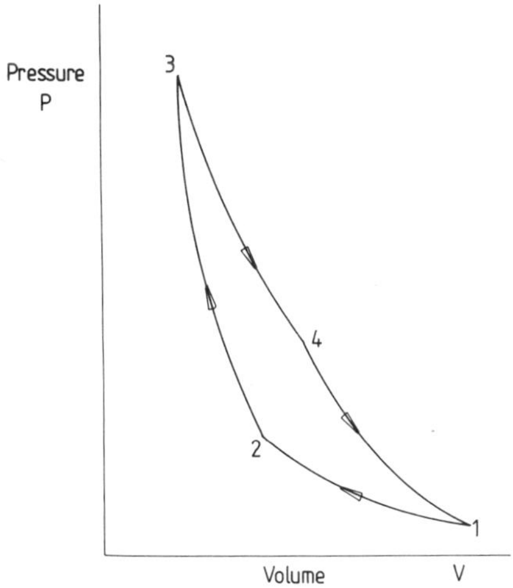

Fig. 3.8 *Carnot cycle using a perfect gas as the working fluid*

3.5 Temperature entropy diagrams

3.5.1 Availability

A process, such as the compression or expansion of a gas, in which no heat is transferred to the surroundings is called an adiabatic process.[3,4] If there are no internal losses, such as turbulence in the gas, then the process is reversible. A reversible adiabatic is called an isentropic process, or one in which entropy is unchanged. Entropy, S, is an important thermodynamic concept. It is not necessary to discuss it in detail here; a clear introduction is given in Reference 2.* For our purposes entropy merely provides a convenient co-ordinate for plotting thermodynamic cycles. If heat Q is added at constant absolute temperature T then the entropy S will increase by an amount.

$$S_2 - S_1 = \frac{Q}{T}$$

* Entropy is a measure of the disorder or randomness in a system. Examples of increasing disorder are the shuffling of a pack of cards and the mixing of hot and cold water. The second Law can be expressed as 'The entropy of a closed system cannot decrease but will increase if irreversibilities are present'.

The Carnot cycle using a perfect gas as the working fluid can be plotted on T,S co-ordinates as in Fig. 3.9*a*. Then

$$S_2 - S_1 = \frac{Q_h}{T_h}$$

$$= \frac{Q_c}{T_c}$$

$$Q_h - Q_c = (S_2 - S_1)(T_h - T_c)$$

$$= W$$

The maximum work W Fig. 3.7*a*, which can be extracted between the two temperatures T_h, T_c is called the availability and is equal to the closed area.

Q_h, Fig. 3.9*b*, is the heat taken in at the source, temperature T_h, and Q_c, Fig. 3.9*c*, the heat rejected to the sink at temperature T_c. Hence

$$W = Q_h - Q_c$$

Fig. 3.9 applies to any Carnot cycle.

3.5.2. Availability of flames and hot gases

In Fig. 3.9 the heat source was assumed to be at constant temperature. Most practical heat sources are in the form of a hot gas;[3] heat can only be extracted from such a source by allowing it to cool from its initial to a lower temperature. It is important to know what is the maximum work energy (availability) which can be extracted from such a source when coupled to a heat engine. The maximum availability occurs when the gas is cooled down to ambient or sink temperature T_c is shown in Fig. 3.10 and could be extracted by means of a series of Carnot engines each operating at successively lower peak temperature. However, if a single Carnot engine is used then there will be an optimum peak temperature in order to give maximum availability or maximum overall conversion efficiency, Fig. 3.10*b*, *c* and *d*. Referring to Fig. 3.10*a*, if an engine is used which follows the cycle 1–2–3, that is, an isothermal compression 1–2 followed by a constant pressure heating process 2–3 and an isentropic expansion 3–1 to close the cyclic, Fig. 3.11*a*, then the full availability will be realised. This engine will have a higher efficiency than the optimum Carnot engine of Fig. 3.10*d*. What this means is simply that the cycle 1–2–3 has a higher efficiency than a Carnot cycle operating from a lower fixed temperature level than T_3, and having the same sink temperature level T_c. However, if the single Carnot engine were to be replaced by a series of Carnot engines operating at successively lower temperatures from T_3 to T_a then the overall efficiency would of course be the same as for the cycle 1–2–3. A similar cycle would be used in the example discussed in Section 3.4 but in this case the optimum cycle would consist of an isothermal compression, a constant volume heating process and an isentropic expansion, see Fig. 3.11*b*.

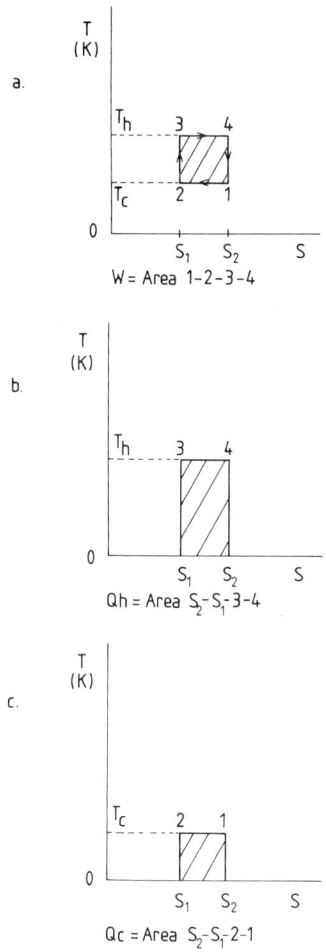

Fig. 3.9 *Carnot cycle plotted on temperature entropy co-ordinates*
a W = area 1–2–3–4
b Q_h = area S_2–S_1–3–4
c Q_c = area S_2–S_1–2–1

The overall efficiency of the cycle such as that in Fig. 3.10*d* may be improved by the use of a recuperator, see Fig. 3.12. In a recuperator the incoming air to the combustor is preheated by the hot exhaust gas and with a well designed recuperator most of the exhaust heat will be recovered.

The matching of the cycle to the heat source is of great importance in selecting an engine/source combination. For a constant temperature source and sink the Carnot cycle will give the highest availability conversion (100%), Fig. 3.9. However, where the source temperature changes then the cycle 1–2–3 shown in Fig. 3.10*a* would give the highest availability conversion (100%).

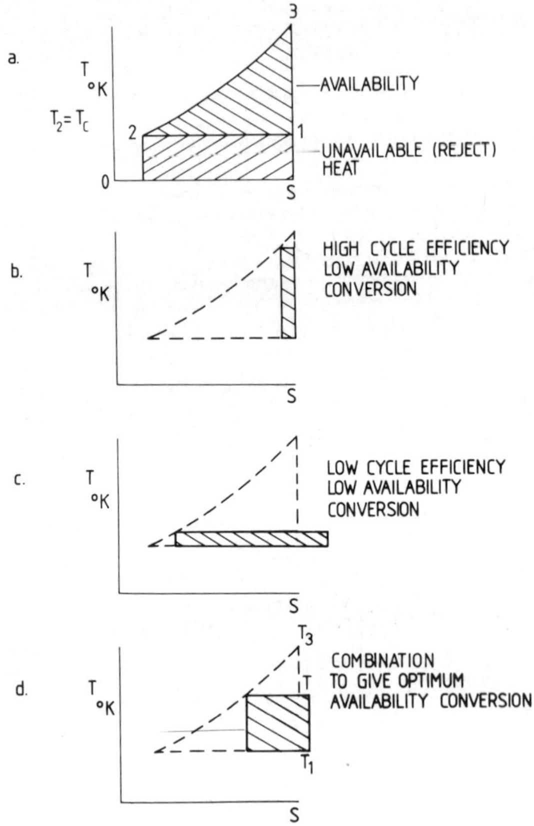

Fig. 3.10 *Carnot engine with variable temperature heat source*

3.6 Heat engines, refrigerators and heat pumps

Fig. 3.13 depicts the four roles of a heat engine. It is most important to understand the configurations given in this figure. Fig. 3.13*a* shows the normal heat engine. Fig. 3.13*c* shows an engine operating between ambient temperature and a cold reservoir at temperature, T_c. This 'cold' engine is, of

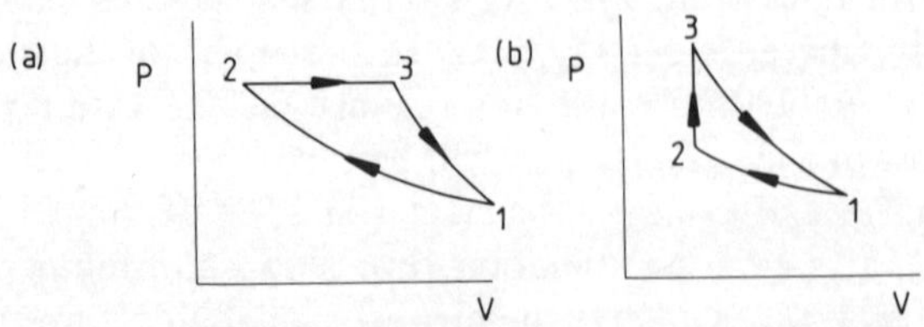

Fig. 3.11 *Cycles which achieve maximum availability conversion from a variable temperature heat source*

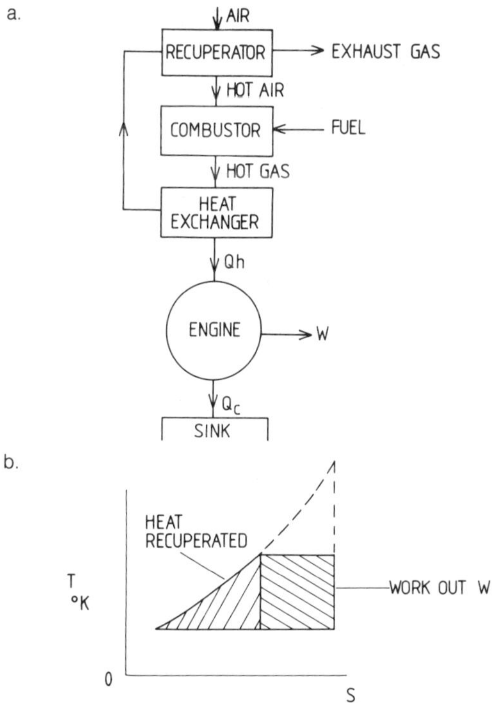

Fig. 3.12 *Exhaust heat recuperation*

course, a heat engine; it is seldom used since only under very rare circumstances is a cold reservoir at a temperature less than ambient available.

As has already been stated, a mechanically reversible engine can be made to run backwards and can then be used to reject heat at ambient temperature, in other words the engine will work as a refrigerator. In a refrigerator we are interested in how much energy in the form of work must be put in to remove heat at temperature T_c, see Fig. 3.13*d*. This is expressed as a coefficient of performance, COP, where:

$$\mathrm{COP} = \frac{Q_c}{W}$$

Alternatively, heat can be abstracted at ambient temperature and rejected at a higher temperature; such a device is called a heat pump. In the case of a heat pump we are interested in the work $W = Q_h - Q_c$ needed to provide heat Q_h at T_h, Fig. 3.13*b*. The ratio of the heat output to the work required is called the figure of merit, M, of the heat pump where:

$$M = \frac{Q_h}{W}$$

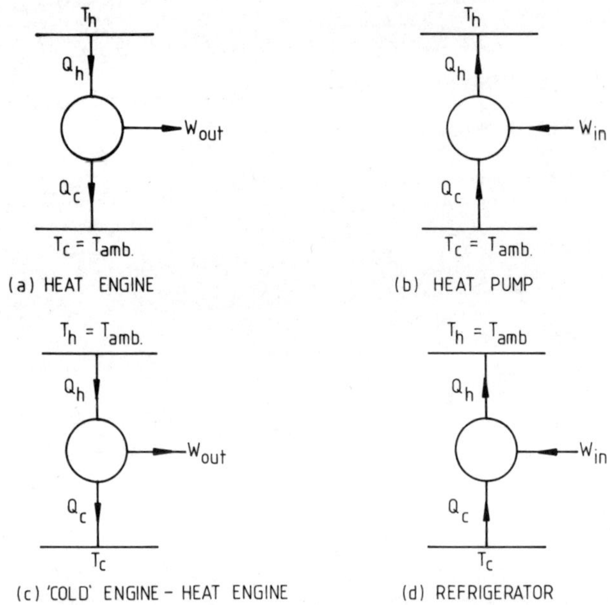

Fig. 3.13 *Four heat engine roles*
a Heat engine
b Heat pump
c 'Cold' engine – heat engine working with reservoir below ambient temperature
d Refrigerator

T_h Temperature of *hot* reservoir
T_c Temperature of *cold* reservoir
T_{amb} Ambient temperature
Q_h Heat exchanged with *hot* reservoir
Q_c Heat exchanged with *cold* reservoir
W_{out} Work output
W_{in} Work input

The maximum figure of merit of a thermodynamically perfect heat pump is:

$$M = \frac{Q_h}{Q_h - Q_c} = \frac{T_h}{T_h - T_c} = \frac{1}{\eta_c}$$

3.7 Practical Limitations

In practice no mechanism is perfect; all suffer from frictional losses to some extent. This means that the flywheel will eventually cease to rotate and the pendulum will cease to swing unless energy is added to replace these losses. By the First Law the energy is not lost but is converted to heat, even if it was possible for this heat to be used to drive an engine it could not, by the Second Law, all be converted back to the original mechanical energy. There is, therefore, in practical devices a loss of *availability* of energy and such losses are known as *irreversible losses*. In electrical circuits similar losses occur owing to the electrical resistance of the electrical conductors.

When heat energy is transferred a temperature gradient will be required to cause the heat energy to flow through thermal conductors having finite thermal conductivity. The temperature will thus be reduced and the thermal energy degraded. This is another example of an irreversible loss. Several sources of irreversibility are shown in Fig. 3.14.

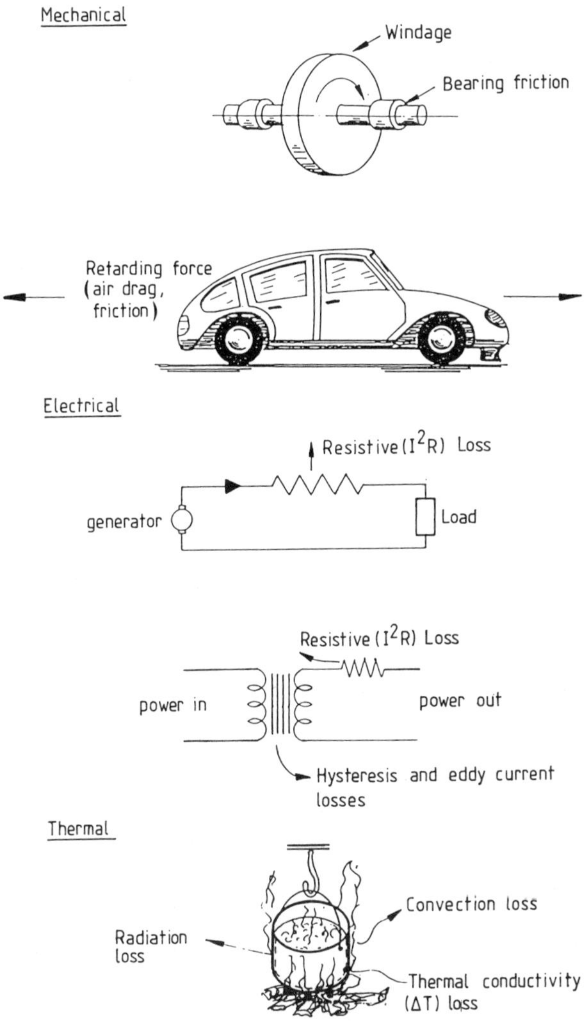

Fig. 3.14 *Some sources of irreversibility*

3.8 Matching

Consider a petrol engine rated at 40 hp and connected to the wheels of a vehicle which requires this amount of power for its propulsion. The result may be quite unsatisfactory, if, for example, the petrol engine shaft is rotating at a high speed, much higher than that required by the vehicle wheels. Clearly then power rating is not in itself sufficient, it is necessary in addition to arrange to 'match' the power source to the load. In this case, since power is torque × speed, it is necessary to interpose a reduction gear which will reduce the speed but increase the torque to be applied to the wheels. The power transmitted will be unchanged (neglecting frictional losses). This is called 'matching'. The earliest reported appreciation of this problem appears to be by Archimedes who is quoted as saying: 'Give me a lever long enough, and a fulcrum strong enough, and single handed I can move the world'. (but presumably not much). The electrical equivalent of the gear box is the transformer which enables electrical systems to be matched by changing the voltage and current from the source but at the same time keeping power (volts × current) constant, see Fig. 3.15.

3.9 Practical considerations in selecting energy convertors

So far we have seen that apart from heat energy it is possible under ideal circumstances to convert one form of energy completely to a different energy form. In the case of thermal energy, if a source and sink of constant temperature are available, the efficiency of conversion is limited to a maximum value set by the ratio of the temperature of the heat source to that of the heat sink. In addition, in practical, non-ideal situations, irreversible losses occur which reduce the efficiency of conversion to a value less than that theoretically possible. As an example, consider the use of coal for domestic heating. The simple direct method, Fig. 3.16*a*, would be to burn the coal in a domestic boiler. Almost 60–65% of the energy content would be used to heat the water in the central heating system and the remaining 40–35% lost to the atmosphere. Alternatively, the coal could be burnt in a central power station and the resulting electricity used in an electric fire. Fig. 3.16*b* shows the processes involved and gives an overall efficiency of 32% or about half that for the direct system. It is not, however, clear that we are comparing like with like and it could be argued that less energy will be used by the electric fire, which, it is assumed, will be placed more conveniently and which can be switched on and off as required. A useful way of showing energy flow is the Sankey diagram. In the Sankey diagram the width of the flow line is proportional to the energy flowing down it. Fig. 3.17 shows the above example in Sankey form.

In selecting an energy convertor for a particular purpose efficiency as such is not usually an explicit criterion. For example, in selecting a central heating

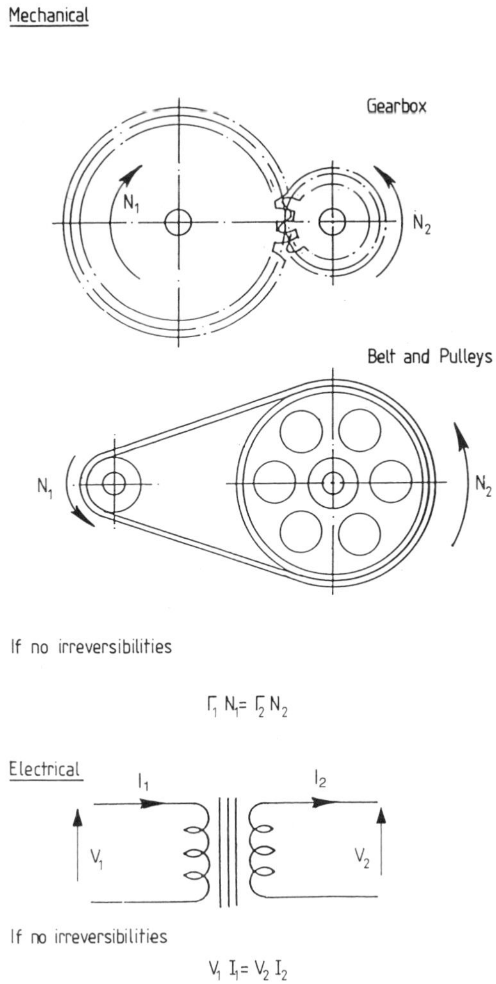

Fig. 3.15 *Matching*

system for a house the lower efficiency of a coal burning system may well be more than offset by the higher price of oil. Other factors, such as capital cost and convenience, will also need to be taken into account. Not all energy forms are equally suitable; for example, as mentioned previously, it is difficult to find an acceptable alternative to electric light. It is often said that electricity should no longer be used for domestic heating. This assumes that the heat loads are the same for all forms of heat; this is no doubt roughly true in a conventional house. However, if we were to adopt electrically heated clothing as used in some military aircraft (not a serious suggestion) this would no longer be the case at all.

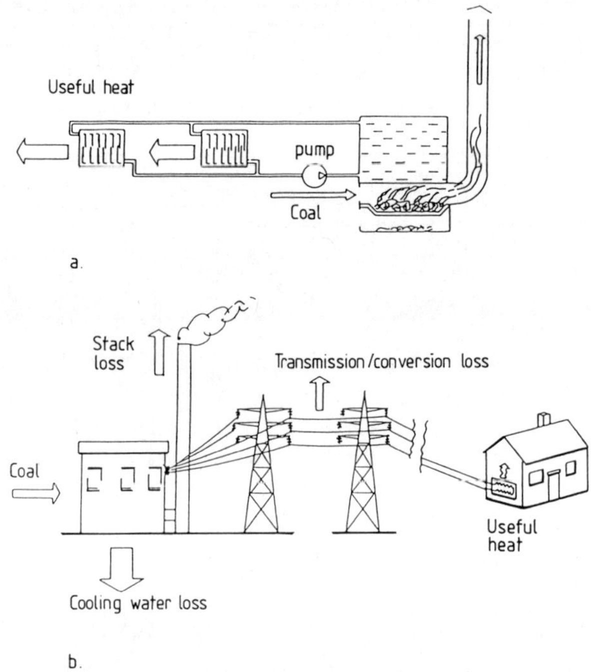

Fig. 3.16 *Comparison between domestic heating methods*
a Domestic boiler
b Simplified system for grid supply to domestic electric fire

The fact that wind and sun are free often causes enthusiasts to overlook such facts as the unreliability and lack of continuity of supply and the high capital cost of plant required to convert these relatively diffuse sources of energy.

The selection of an energy conversion system for a particular application is a complex business. Some factors which need to be taken into account are:

Capital cost: this includes the initial cost, together with the cost amortised (spread over) the lifetime.

Running cost: including cost of fuel (if any) and efficiency of conversion, and cost of servicing and replacement parts.

Maintenance requirement: complex plant may be quite satisfactory in a large installation, e.g. in a power station where skilled maintenance can be provided, but quite unsuitable in a domestic situation or a remote area where servicing skills are not available.

Reliability and continuity of supply: for irrigation purposes all that is necessary is that the plant should provide the average energy required to pump the water. A windmill, for example, would be satisfactory in this respect. On the other hand, if the power is required for the lights in a hospital operating theatre, or even to power the domestic television, the continuity of supply is essential.

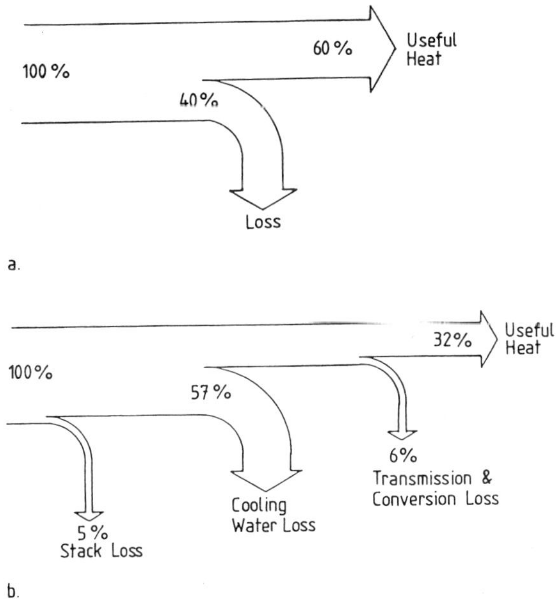

Fig. 3.17 *Comparison between domestic heating methods–Sankey diagram*
a Domestic boiler
b Electric fire

Safety: this is an important consideration and one sometimes overlooked in discussion of small plant of the alternative technology type. The shedding of a blade by a high speed windmill or the explosion of a methane generator can be dangerous.

At government level a much broader view is required when considering energy conversion policy matters. Such factors as security of supply, import substitution, pricing policies, long term energy policy, effect on employment, environmental pollution and safety are all of importance.

References

1 ROGERS, G. F. C. and MAYHEW, Y. R.: 'Engineering thermodynamics, work and heat transfer' 3rd edn (Longmans, London, 1980)
2 RAMSEY, J. A.: 'A guide to thermodynamics' (Chapman and Hall, London, 1971)
3 BRUGES, E. A.: 'Available energy and the Second Law analysis' (Butterworth, London, 1959)
4 RICE, G.: 'Stirling engine availability criteria', 18 IECEC Conference, Orlando, Florida, USA, August 1983

Chapter 4

Heat engines – the conversion of heat to mechanical energy. Reversed Heat Engines – Refrigerators

4.1 Introduction

The Industrial Revolution was based on power provided by reciprocating engines which made use of steam generated in coal-fired boilers. In modern steam plant the steam engine has now been displaced by the steam turbine which is used in power stations for the large scale generation of electricity and also for ship propulsion. Nevertheless, steam engines are still to be found in many parts of the world where they can be seen operating machinery in sugar mills and hauling trains on the railways. Some 8000 steam locomotives are still in service in China. For railway use steam is raised by coal or wood, and in the factories waste biomass, such as bagasse, is employed as fuel. There may well now be a market for small reciprocating steam engines in the 1 to 20 hp range in the rural areas of the developing countries. Coal, and to a less extent wood, provided the energy for the early steam engines and coal has continued to the present to make a major contribution to central electricity generation. Over the past three decades the use of oil to raise steam has increased considerably and nuclear energy has begun to make a significant and increasing contribution.

In the early days of the steam engine the possibility of a boiler explosion was an ever-present hazard. There was little control over the materials from which the boilers were constructed, there was no indication of water level and pressure gauges were not used. Safety valves, where fitted, were often incorrectly adjusted, either deliberately, in order to increase power, or by carelessness. Because of this safety problem there was much interest in developing alternative, safer, engines in which a gas rather than steam was used as the working fluid. An American, Ericcson,[1] was particularly prolific in his inventions. Another inventor was Revd Robert Stirling, a Scottish clergyman, who, in 1816, patented a novel hot air engine which bears his name. At first the Stirling engine received considerable attention but interest waned with the improvements in boilers and later the development of internal combustion engines. More recently, interest in the engine has revived, and it has possible applications in association with fossil, biomass and solar powered heat sources.

The physical principles underlying the operation of vapour reciprocating and turbine convertors and also the reciprocating Stirling engine and closed cycle gas turbine are discussed in this chapter. All these engines are *heat engines* as defined in Chapter 3. Finally, refrigeration is described.

4.2 Rankine cycle engines (vapour engines)

4.2.1 Introduction

The basic components of a simple vapour engine system are shown in Fig. 4.1. Liquid is evaporated in a boiler and is then allowed to expand in an expander, hence performing mechanical work. The expander is usually either a

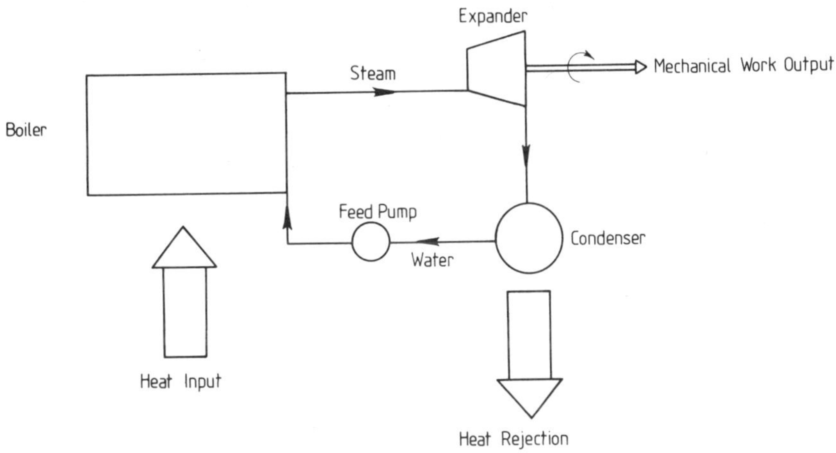

Fig. 4.1 *Basic components of a simple Rankine engine*

reciprocating engine or turbine. After expansion the vapour is condensed and returned to the boiler by means of a pump, the feed pump. A small fraction of the total work output of the expander is used to drive the feed pump. Since this is a very small fraction of the total work produced it can usually be ignored in calculations. Although the first vapour engines used steam as the working fluid, any other liquid could in principle be used and the next section considers the desirable properties for a working fluid and discusses how fluids are selected for a particular application.

4.2.2 *Properties of vapours*

4.2.2.1 Saturated vapour: When a quantity of water is placed in a closed container from which the air has been removed, the space not occupied with water will become filled with vapour. The final steady pressure attained by this vapour will depend only on the temperature of the water. The water vapour is said to be 'saturated' and the corresponding pressure is called the 'saturated vapour pressure' or SVP. At room temperature the SVP of water is low, for example at a temperature of 15 °C the SVP for water is only about 1705 N/m^2 or 0·017 atm. The low pressure of water vapour at low temperatures, and hence its large volume, causes difficulty in the design of low temperature solar powered engines and has led to the use of other liquids as the working fluid. When the container temperature is raised the SVP of the water vapour rises rapidly until it reaches a value of $1{\cdot}01 \times 10^5$ N/m^2 (1 atm) at 100 °C. If the container is open to the atmosphere then the temperature will cease to rise and the water is said to 'boil'. The temperature of the boiling point will depend on the value of the atmospheric pressure, and, for example, on top of Mount Everest (which is at a height of 8900 m above sea level) water will boil at 71 °C. If the container is sealed, then, as more heat is added, the pressure will continue to rise.

As the saturated vapour pressure rises the vapour density increases, or, in other words, the volume of a given mass of vapour decreases. Table 4.1 shows how the vapour pressure and vapour density of water change with temperature. We see that the vapour pressure and density at 20 °C are 2300 N/m^2 and 0·0175 kg/m^3 but at 100 °C they have changed to 1 atm and 0·62 kg/m^3. This gives a ratio of 44 for pressure change and 34 for density change between these two temperatures. In order to allow for the full expansion of water vapour over this temperature range a piston in a cylinder would require to expand over a volume ratio of 34.

4.2.2.2 *Superheated vapour*

If a quantity of saturated steam is isolated from its water surface and its temperature is raised, it will behave as a gas. The temperature will no longer solely determine the pressure, but the latter will now also depend on the volume of the container. Steam heated in this way is called 'superheated steam' and other vapours are similarly referred to as 'superheated'.

4.2.2.3 *Wet vapours*

If an isolated quantity of saturated steam is cooled then partial condensation will occur and the pressure will drop. The resulting mixture is known as 'wet steam' and the fraction of steam in vapour form, by mass, is known as the 'dryness fraction'. The same terminology is employed for other vapours.

The heating of a liquid at constant pressure is of interest in vapour engines. For example, if ice, at atmospheric pressure, is heated it will first melt and the resulting liquid will rise in temperature until it reaches 100 °C when it will turn

Table 4.1 *Properties of saturated steam*

Temperature °C	Pressure bar	Vapour density kg/m^3	Vapour specific volume, m^3/kg
0·01	0·00611	0·00485	206
20	0·023	0·0175	57·8
30	0·0424	0·0304	32·9
40	0·0737	0·053	19
60	0·199	0·13	7·7
80	0·474	0·29	3·4
100	1·01	0·625	1·6
120	2·0	1·14	0·88
150	4·8	2.56	0·39
200	16·0	8·13	0·123
250	40·0	20·4	0·049
300	86·0	46·3	0·0216
350	166	115	0·0087
374·15*	221·20	315	0·00317

*Critical Point
1 bar = 10^5 Pa
1 atm = 1·013 bar

to vapour. The vapour temperature will then rise. The process is shown in Fig. 4.2*a*. The heat supplied is plotted on the horizontal axis and the temperature on the vertical axis. Heat supplied at constant pressure is called enthalpy h. The quantity of heat necessary to cause the solid–liquid phase change and the liquid–vapour phase change are called the enthalpy of fusion and the enthalpy of evaporation, respectively. In the case of water–steam the enthalpy change is large, approximately $2{\cdot}2 \times 10^6$ J/kg. The temperature/enthalpy relation depends to some extent on the pressure at which the process is carried out and the values are given for water in standard steam tables.[2] Fig. 4.2*b* and *c* shows the relationship, for water, when plotted on temperature enthalpy and temperature entropy co-ordinates.

4.2.3 Reciprocating engines

Since the principle of the reciprocating steam engine[2,3,7] is well known it will be described only briefly. This description will, however, serve to introduce the general class of vapour engines.

Suppose that water is boiled in a container and the steam produced is allowed to operate an engine. Assume that the pressure of the steam is maintained constant by controlling the heat input to balance the steam requirement of the engine. If the steam is admitted by a valve to a piston cylinder arrangement it can be allowed to expand in the cylinder, thus doing work on the piston. When the inlet valve is closed the steam in cylinder will still be at a pressure greater than that of the surroundings and can do further work by expanding until its pressure has fallen almost to that of the surroundings. A second valve now opens and allows the steam to escape. The piston can be returned and the process repeated. In the engine which has just been described the steam is released at atmospheric pressure and therefore at a temperature

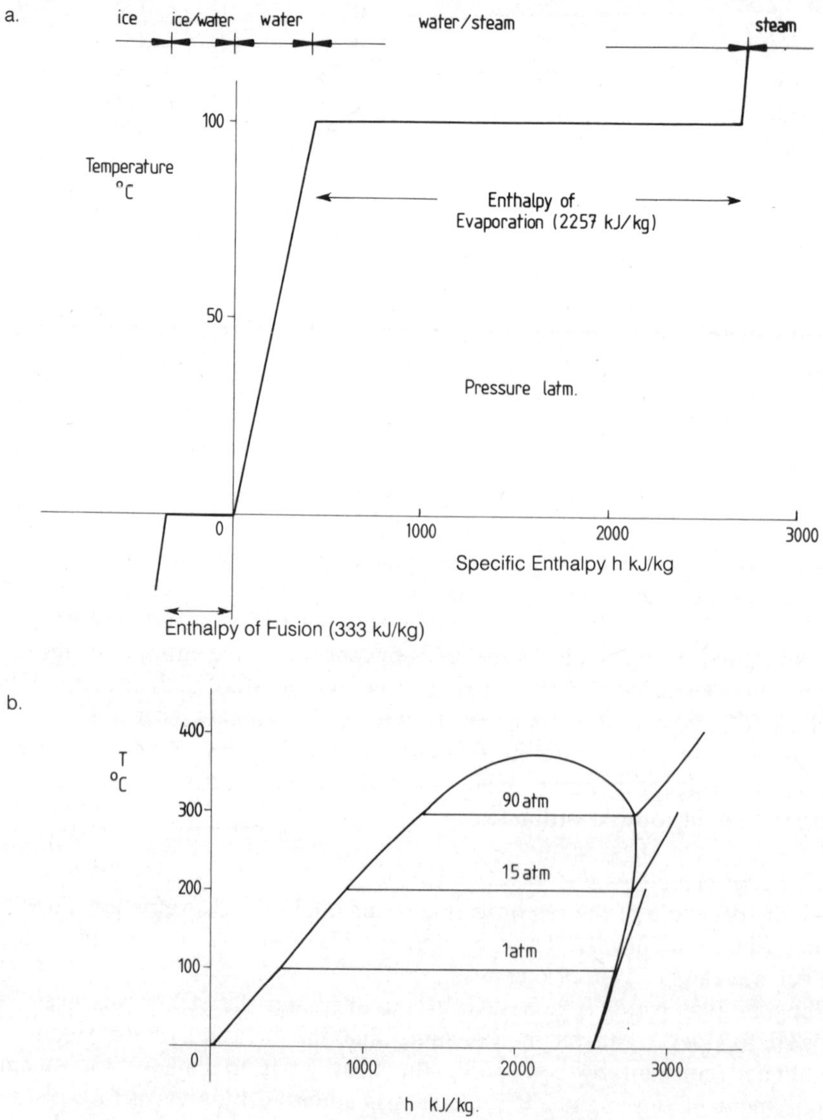

c.

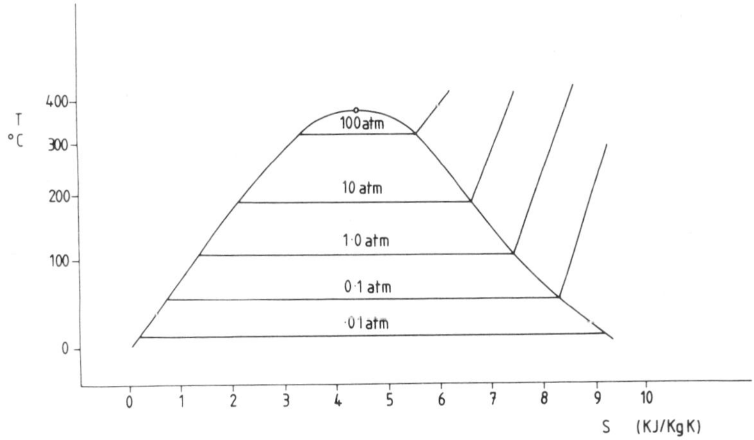

Fig. 4.2 *Properties of water*
a Temperature versus heat supplied (enthalpy) for water at 1 atm pressure
b Temperature versus *h* for water at different pressures
c Temperature versus entropy for water at different pressures

corresponding to the boiling point of 100 °C. Since ambient temperature is around 20 °C, further energy could have been extracted by allowing the steam to condense in a second container, the condenser. In early steam engines condensation was achieved in the working cylinder by injecting a spray of cold water. Watt appreciated that this cooling of the cylinder walls led to some condensation of the next charge of steam and to avoid this loss of energy suggested the use of a separate condenser. During condensation the pressure in the cylinder is less than atmospheric pressure and this higher pressure of the atmosphere causes the piston to return, thus doing further work. This work is not of course obtained from the atmosphere, it has already been performed by the piston when it pushed away the atmosphere on its original expansion stroke. The pressure in the condenser will be very much less than 1 atm and it is necessary both to maintain a good seal and also to extract any air that may leak in so that an air pump is required. What has been described above is the single acting expander, a more usual arrangement is the double acting expander, see Fig. 4.3.

In the early days of the steam engine, variability of the quality of materials and also difficulty in controlling standards in manufacture meant that the strength of the boiler was uncertain and pressures had to be kept low. Even so, boiler explosions were frequent. In fact, because of the danger of explosion, Watt always employed low steam pressure in engines in which the boiler temperature was around 100 °C and steam pressure had a maximum value of little more than 1 atm. Because of the low boiler temperature the efficiency of the early steam engines was low and a Watt atmospheric engine of *circa* 1796 would have an efficiency of around 3%. Later Trevithick pioneered the use of higher boiler pressures for traction use; his first locomotive ran on the Penydarren tramway in Wales in 1803.

The reciprocating steam engine is a form of heat engine. We have seen that a heat engine takes in heat from a high temperature source, converts some of this energy to mechanical or other energy form and rejects the remainder of the heat to a cold reservoir. The engine works in cycles, that is, without permanently altering the working fluid, in this case water. For given temperatures of the heat source and sink the maximum efficiency of conversion will occur when the heat transfer processes take place at constant temperature. This is the case in evaporation and condensation since heat is absorbed or rejected at constant temperature in the liquid to vapour phase change in the boiler and also in the phase change vapour liquid in the condenser.

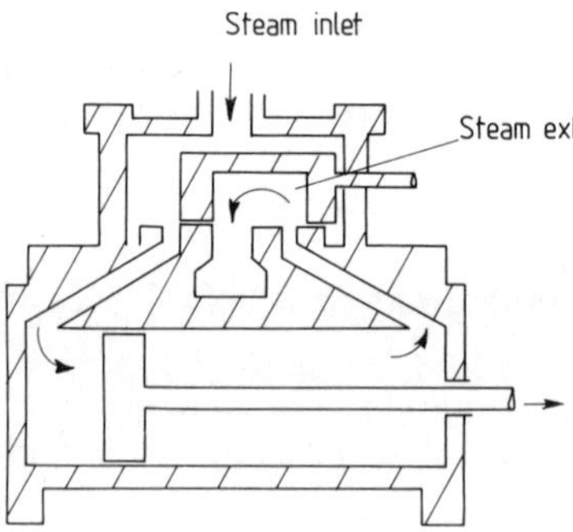

Fig. 4.3 *Double acting expander*

The theoretical efficiency of the steam engine is thus very close to the theoretical limit of Carnot efficiency but does not quite reach this value. These engines work on a thermodynamic cycle called the Rankine cycle and are sometimes referred to in the literature as Rankine engines.

The Carnot and Rankine cycles are shown in Fig. 4.4*a* and *b*. The Rankine differs from the Carnot in that the steam is completely condensed at low pressure and then raised to boiler pressure by means of the feed pump. This results in a slightly lower theoretical efficiency. However, in practice this theoretical loss is more than compensated for by practical inefficiencies arising from the need to compress wet steam if the Carnot cycle was to be followed. Practical steam engines usually employ superheat as shown in Fig. 4.4*c*. This has two advantages: it raises the average temperature at which the working fluid is heated, and hence the cycle efficiency; superheat also enables the steam to be expanded without partial condensation, a considerable advantage when used in steam turbines since the water droplets can cause severe blade erosion. Vapour cycles are often plotted on enthalpy/entropy axes, this is known as a Mollier diagram. Fig. 4.5 shows a Rankine cycle with superheat, plotted on enthalpy/entropy co-ordinates.

In order to achieve a good thermodynamic conversion efficiency it is necessary to have a large temperature difference between the boiling and the condensing temperatures. The upper temperature is determined by either the nature of the heat source or by maximum pressure considerations, whereas the

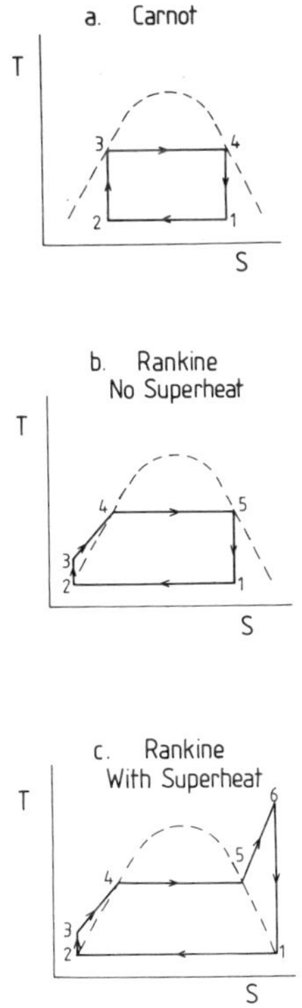

Fig. 4.4 *Carnot and Rankine cycles*

a Carnot

1–2 Constant temperature and pressure compression of water steam mixture

2–3 Isentropic compression of water steam mixture

3–4 Constant temperature expansion owing to water evaporating to steam at constant pressure

4–1 Isentropic expansion of dry steam at high pressure to wet steam at low pressure

b Rankine – no superheat

1–2 Constant pressure and temperature compression until the wet steam is completely condensed to water

2–3 Pressurisation of water in feed pump to raise pressure to high pressure level

3–4 Heating of water at constant high pressure

4–5 Vapourisation of water at constant temperature and pressure

5–1 Isentropic expansion of initially dry steam 5 to wet steam 1

c Rankine – with superheat

Processes 1–5 as the Rankine without superheat

5–6 dry steam further heated at constant pressure to give superheat

6–1 Isentropic expansion to dry steam at point 1

lower or condensing temperature is fixed by the method of cooling. If cold water is available then it may be possible to cool the steam to a temperature of 30 °C or so. This is difficult to arrange in a mobile application such as a steam locomotive and here the spent steam is allowed to escape to the atmosphere after expansion in the cylinder. In order that the steam can escape into the atmosphere it must obviously be at a pressure somewhat greater than atmospheric pressure, and hence an effective condensing temperature of around 100 °C. Because of this relatively high condensing temperature the efficiency of the best railway locomotive was not greater than about 9%.[4] This figure for efficiency refers to the drawbar horsepower or the net energy produced by the locomotive in pulling the train. The gross energy produced by the locomotive includes the losses in transmission through to the wheels and rails and also the energy required to move the locomotive itself. The corresponding gross efficiency is nearer to 14–16%.[5] A further disadvantage of non-condensing engines, as these are known, is that steam is lost after each expansion instead of being condensed and returned to the boiler, and a

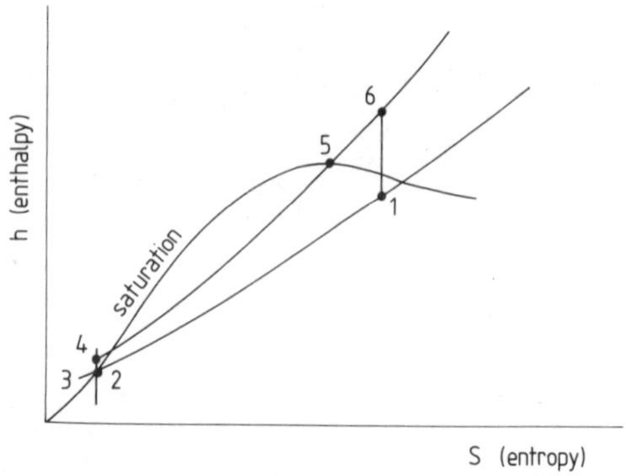

Fig. 4.5 *Rankine cycle plotted on a Mollier diagram*
Heat in = $h_6 - h_2$
Work out = $h_6 - h_1$

$$\text{Efficiency} = \frac{h_6 - h_1}{h_6 - h_2}$$

continuous feed of fresh water is thus required by the boiler. The boiler feed water must be treated to remove dissolved salts (softened) otherwise scale will be deposited on the boiler surfaces which will lower the heating efficiency. There are, however, some compensating advantages which arise from using non-condensing engines. Since the steam is always at a pressure greater than atmosphere, and is rejected on each cycle, an air pump is not necessary. Also the practice of exhausting the steam through the chimney to provide a draught through the furnace eliminates the need for a blower. Condensing engines achieve a higher efficiency than the non-condensing type, but, as has been seen, one problem experienced when water is used as the working fluid is the very large expansion in volume of the steam. This expansion ratio is more than can be achieved in a single cylinder and it is usual to arrange for several cylinders of increasing size to be connected in series. The steam expands in the first cylinder to a lower pressure and is then fed into a larger cylinder into which it again expands. This practice is called compounding. A typical triple compounded reciprocating engine of the 1880s would have an efficiency of around 18%. Fig. 4.6*a* and *b* shows how the use of higher boiler pressure led to improved efficiency.[6]

An important practical problem which arises in the design and operation of

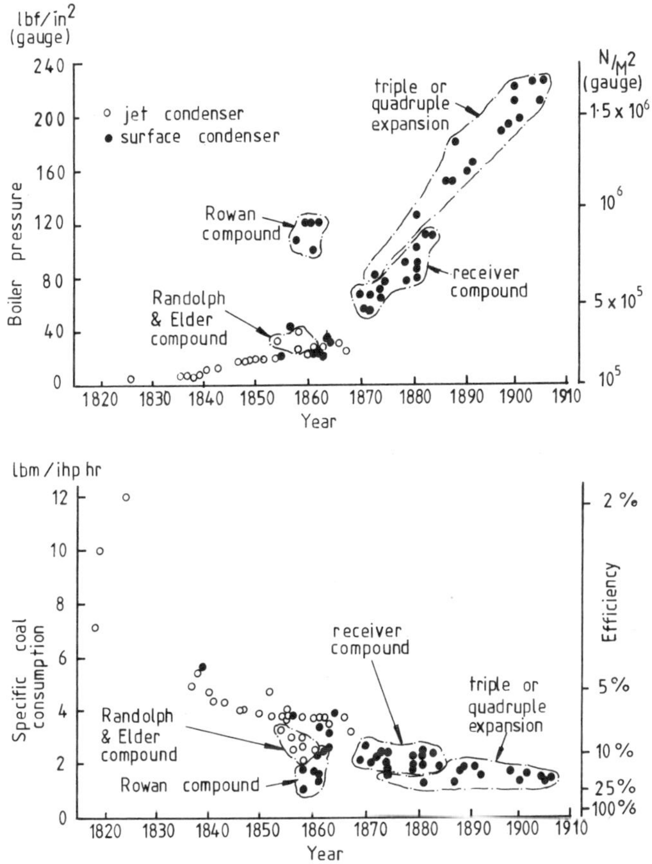

Fig. 4.6 *Boiler pressure and specific coal consumption of marine steam engines: 1815–1910 (Reference 4)*
○ jet condenser
● surface condenser

reciprocating engines is the provision of lubrication for the piston.[3] An arrangement to feed oil to the moving piston is required but this oil must be prevented from entering the boiler.

In condensing engines the working fluid is conserved and fluids other than water may be used. The properties of some of these fluids and the criteria for their choice is discussed in a later section.

4.2.4 Turbine expanders

As an alternative to expanding steam or other working fluid in a piston cylinder arrangement it is possible to allow the steam to escape from the boiler in the form of a high velocity jet and to use the kinetic energy in this jet to rotate a wheel or turbine. This method is also used in a form of hydraulic turbine, the Pelton turbine (Chapter 10). Turbines of this type are called *impulse* turbines. In the impulse turbine all the available energy is converted to kinetic energy in the jet at the nozzle.

For a single stage, in order to transfer all the kinetic energy from the jet to the blade, the blade velocity should be half that of the jet. A simple, single stage steam turbine of this type is called a de Laval turbine. Since steam leaves the nozzle with a very high velocity (typically 500 m/s) then the blade velocity will be high and hence the turbine rotation speed. De Laval turbines rotate at speeds of 20 000 to 25 000 rev/min.

Modern large impulse steam turbines employ multiple stages; in each stage the steam is expanded to a lower pressure and the resulting jet of steam is made to apply a force to the blades of a rotor mounted in a common shaft. This arrangement is called a Rateau turbine. Alternatively, the steam can be expanded to final pressure in one nozzle stage but the resulting jet passed successively through a series of rotor blade stages in each of which it is decelerated (Curtis turbine).

The other general turbine type is the reaction turbine, which operates in the same manner as a rocket, employing the reaction on the blades as the steam is expanded within the blade stage. The garden water sprinkler works in this way and can be regarded as having 100% reaction. Modern reaction turbines are also built in multiple stages; they are not pure reaction but normally employ 50% impulse, 50% reaction in each stage. This means that 50% of the available energy is converted to kinetic energy in the nozzle and the remaining 50% change occurs on passing through the turbine rotor. Nothing is new, and the first reported use of the reaction turbine is as long ago as around 300 BC when a Greek engineer and mathematician named Hero designed a small demonstration model, see Fig. 4.7.

Large steam turbines are normally built up from an impulse stage at the high pressure end and a reaction stage at the low pressure end. Because of the large volume increase at low pressure there are usually two reaction turbines in parallel, each accepting half the steam flow.

Although the Greeks were aware of the steam turbine it was not until 1884

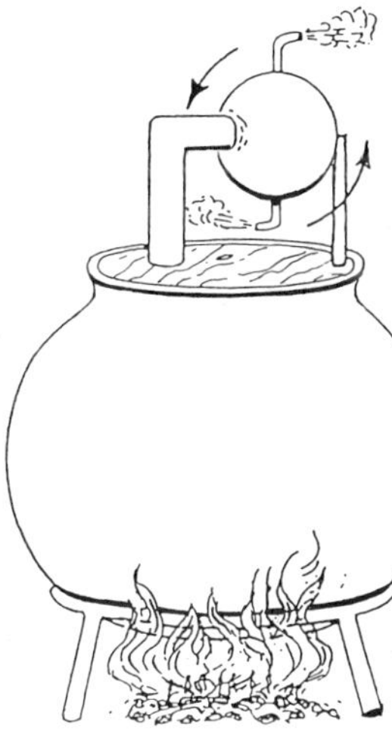

Fig. 4.7 *Hero's turbine*

that the first practical steam turbine was built by Sir Charles Parsons. The main advantages of the turbine over the reciprocating steam engine are:

1. Higher efficiency
2. High power/weight (because of high speed)
3. High power/volume
4. Low noise and vibration
5. Low maintenance cost and low reliability

The steam turbine has almost entirely displaced the reciprocating engine in industry for high power applications. The steam turbine now supplies the greater part of the world's electricity supply, and turbines as large as 660 MW are in service. Small turbines tend to have very high shaft rotation speeds and also require very tight mechanical tolerances, for these reasons reciprocating expanders are preferred for small power output. The size at which the transition from reciprocating to turbine occurs will depend on the working fluid chosen and the application for which it is intended, but typical ranges for steam will be greater than 100 kW and for organic fluids 1 kW to 250 kW.

4.2.5 Choice of working fluid

Water is not the only fluid which can be used as the working fluid for a heat engine and there is, in fact, a number of choices, some of which have advantages over water for particular applications. What then are the criteria to be considered in the selection of a working fluid? Clearly it should, if possible, be cheap, non-corrosive and non-toxic. In addition it should enable the engine to operate over the temperature range available without either too high a pressure at the highest temperature nor too large a volume at the lowest temperature. For the temperature readily achieved with solar heat sources, say less than 200 °C, there are a number of organic liquids which have suitable properties. These organic fluids have the odd property of 'dry expansion'. When steam is expanded some of it will condense to give droplets; this is termed 'wet expansion' and is also exhibited by liquid metals. These liquid droplets can be a nuisance by causing blade erosion in turbines but helpful in providing a liquid for lubrication in reciprocating expanders. For those fluids which show 'dry expansion' when a saturated vapour is expanded it does not partially condense but becomes drier, that is it becomes superheated Fig. 4.8. If a very high temperature heat source is available then liquid metals such as mercury, potassium and sodium have advantages. Fig. 4.8.

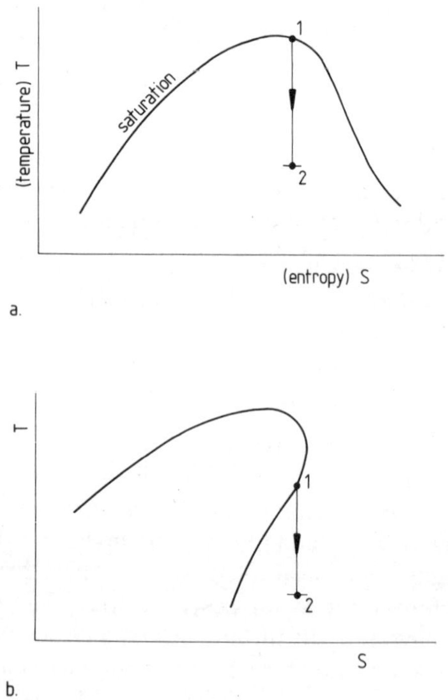

Fig. 4.8 *Vapour expansion modes*
a Wet expansion
b Dry expansion

Water obviously has the major advantages of cheapness, non-toxicity, availability, safety and ease of containment. It does, however, give rise to rather high pressures as the temperature is raised, also at low pressure the large volume can cause design problems. It has been pointed out earlier that for steam to expand from a pressure of 10^5 N/m^2, corresponding to a temperature of 100 °C, to a pressure of $2{\cdot}3 \times 10^3$ N/m^2, corresponding to a temperature of 20 °C, a volume expansion ratio of 34 is needed. If an organic fluid freon 113 is used between the same temperature the corresponding pressures are $9{\cdot}53 \times 10^5$ N/m^2 and $9{\cdot}3 \times 10^4$ N/m^2 with a volume expansion of 9·1.

The properties of the two fluids are given in Table 4.2. Since the latent heat, or enthalpy of vapourisation, is less for freon 113 than water, a mass 15 times greater is needed if the same amount of latent heat is to be removed. However, since the density of freon is much higher that of steam, the relative initial volumes of steam to freon have the ratio 5.3. Fig. 4.9 shows the relative cylinder sizes for the two fluids to give the same energy/expansion, a ratio in total volume of 34/1·71 or 20.

	STEAM	FREON 11
MASS	1	15·4
INITIAL VOLUME (100°C)	1	0·188
FINAL VOLUME (20°C)	34	1·71
RATIO FINAL VOLUME / INITIAL VOLUME	34	9·12

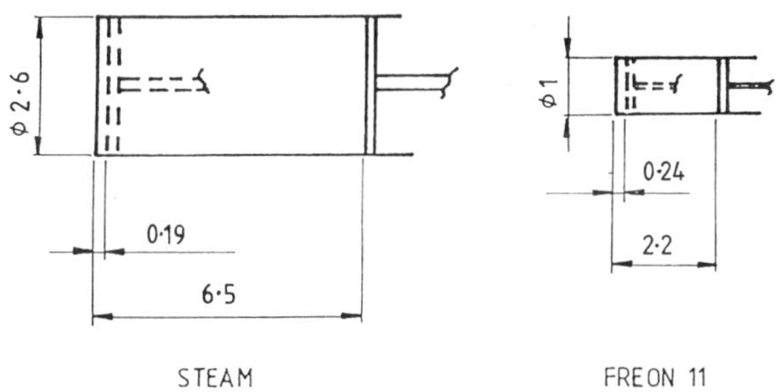

Fig. 4.9 *Relative cylinder sizes for water and freon 11*

Table 4.2 *Comparison of some properties of freon II with water*

Temperature °C	Material	Vapour pressure, bar	Vapour density, kg/m^3	Latent heat kJ/kg
100	H_2O	1·01	0·60	2258
	freon II ($CFCl_3$)	9·53	49·04	146·9
20	H_2O	0·023	0·0175	2448
	froen II ($CFCl_3$)	0·93	5·38	183·4

Organic fluids such as freons are also used in small vapour turbines, but for rather different reasons than those dictating the choice of working fluid in reciprocating engines. The blade speed in a turbine is related to the velocity of the vapour on expansion through the jet. In the case of steam this velocity is high and leads to high blade velocity and hence high rotational speeds, particularly for small turbines. The vapour velocity is found to be inversely proportional to the square root of the molecular weight of the vapour. Water has a molecular weight of 18 whereas freons are around 360, giving a factor of $\sqrt{20}$ or 4·5 reduction in velocity for the same pressure ratio. Also, owing to the higher mass flow, for a given power, and lower vapour velocity, the blade sizes are larger, thus the tolerances on small power turbines are less critical than is the case for steam.

If the engine is designed to produce electricity then it is possible to place both the engine and generator in the same container which can be sealed, the electrical output being fed out through insulators. If, however, the engine is intended to produce mechanical power then it is necessary to arrange for a rotating or sliding seal so that a shaft can be coupled to the external load. The working fluid will be on one side of the shaft seal and the atmosphere on the other. The seal will be required to prevent loss of the working fluid to the atmosphere. If, in addition, the lowest pressure, the condenser pressure, of the working fluid is below atmospheric pressure then the seal must also be arranged to prevent the ingress of air at this pressure. From Table 4.1 it is seen that water will present this air leakage problem but this is not the case with some of the organic fluids whose vapour pressure will be greater than atmospheric pressure at all working temperatures.

4.3 Engines using gas as the working fluid

In the previous section the performance of engines employing vapour as a working fluid was discussed. This section describes engines in which the working fluid is a gas. The first engine described, the Stirling engine, is a reciprocating engine and the second, the closed cycle gas turbine, is a rotating engine.

4.3.1 Stirling engine

The Stirling engine is an old concept first proposed by a Scottish Clergyman, the Revd. Robert Stirling, in 1816. The Revd. Stirling was born in 1790 and ordained Minister in September 1816. Two months later he took out the patent for the Stirling air engine and heat regenerator, or economiser, as he described it. Further patents were taken out in 1827 and 1840, in association with his brother James. It is not entirely clear what was the contribution of James. Kolin[1] suggests that it was James who was responsible for the design and construction of the early Stirling engines although the original idea was

Robert's. Stirling was not only a gifted inventor but was also a popular and effective Minister and a respected theologian; he was in fact presented with an Honorary Doctor of Divinity Degree at St Andrews in 1840. He had a long life, retiring in 1876 and dying two years later. Other members of the family made important engineering contributions. His brother studied for the church but did not complete his theological studies and became a mechanical engineer. Stirling's four sons were well-known mechanical engineers, perhaps the most notable being Patrick, who was the designer of the Great Northern Railway's 8 ft single drive wheel locomotive.

It is interesting to note that at the time of his invention Stirling was unaware of both the First and Second Laws of Thermodynamics. In fact, he believed that heat was only necessary to replace work lost by friction. For this reason, he did not make proper provision for heating areas and his engines suffered from a tendency to burn out the heating surfaces. In spite of his lack of theoretical understanding, Stirling clearly had an intuitive appreciation of engine requirements and he designed an engine capable of achieving the maximum theoretical thermodynamic efficiency.

In its original form the Stirling engine was slow, heavy and inefficient, but in spite of these limitations a number was made and successfully operated. Small engines of power output less than 1 hp were produced in considerable numbers for such applications as domestic and office fans and dentist drills. These engines, which were fuelled by kerosene burners, were produced until the 1920s (other applications included pumping in mines).

The real advance in the Stirling engine development started in 1938 when the Philips Company took up the idea as a possible solution to the problem of providing power supplies for remote radio receivers and transmitters.[8] Power levels of up to a few hundred watts were envisaged. Work continued through the Second World War and the Philips' results aroused considerable interest in the USA at the end of the war. The advent of the transistor removed the need for such power supplies and Philips turned their attention to the design of a larger engine. By using modern materials they were able to design an engine, having an efficiency and weight comparable to those of a modern automotive diesel. (A very complete account of the development of early gas engines is given by Finkelstein in Reference 10 and a good summary of the Philips work is given in References 8, 9 and 11). Later, United Stirling in Sweden set up a major programme in engine design and development[9] and development work has been carried out in both universities and industry in other countries, including China, Japan, France, Germany, Sweden, UK, USA and USSR.[12]

In a recent development in the UK a consortium[13] has been set up between three universities (Bath, Royal Naval Engineering College, Manadon, and Reading) and three industrial firms (Petters, AED and BP) to develop a Stirling engine. The engine is filled with helium at a mean pressure of 150 bar and a peak temperature of 700°C and is designed to give a power output of 20 kW, see Fig. 4.10. The consortium, which was set up in 1978, represents a novel and potentially important experiment in university/industry collaboration.

The Stirling engine is a heat engine and has the characteristics, both advantages and disadvantages, of all heat engines, as described in Chapter 3.

Fig. 4.10 *20 kW U.K. Consortium Stirling engine*

Thus the Stirling engine can be used with any heat source including direct flame heating from oil, gas, wood or coal combustors, by solar and by nuclear energy. As an alternative to conventional combustion fuels such as coal, oil and gas, vegetable waste can be combusted in a fluidised bed. The engine can be heated by coupling it directly to one of these sources of heat or it can be separated from the heat source and the heat transported by a heat pipe. There is clearly considerable flexibility in the choice of heat source. A major economic penalty is the need for a high temperature heat exchanger to transfer the heat to the engine working fluid from the heat source. In order to achieve good heat transfer, a large surface area is needed and hence a complicated arrangement of small bore piping. Since the working fluid is not consumed, an expensive substance such as helium can be used; however, if the power is to be extracted by a mechanical shaft it is necessary to design a seal between the engine body and the output shaft which will not allow any significant loss of helium. The seal problem is still one of the major technical difficulties in the development of Stirling engines using helium or hydrogen as the working fluid. For this reason interest in using air as the working fluid in lower speed engines has revived. In all Stirling engines two pistons are required to move with a phase shift between them and output is taken from one of the pistons. There is a number of possible designs of piston linkage and drive to achieve this requirement, including the bell crank, rhombic drive, Rinia and others. Finally, the rejected heat must be transferred from the working fluid to the low temperature sink. If the latter is water, as in a marine application, the problem is relatively straightforward; if, however, ambient air is to be used then a large heat exchanger, several times the size of radiator of a conventional internal combustion engine of the same speed and power, will be required.

Since there are various solutions to each of the fundamental engine component problems, the Stirling engine can take many forms. Some of the more popular of these designs are described in the next sections.

However, before considering these detailed engine problems a further Stirling development by Philips should be mentioned. The Stirling engine may be driven in reverse using an external source of mechanical power such as an electric motor. The engine can now be used either as a refrigerator or heat pump depending on whether ambient temperature is used as the high or the low temperature reservoir. Philips manufacture a small Stirling engine which is used as a low temperature refrigerator and operates with the cold reservoir at liquid air temperature.[14,15]

4.3.2 Stirling Engine Cycle

The ideal Stirling engine cycle is indicated in Fig. 4.11. A gas is heated at constant volume (2–3) and then allowed to expand; it is this expansion (3–4) which may be arranged to move a piston and hence perform external work. If the gas is compressed at the same temperature then it will require just as much work to return it to its original volume so that no net work will be extracted. However, if the expanded gas is allowed to cool (4–1) then less work will be required in order to restore it to its original volume (1–2). This is the principle of operation of the Stirling engine cycle. Gas is allowed to expand, doing work, at a high temperature, and then is cooled before compressing to its original volume. Since the gas does not change in temperature in the expansion the

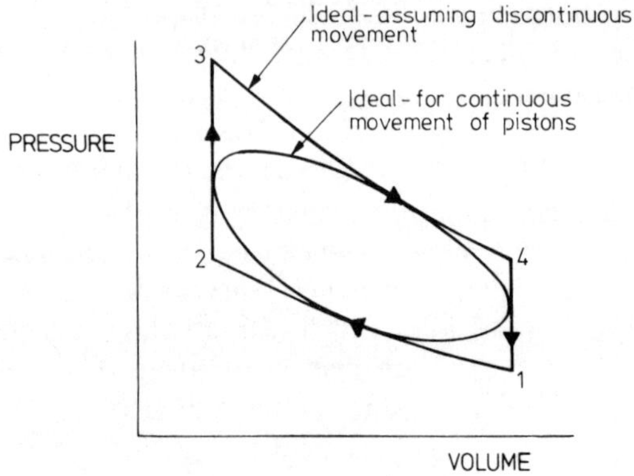

Fig. 4.11 *Pressure–volume diagram of Stirling engine*
1–2 Isothermal compression
2–3 Regeneration
3–4 Isothermal expansion
4–1 Regeneration

work out is equal to the heat taken in. Similarly, the compression work is equal to the heat given out at the cold temperature. Hence:

$$
\begin{aligned}
Q_h &= \text{Heat taken from hot source} \\
Q_c &= \text{Heat given to cold sink} \\
W &= Q_h - Q_c \\
&= \text{work from cycle}
\end{aligned}
$$

$$
\begin{aligned}
\text{Efficiency } \eta &= \frac{W}{Q_h} \\
&= \text{Carnot efficiency}
\end{aligned}
$$

However, in the above description an important process has been omitted; the compressed gas at the end of the compression stroke is at the sink temperature and will require to be heated up to the source temperature before further expansion can take place. This will require a considerable quantity of heat which must be added to the heat taken in during the expansion stroke Q_h and which will significantly reduce the cycle efficiency below that of Carnot. In the Stirling engine the gas is cooled at constant volume at the end of the expansion stroke by passing it through a mass of wire gauze, the regenerator. One end of the gauze mass is at the source temperature and the other end at the sink temperature. In this way the gas is cooled as it passes through the regenerator in a gradual way. On passing back through the gauze to the heat source it retrieves the heat which 'parked' on the way down. Hence no net heat is required for the two constant volume operations. In this way the use of the regenerator enables the Stirling cycle to, theoretically, achieve the Carnot efficiency. The heat exchanged in the constant volume processes is typically several times larger than the heat taken from the hot source Q_h. Hence, in a real engine it is most important that the regenerator should operate with high efficiency, and a practical regenerator can recover as much as 98% of the heat given up to the regenerator.

The energy/cycle is given by the area of the indicator diagram Fig. 4.11. In practical engines energy per cycle can be increased considerably by raising the mean pressure, see Fig. 4.12. Modern engines operated at a mean pressure as

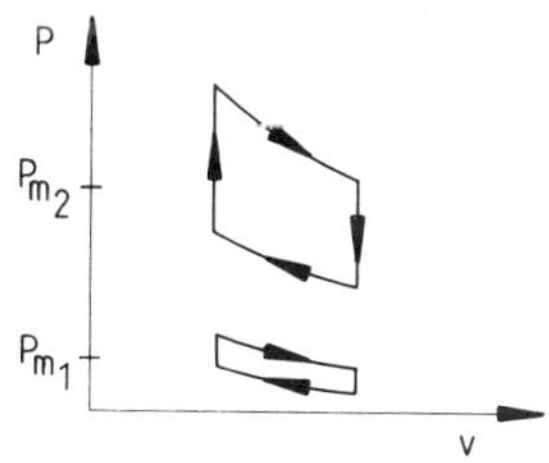

Fig. 4.12 *Effect of mean pressure on energy per cycle*

high as 100 atm. Ideally, as engine speed is increased the power output will rise linearly with speed, since the energy per cycle is constant. In practice, heat transfer limits the power into and out of the engine, thus for a given mean pressure the power will rise less rapidly than a linear relation; as the speed is further increased the internal pressure loss becomes increasingly important and the power output of the engine reaches a maximum value and then falls. The speed at which the maximum power is developed depends on the mean pressure, see Fig. 4.13. The heat transfer from the heater wall to the working

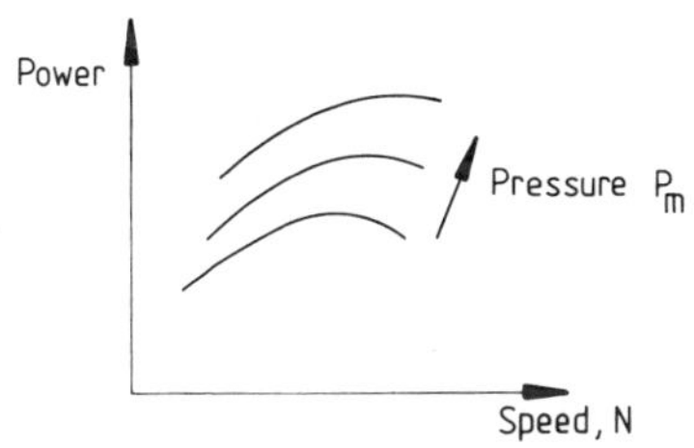

Fig. 4.13 *Effect of mean pressure on power output*

fluid and from the working fluid to the cooler wall can be improved by selecting a gas of high thermal conductivity, such as hydrogen or helium, rather than air, see Fig. 4.14. The use of expensive gases results in the practical problem of developing a seal on the output shaft to prevent gas loss. In the case of air as the working fluid a simple compressor can be used to replace any leakage losses.

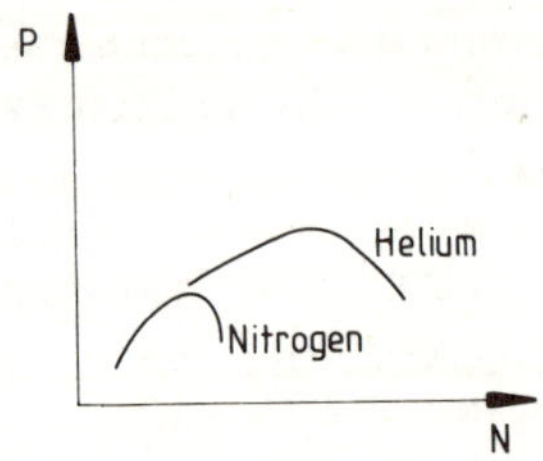

Fig. 4.14 *Effect of working fluid on power output*

The next two sections discuss piston and drive arrangements necessary to perform the thermodynamic cycle and methods of transferring heat to the engine.

4.3.3 Stirling engine geometries

There are two basic Stirling engine geometries

1 Two piston engine
2 Displacer engine

For the *two piston engine*, Fig. 4.15*a*, the sequence of operation is as shown in Fig. 4.16*a*. In this engine, the gas is initially contained in the cold section corresponding to point 1 on the diagram in Fig. 4.11. Piston A is moved to compress the gas at constant temperature to a state corresponding to point 2 on Fig. 4.11. Then both pistons are moved to transfer the gas at constant volume through the regenerator from the cold region to the hot region. The gas is now

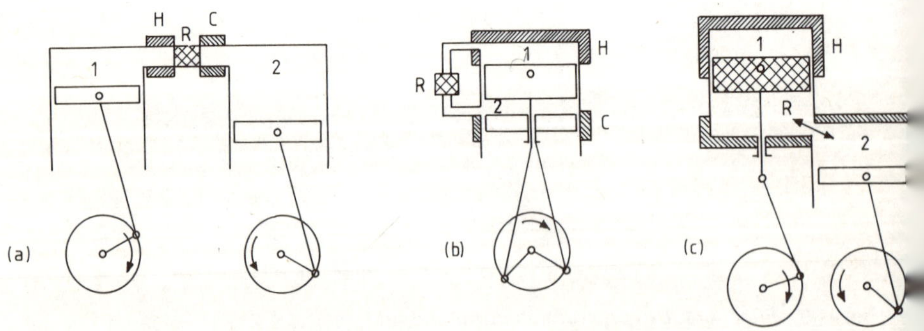

Fig. 4.15 *Main types of Stirling engine*
H Heater
R Regenerator
C Cooler
1 Expansion space
2 Compression space
a Two piston engine
b Displacer and piston in same cylinder – displacer engine
c Displacer and piston in separate cylinder

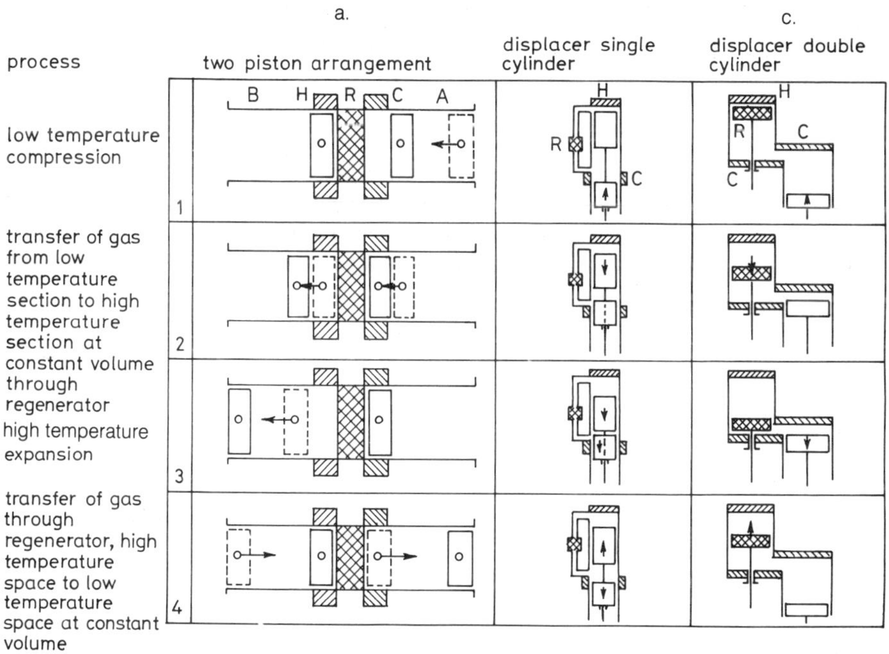

Fig. 4.16 *Sequence of operation of three basic types of engines in the discontinuous motion*
H Heater
R Regenerator
C Cooler

at point 3 in the diagram. piston B is allowed to expand at constant temperature, to point 4 on the diagram, and once more both pistons are moved at constant volume to restore the gas to its original state, point 1.

An obvious disadvantage of this geometry is that of the need to provide a seal for the piston B which operates in the hot region; this problem can be overcome by extending the piston above the piston rings by using a heat resisting material such as ceramic or stainless steel. Fig. 4.17 shows the use of a double acting piston in a multicylinder arrangement.[8]

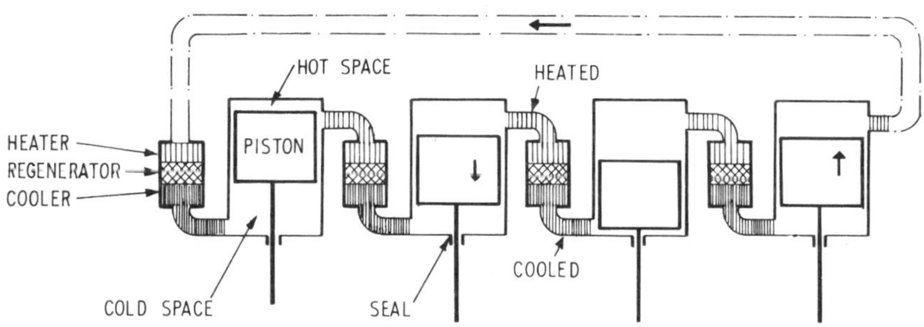

Fig. 4.17 *Multicylinder double-acting Stirling engine*

The *displacer type engine*, Fig. 4.15*b*, has a single power piston working in a cylinder closed at one end and the gas is moved from the hot to the cold region by moving a solid mass of material from one region to the other, thus displacing the gas; this piston is known as the displacer and does not perform any work. The regenerator may be either situated externally to the engine, as in Fig. 4.15*b*, or may be incorporated into the displacer. The sequence of operation is given in Fig. 4.16*b*. The displacer geometry was adopted by the Philips Company and used with the rhombic drive which is described below.

A further variant of the displacer engine is shown in Fig. 4.15*c*. and in this engine a double cylinder configuration is used. The sequence of operation is given in Fig 4.16*c*.

4.3.4 Drive mechanisms

It is difficult in practice to achieve the discontinuous motion of the piston necessary to achieve the theoretical indicator diagram. Normally both pistons are moved with a sinusoidal motion and the displacer piston leads the power piston by an angle of about 90°. The effect of this is to remove the corners of the theoretical indicator diagram shown in Fig. 4.11 and hence reduce the power output from that of the ideal Stirling cycle.

A further deviation from the theoretical cycle arises because the expansion and compression processes are carried out rapidly. This results in the processes being more nearly adiabatic than isothermal. For these reasons the Stirling engines have a theoretical cycle efficiency somewhat lower then the Carnot value.

4.3.4.1 Constrained pistons In these engines the two pistons are connected together by a mechanical linkage. In the early Stirling engines a bell crank was used to achieve the phase difference between the two pistons. The Philips Company have pioneered the use of the rhombic drive, see Fig. 4.18. This drive has a number of advantages over the simple bell crank. It is balanced, has no side forces, and, owing to the linear movement of the piston, simplifies the sealing problem. The rhombic drive is used with the single cylinder arrangement see Fig. 4.15. Unfortunately it is expensive, and has been replaced in recent multicylinder engines by the Rinia arrangement, see Fig. 4.17.

Many other drive mechanisms are described in the literature on Stirling engines.[9]

4.3.4.2 non-constrained pistons. All these engines are of the displacer type. They differ in the manner by which the displace motion and phasing are achieved. Since the displacer operates with the same pressure on each side, very little force is required to move it. This movement can be arranged in many ways and some of these are described in the following:

1 Displacer piston moved by separate (electric) motor, Ringbohm engine.[9] This method is particularly suitable for large engines since the

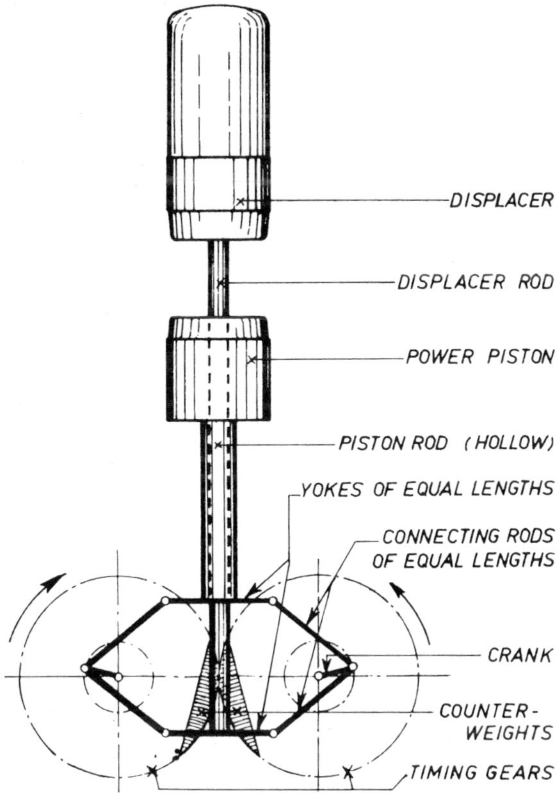

Fig. 4.18 *Rhombic drive*

relative cost of the displacer drive is then small. An advantage of the Ringbohm system is that the displacer can be moved such that the working fluid is in the heated section of the cylinder to give self-starting, see Fig. 4.19.

2 Beale free piston engine.[16] The power piston is not connected to any mechanical linkage to constrain its motion and oscillates freely in the cylinder. Power take off can be electrical, hydraulic or mechanical. The displacer and power pistons are arranged to have different mass, the power piston being considerably heavier. The displacer is connected to the power piston by means of a gas reservoir in the latter, see Fig. 4.20. The operation can be understood by reference to Fig. 4.16*b*. Assume that initially the engine is in state 1. The power piston rises, compressing the gas until the pressure exceeds that in the reservoir. The displacer is then moved down by this pressure difference (3). Displacer and piston move down together. The pressure in the engine falls and the displacer is moved since the reservoir pressure is now higher than the

gas pressure. The full theoretical solution of the free piston engine operation is complex because of the coupled mechanical/fluid motion involved. The power piston can be connected directly to an electrical generator so that the engine does not require a seal for the output shaft. Another application, also for a sealed unit, is a combined engine/heat pump,[17] in which the heat pump is simply a reversed Stirling engine, Fig. 4.21. Free piston engines have also been coupled directly to water pumps for irrigation applications.

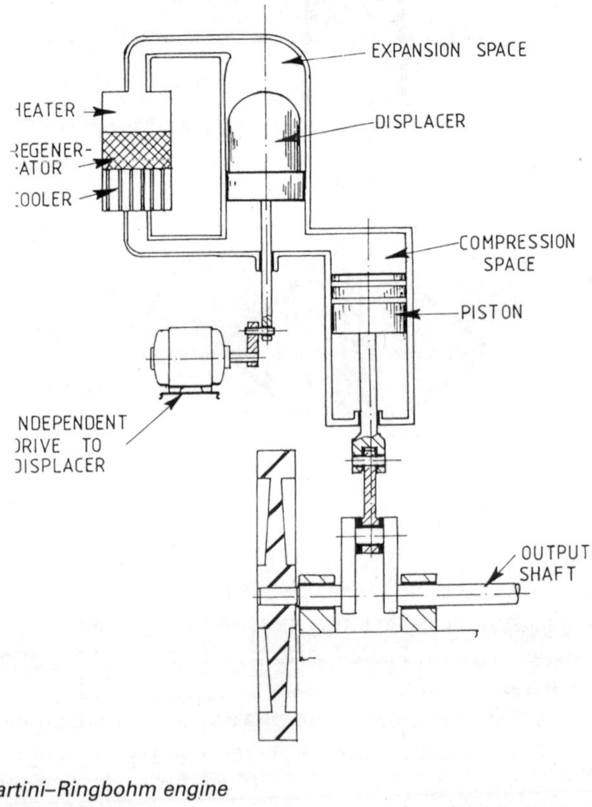

Fig. 4.19 *Martini–Ringbohm engine*

3 Liquid piston engines – fluidyne,[18,19] in this design both the power piston and the displacer piston are liquid (usually water) see Fig. 4.22*a*. The U-tube contains the displacer in the form of a water column which oscillates to move the gas from the hot to the cold cylinder and back through the regenerator. This displacer motion is achieved by the asymetric jet stream effect of water from the power piston, see Fig. 4.22*b*.

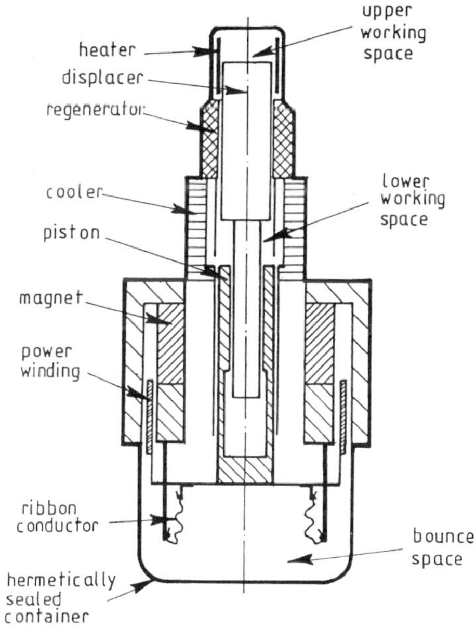

Fig. 4.20 *Beale-type free piston Stirling engine with Integral alternator*

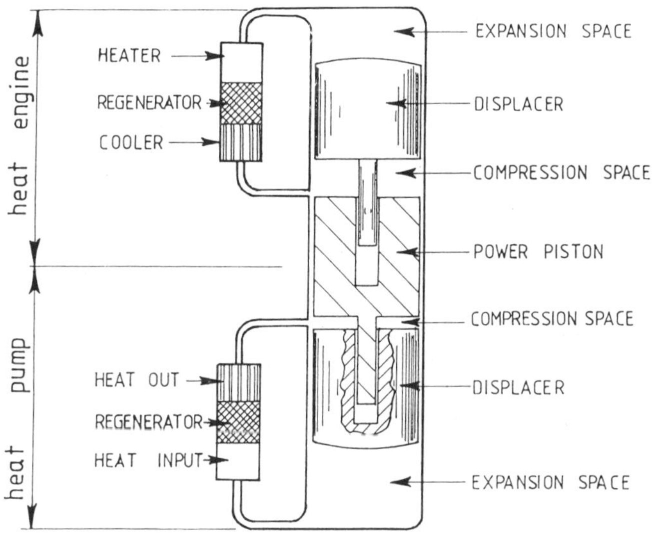

Fig. 4.21 *Free piston Stirling heat pump*

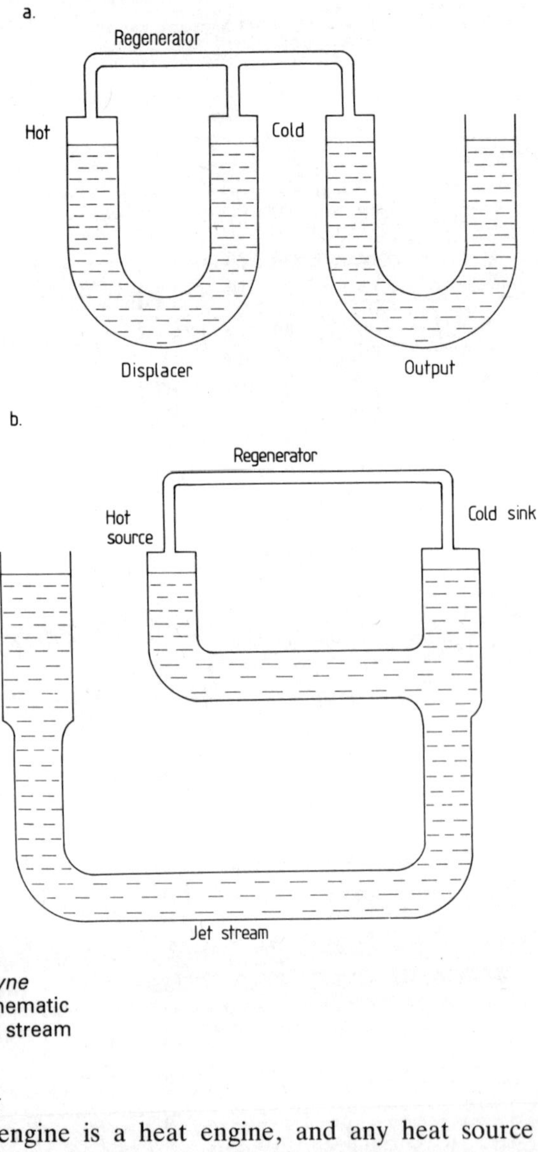

Fig. 4.22 *Fluidyne*
a Schematic
b Jet stream

4.3.5 Heating

The Stirling engine is a heat engine, and any heat source may be used, including flames, solar energy, nuclear heat sources or energy stored in fused salts.

When a hot gas is used as the source of heat the availability restriction on energy conversion discussed in Chapter 3 applies. It is important to incorporate an air preheater to recover some of the heat corresponding to the temperature at which the gas leaves the engine.

A conventional method of combustion and heating of the Stirling engine is illustrated by Fig. 4.23. Heater tubes are expensive in material and fabrication and low power air charged engines use only the cylinder head for heat transfer. At Reading University a fluidised bed has been used for the heating of the Stirling engine, Fig. 4.24.[20]

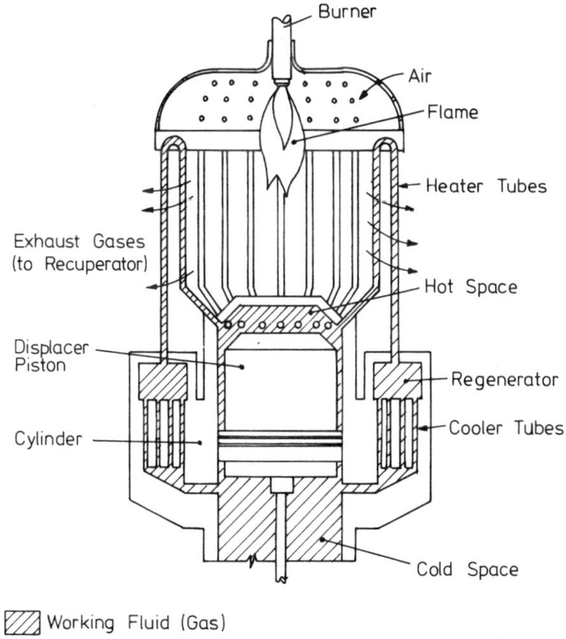

Fig. 4.23 *Conventional Stirling engine heater*

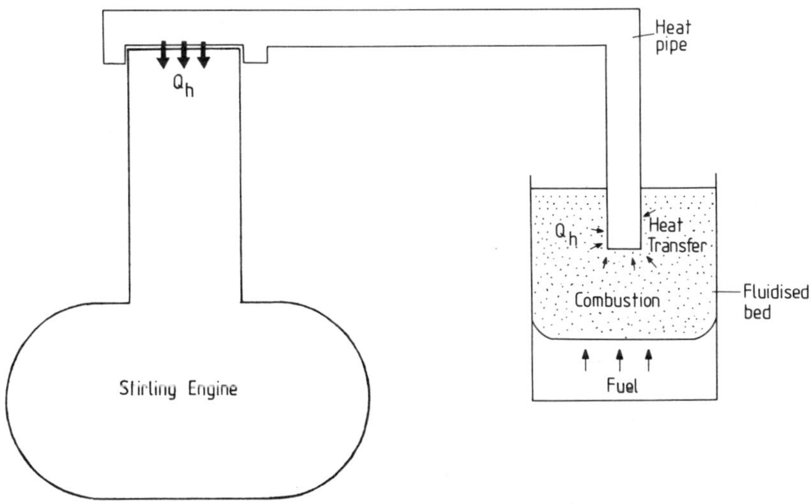

Fig. 4.24 *Application of fluidised bed and heat pipe to Stirling engine*

4.3.5 Fluidised bed: If a gas is passed through a packed bed of particles, such as coarse sand, in a vertical direction then pressure drop across the bed increases as the gas velocity is increased. However, as the fluid velocity reaches a certain minimum value the bed of solid particles expands slightly and the individual particles become supported in the gas stream with freedom of movement relative to one another. The bed is then said to be fluidised and has the appearance of a boiling liquid with a well-defined free surface. For a limited range of gas velocity the fluidised bed will remain in this apparently liquid state and the pressure drop across it is largely independent of gas velocity. At higher gas velocities entrainment of the solid particles will occur and the surface loses its well-defined interface. Combustion of a gas or vapour in such a fluidised bed may be achieved by arranging to feed a combustible gas or vapour together with oxygen or air into the bed. Solid fuel, such as coal can also be burnt in a fluidised bed. The flow of reactants may be initially insufficient to cause fluidisation until the bed is ignited. Fluidised bed combustion is characterised by uniform temperature and high heat transfer to surfaces placed in the bed. Surface catalytic effects at the surface of the solid particles enables the bed to be operated at temperatures which can be considerably below the adiabatic flame temperature necessary for the desired reaction.

The fluidised bed has a number of advantages over direct flame heating. The major advantages are:

1 Uniform temperature, thus avoiding hot spots.
2 High heat transfer.
3 Reduced oxides of nitrogen in the exhaust gas due to the lower temperature of the bed.

4.3.5.2 Heat pipe. A further development which enables the heat to be transported from the single heat source to a number of engines is the heat pipe,[21] which has been developed for this application both at Philips and at Reading University,[20] Fig. 4.24.

The heat pipe[21,22] is, as the name suggests, a device of high thermal conductance, that is, it will transport thermal energy without an appreciable drop in temperature. The heat pipe may be regarded as a development of the thermal syphon. The thermal syphon, Fig. 4.25*a*, consists of an evacuated sealed tube containing a small quantity of liquid. If this liquid at the bottom of the tube is heated, it will evaporate and the resulting vapour will flow to the top of the tube, where it will condense, returning to the evaporator region by gravitational forces. The net effect is for thermal energy to be transported from the evaporator to the condenser end of the tube. Since the energy exchange is by latent heat, the temperature difference between the two ends is small. Thus the effective thermal conductance of the tube is high. An important restriction on the use of the thermal syphon is the requirement that the evaporator section should be placed in a position lower than the condensing section, in order that the condensate should be returned. In the heat pipe, Fig. 4.25*b*, capillary forces are used to return the condensate to the evaporator region, hence the restriction on relative position of the evaporator and condenser sections is removed.

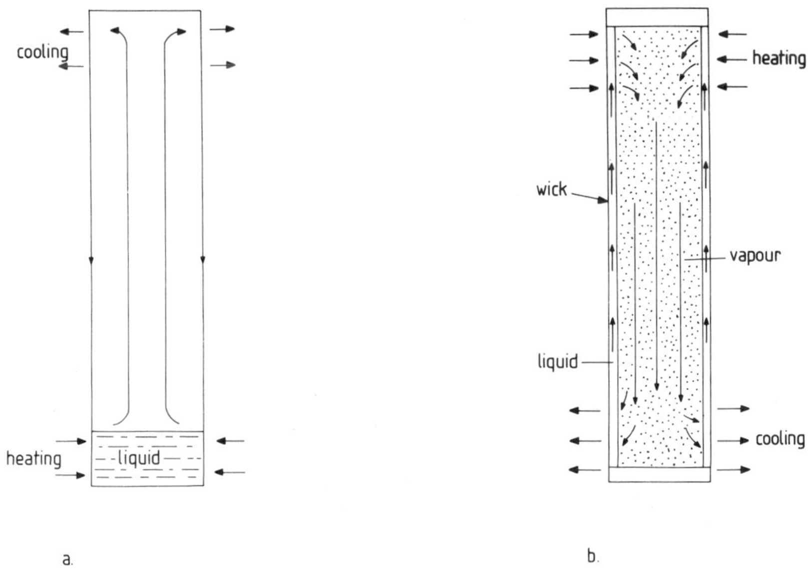

Fig. 4.25 *a* Thermosyphon
b Heat pipe

Various forms of construction may be used to give capillary return. One of the simplest is the use of a few layers of fine wire mesh welded to the inside surface of the tube. Heat pipes can be made in several geometries; flat plate, flexible and annular versions are all possible.

For the Stirling engine application, stainless-steel tubes with stainless-steel mesh are used for the containment and wick structure and the working fluid is sodium. A typical operating temperature is around 700 °C and axial heat fluxes up to 15 kW per square centimetre of pipe cross-section are possible.

Fig. 4.26 shows such a heat pipe operating in the laboratory. Fig. 4.27 shows a complete heat pipe fluidised bed Stirling engine arrangement.

4.3.6 Characteristics of Stirling engines

Stirling engines have been constructed over the power range from a few watts to 400 or 500 hp. The high performance, high pressure, helium charged, high temperature (700 °C) engines have a power to weight, power to size, and efficiency performance similar to that of a diesel engine of the same power and

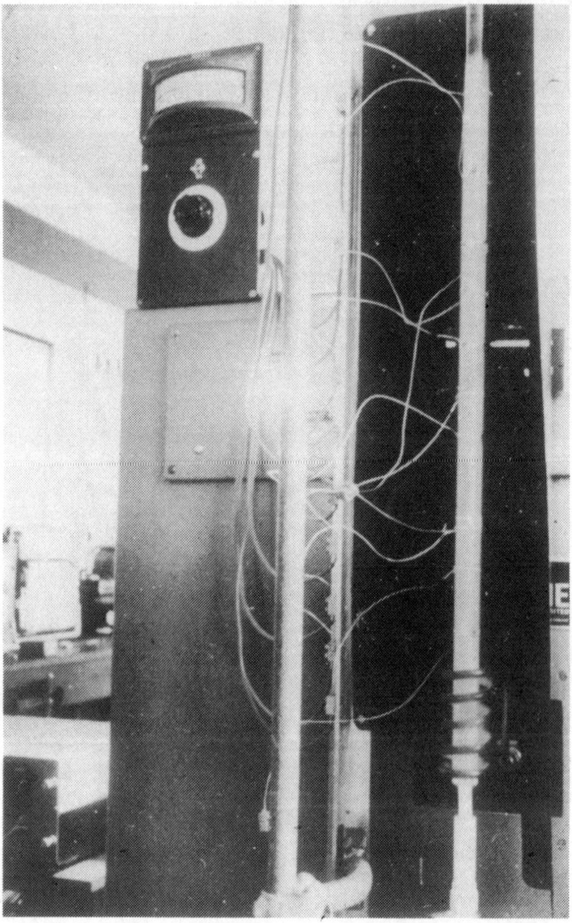

Fig. 4.26 *Sodium in stainless steel heat pipe*

speed. Fig. 4.28 shows the efficiency and power of a typical engine as a function of speed, these curves are plotted for a range of mean gas pressures. Fig. 4.29 shows the corresponding torque/speed characteristics.[11] Cheaper engines employ air as the working fluid, are operated at lower speeds and are heavier and more bulky, see Fig. 4.30. For some applications these disadvantages are not severe and are more than offset by the simplicity of sealing and other constructional factors.[23]

General features of the Stirling engine include:

1 A sealed unit, which simplifies lubrication problems, reduces maintenance requirements and contributes to long life. Another helpful factor in achieving a long life is the absence of valves and valve gear.

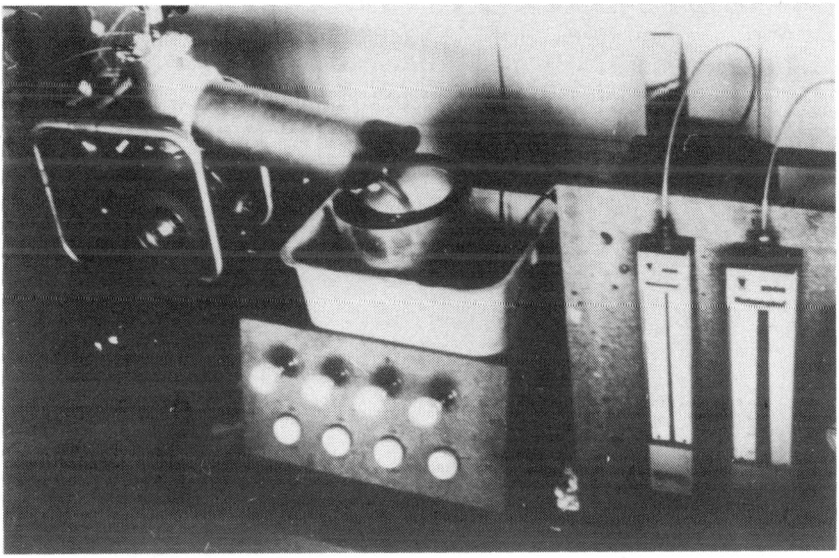

Fig. 4.27 *Fluidised bed, heat pipe, Stirling engine system at Reading University*

2 The ability to operate on different fuels.
3 Low exhaust emission. Unlike the internal combustion engine the combustion process is continuous and under these conditions a more complete combustion is achieved. Anti-knock additives such as lead are not required.
4 The torque does not vary greatly with speed and maintains its value at low speeds (Fig. 4.29). In a traction application this allows the use of a lower power engine than the equivalent internal combustion engine.
5 The response time achieved by the use of a pressurised accumulator is comparable to that of a diesel.
6 Low noise and vibration. One contributory factor is the absence of an explosion in the cylinder as is the case in the internal combustion engine, and also less exhaust noise than the internal combustion engine. In particular, when the rhombic drive is adopted vibrations due to lack of balance are eliminated. Other contributing factors to low noise include the absence of valves and valve gear.

4.3.7 Applications of Stirling engines

Although an old idea, it is only recently that Stirling engines[24–26] have received significant design attention, and apart from its use in reversed form as cryogenic refrigerators there has been no commercial production. There are, nevertheless, several applications which show promise.

The high temperature, high pressure, helium filled engines compare favourably in performance with a diesel of the same power and speed in terms of size, weight and efficiency. They also have the other advantages set out in the previous section. The cost will inevitably be somewhat higher than that of the diesel but this may be more than offset by low noise and pollution levels.

The possibility of replacing the gasolene SI car engine has interested several companies, most recently MTI in the USA who have claimed important advances in the performance of their engines.[24]

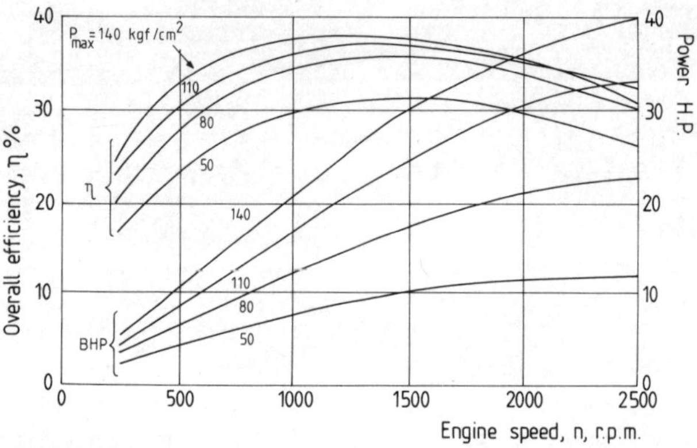

Fig. 4.28 *Efficiency and power variation with speed for different values of charge pressure (Ref. 11)*

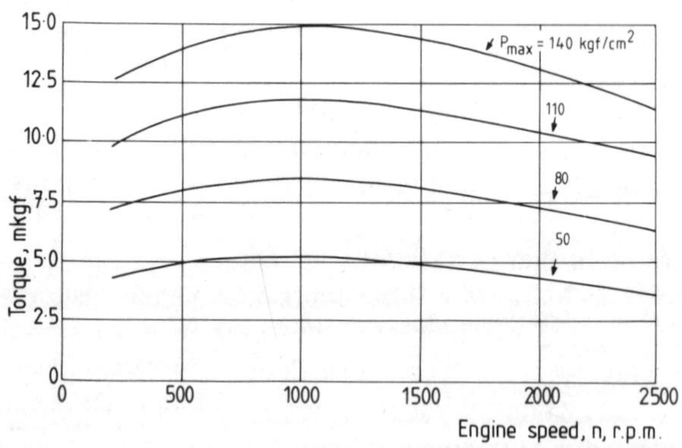

Fig. 4.29 *Torque variation with speed for different values of charge pressure (Ref. 11)*

Recently there has been interest in the design of large, slow running Stirling engines in the 1000 hp range for use in rail and marine transport applications and also for isolated electrical generation. These large engines will be coal fired, using a fluidised bed as the combustor and air as the working fluid. Air is not at such a disadvantage with respect to helium in low speed engines and has the considerable advanatge of requiring a much less gas-tight seal.

There is also a possible application for small air charged engines in the 15 kW power range for electricity generation in the developing countries. Coal or biomass would be used as fuel.[27,28]

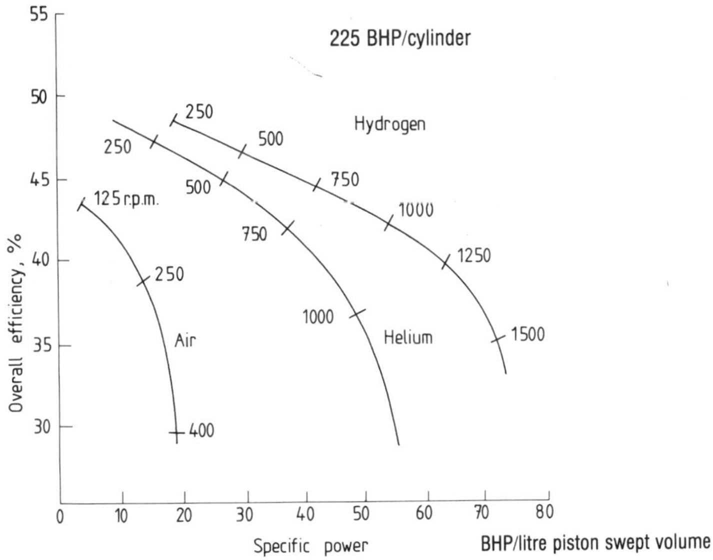

Fig. 4.30 *Efficiency versus specific power for three different gases (Ref. 11)*

The free piston engine has been developed to an advanced stage and shows promise for domestic heat pumps, small scale electricity generation and water pumping.[26]

4.4 Rotary gas engines: the closed cycle gas turbine

Just as steam at high pressure can be allowed to expand through a turbine, so gas at high pressure can also be used to drive a turbine. There is one important difference between a gas and a vapour, however, and it is this difference that held up the development of the gas turbine for many years. In the case of steam, after expansion through the turbine the steam is condensed to water in the condenser, at low pressure, and then returned to the boiler at high pressure by means of a pump, the feed pump. The power required to drive this pump is very small compared with the total power output of the steam turbine. In the case of expansion of a gas through a gas turbine to a low pressure the gas is cooled to a low temperature and pressure and then must be returned to the heater which is at high pressure. The operation of compressing a gas requires considerable work. Typically around two-thirds of the work extracted from the gas by the turbine is required to be fed back to drive the compressor. Since such a large fraction of the turbine output is required to drive the compressor, the gas turbine is extremely sensitive to the efficiency of both the turbine and the compressor.*

The principle of operation of the gas turbine is as follows: if a gas is compressed by means of a fan or compressor and allowed to flow through a pipe, the stream of gas may be used to drive a second fan or turbine placed downstream of the compressor, Fig. 4.31*a*. Obviously, even when no losses occur, the power produced by the turbine cannot exceed that put into the gas stream by the compressor. However, if heat is added as shown in Fig. 4.31*b* and the gas pressure in the pipe is unchanged, the heated gas will have an increased volume. Hence, a greater volume will flow through the turbine, which can then generate more power than is required by the compressor. It is then possible, in principle, to drive the compressor from the expander turbine and also have, in addition, net external power. This is the principle of the gas turbine cycle. However, as mentioned earlier, the compressor requires typically as much as two-thirds of the output of the expander turbine. This is under ideal, or loss free, operation. If the compressor and turbine both have efficiencies of 80% then the turbine's output will be reduced to 0·8 of the ideal value and the compressor requirement will be increased by 1/0·8. The result is

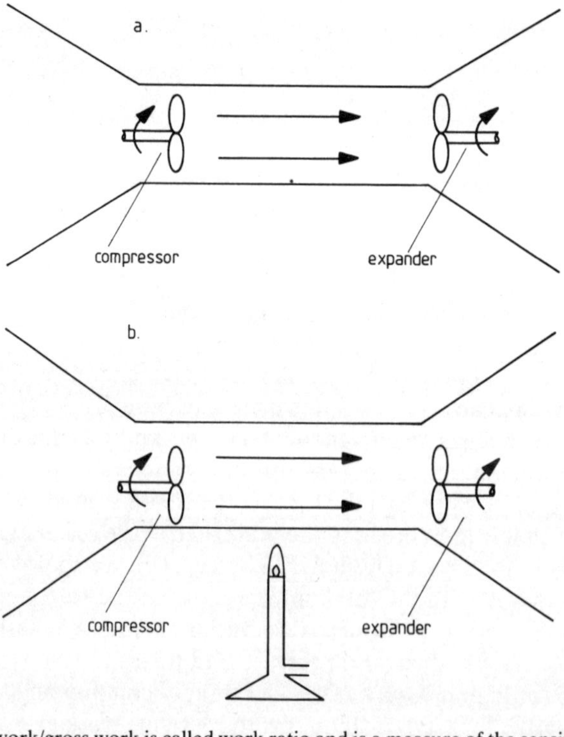

* The ratio net work/gross work is called work ratio and is a measure of the sensitivity of the cycle to irreversibilities. Work ratio for a steam engine is very nearly 1·0 whereas for the gas turbine it is only 1/3.

c.

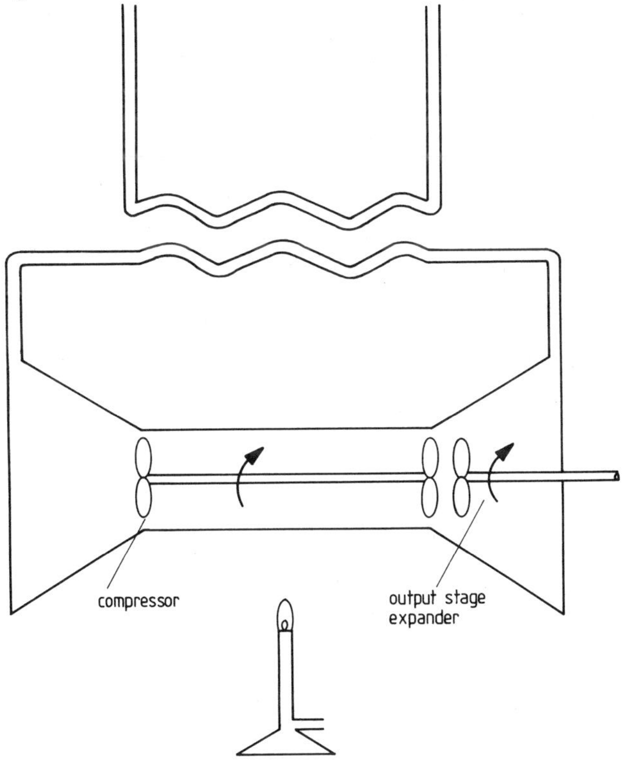

d.

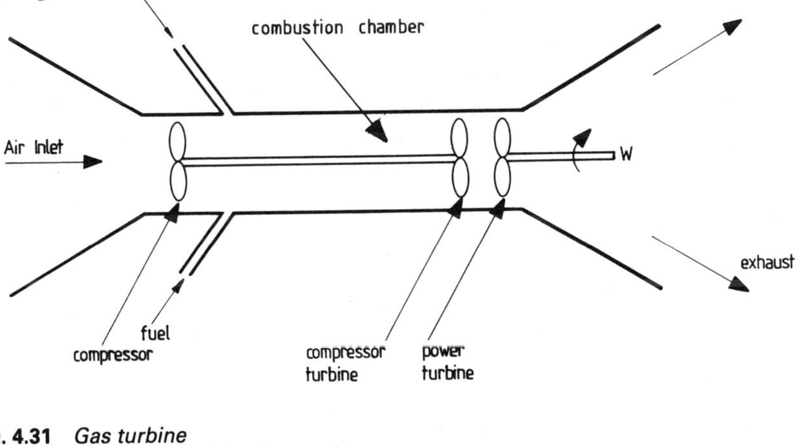

Fig. 4.31 *Gas turbine*
a and *b* principle of operation
c closed cycle
d open cycle

that the expander, when coupled to the compressor, will have an overall efficiency of 0·8 × 0·8, or 64%, barely sufficient power to operate the compressor and leaving no net power output. This is the reason for the early difficulties in constructing successful gas turbines. Advances have been made in two respects; in the development of higher temperature blade materials and in a better understanding of fluid dynamics leading to improved compressor efficiency. In some gas turbines a second, separate, expander is used to provide the power output, Fig. 4.31*c*.

As in the case of the Stirling engine, the thermodynamic efficiency is not affected by the nature of the gas chosen as the working fluid nor its mean pressure. However, in practice efficiency can be improved by choosing a gas such as helium, which has good heat transfer properties, and also the power to weight ratio is improved by operating at high pressure. For small scale applications the gas turbine does not appear to have significant advantages over the vapour reciprocating and turbine engines. In open cycle form, Fig. 4.31*d*, it is extensively used for aircraft propulsion and for industrial power production using oil as the fuel (Chapter 5).

4.5 Refrigeration

4.5.1 Physical principles

Heat is required in order for a liquid to evaporate. The heat necessary to evaporate unit mass of liquid was, in the past, called latent heat; it is now more usual to refer to it as the enthalpy of evaporation. We have already met this quantity in the discussion on vapour engines earlier in this chapter. If no additional heat is supplied the liquid will cool. We are all familiar with this effect which is so noticeable on leaving the shower.

The simplest coolers make use of the direct evaporation of water to give a cooling effect. Water is stored in slightly porous jars which allow a small quantity to seep through the porous walls. The outside of the jars are thus always wet and some of the heat required to evaporate the water is abstracted from the jar.

Modern refrigerators rely on the same mechanism of evaporation but normally use fluids having more suitable refrigerant properties than water and the units are sealed in order to prevent loss of the refrigerant. There are two basic types of vapour refrigerator:

1 Compression refrigerator
2 Absorption refrigerator

4.5.2 Simple vapour compression refrigerator

Fig. 4.32 shows the four principal components of a vapour compression refrigerator. These are:

1 Evaporator
2 Compressor
3 Condenser
4 Throttle or expansion valve

Cooling is achieved by allowing the liquid refrigerant to evaporate in the evaporator chamber. The low pressure vapour is then compressed in a mechanical compressor to a higher pressure and passed to a condenser where it condenses back to a liquid. It is necessary to arrange for the heat to be removed from the condenser. The resulting high pressure liquid is allowed to pass through a restriction or throttle valve, from which it emerges at low pressure and the cycle is repeated. The common domestic refrigerator is based on this principle. It may be asked why the vapour will condense at a surface which is maintained at a higher temperature than the surface from which it originally evaporated? The explanation for this is the higher pressure to which the vapour has been raised by the compressor. For a liquid, at a given temperature, there will be a pressure at which the vapour will be in equilibrium with the liquid. That is where the rate of evaporation of the molecules from the surface of the liquid just balances the rate of condensation. This is known as the saturated vapour pressure. The saturation vapour pressure of a liquid increases rapidly with increase in temperature. If the vapour is maintained at a

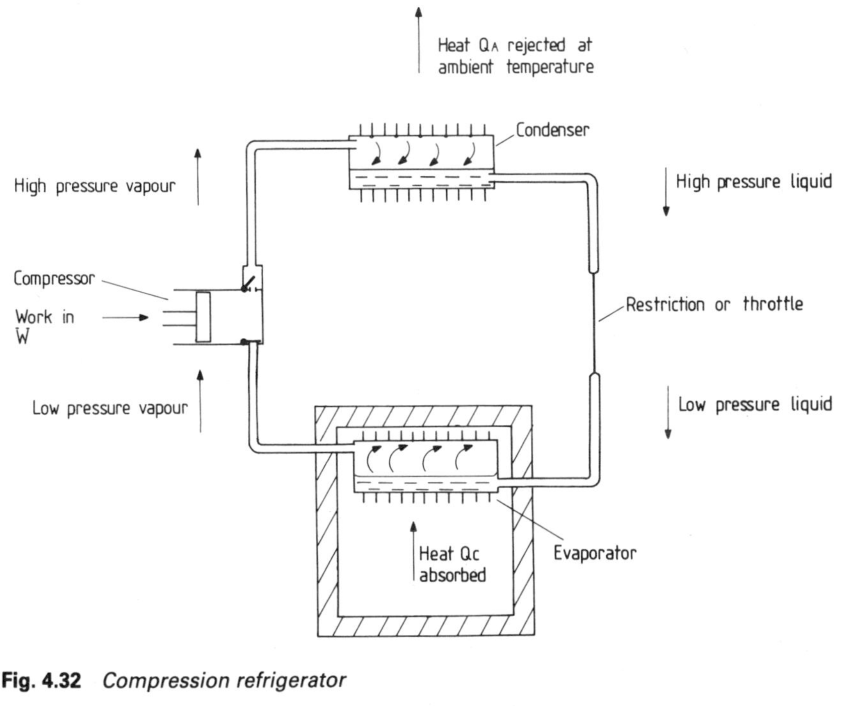

Fig. 4.32 *Compression refrigerator*

pressure higher than the saturation vapour pressure it will condense, and conversely, if the vapour pressure is less than the saturated vapour pressure then the liquid will evaporate. The refrigeration effect will depend on the enthalpy of evaporation at the pressure at which the refrigerant evaporates. This information is available for the more usual refrigerants in the form of tables and of pressure–enthalpy curves. Table 4.2 lists this information for ammonia, freon 12 and water. Water is obviously unsuited as a refrigerant if temperatures below 0 °C are required. It also suffers from the disadvantage of having a high volume of vapour per unit mass at low pressures resulting in a bulky refrigerator. For this reason other fluids, such as ammonia or the freons, are normally employed in commercial equipment; water is, however, used as the refrigerant in solar air conditioning systems.

As has been discussed in Chapter 3, the goodness of a refrigerator is expressed in terms of coefficient of performance, or COP, which is given by the relation:

$$\mathrm{COP} = \frac{\text{heat abstracted from the cold source}}{\text{work input}}$$

$$= \frac{Q_c}{W}$$

In principle it is possible to use an expander in place of the throttle valve; the expansion of the liquid into a mixture of liquid and vapour in the expander would then provide a work output. This work could be used to reduce the external work required to drive the compressor. In practice it is not economic to install an expander, hence practical refrigeration cycles are less efficient than the ideal cycle and have a COP which is, typically, 70% of the maximum theoretical COP. The compressor will also be about 70% of theoretical efficiency owing to friction and other factors, and the result of these two effects is to reduce the practical value of COP to around one-half that for an ideal cycle. Typical COPs for commercial domestic refrigerators are in the range 2.5 to 3.

4.5.3 Absorption refrigerator

Early refrigerators made use of the two fluid or absorption cycle which has the advantage of not requiring a compressor. The absorption cycle is similar to the compression cycle, the only difference being in the manner in which the vapour is compressed. In both systems high pressure vapour is condensed to a liquid, then expanded to a region of low pressure through a throttle valve and the resulting wet mixture evaporated to give a low pressure vapour. Thus the circuit from A–B shown in Fig. 4.33 is common to both the compression and the absorption systems. In the absorption cycle the refrigerant vapour is first absorbed in a second liquid and the resulting solution is then raised to a high pressure by means of a pump, Fig. 4.34. Ammonia is often used as the

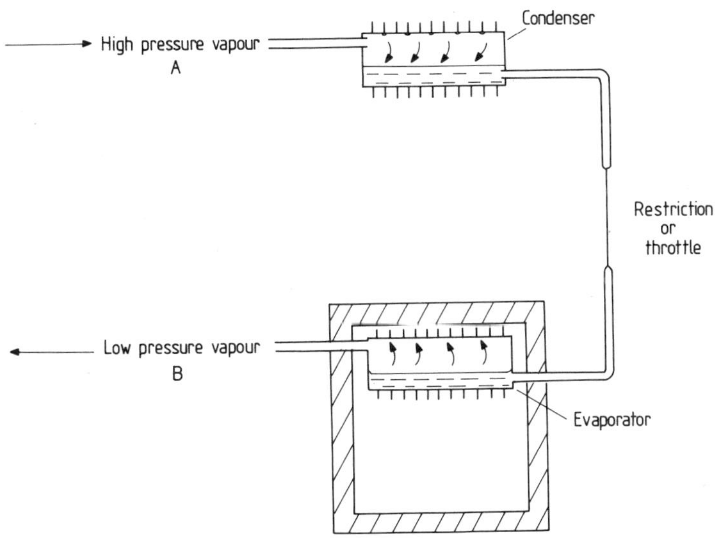

Fig. 4.33 *Common features of compression and absorption refrigerator*

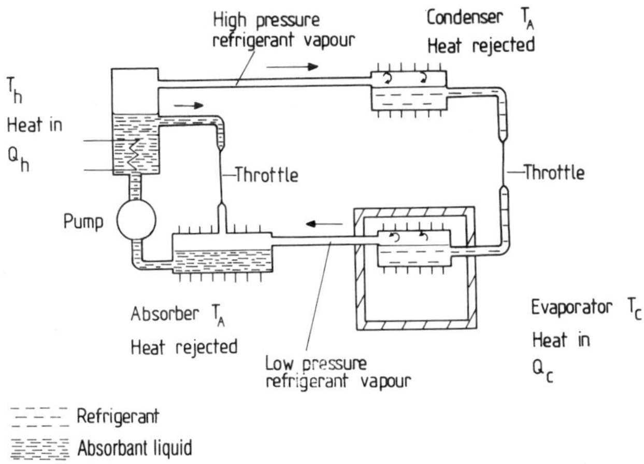

Fig. 4.34 *Absorption refrigerator*

refrigerant together with water as the working fluid absorber*. The refrigerant is separated from the absorber liquid by heating in a heater section and the high pressure refrigerant vapour then passes to the condenser section in the usual manner. The absorber liquid, which is at high pressure, is returned to the heater section by means of a second throttle valve. The absorption cycle just

* Ammonia is extremely soluble in water. As an example, water will absorb roughly its own weight of ammonia at atmospheric pressure and a temperature of 0°C. This value increases with increase in pressure but falls by a factor of six when temperature is raised to 100°C.

Table 4.3 *Properties of some refrigerants*

Substance	Temperature °C	Specific volume m^3/kg	Saturated vapour pressure, bar	Enthalpy liquid–vapour kJ/kg
Ammonia	−10	0·42	2·9	1298
	0	0·29	4·3	1263
	+30	0·11	11·7	1146
Freon 12	−10	0·077	2·2	156
(CF_2Cl_2)	0	0·055	3·1	151
	+30	0·023	7·4	135
Water	+10	106	0·012	2477
	+20	57·8	0·023	2451
	+30	32·9	0·042	2430

described requires the use of a small pump but no compressor. The provision of the mechanical power source to drive this pump can be a problem in the case of solar systems if an electrical power supply is not readily available. The terms high and low pressure have been used to distinguish the two pressure regions in the refrigerator. In practice, as reference to Table 4.3 shows, the pressure ranges involved are sub-atmospheric when lithium chloride is used as the absorber together with water as the refrigerant. In fact it is possible to devise designs for small plant in which a gravitational head can be used as an alternative to the pump, and the circulation is by thermal syphon. With the lithium chloride–water system the temperature cannot be reduced beyond about 4·5°C.

An alternative way to avoid a pump is to operate on an intermittent cycle. The simplest type of intermittant cycle refrigerator, known as the ice-ball, was invented in 1920 and used extensively in homes in USA and Canada and Europe between the two world wars. The principle of operation is shown in Fig. 4.35 and the device was heated by paraffin, gas or even firewood. It is not convenient to use the evaporator and cooler in this dual role in larger plant and a separate boiler and condenser are employed.

The advantage of the intermittent absorption refrigeration when used with solar heating is that no other source of energy is required. Such refrigerators are particularly suited to community cold stores in developing countries and an example is given in Chapter 7. Exell has developed a solar refrigerator of this type, see Fig. 7.49.

An interesting variant of the absorption machine is the three fluid or Electrolux refrigerator. In this refrigerator the mechanical pump is eliminated and only heat is required for its operation. This design is popular for gas heated domestic refrigerators. In the Electrolux system, in addition to the refrigerant and absorber, usually ammonia and water, respectively, a third gas, hydrogen,

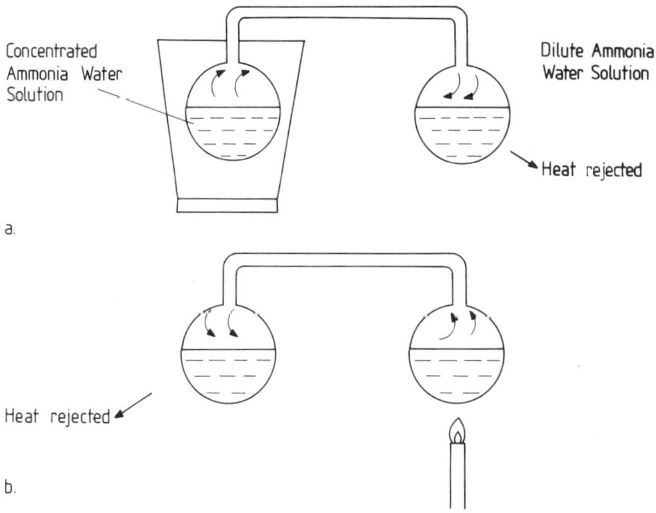

Fig. 4.35 *Ice-ball intermittent refrigerator*
a Cooling
b Regeneration

is added. The effect is to maintain an effectively constant pressure throughout the system. This is achieved as follows. Fig. 4.36 shows two vessels containing a refrigerant in liquid form. Some of the liquid will escape as vapour and the pressure of this vapour will be considerably higher in the hotter of the two vessels. However, if the vessels are connected by a tube and hydrogen gas added then the pressure in both vessels will reach the same value, but in the hotter vessel the vapour pressure of refrigerant will remain higher than that in the cooler vessel and the hydrogen pressure will be somewhat higher in the cooler vessel than in the hotter vessel, to give an overall pressure equalisation.

This is the principle on which the Electrolux refrigerator is based. In addition it is necessary to arranged for a flow of the three components, ammonia, water and hydrogen, around the system. The overall arrangement is

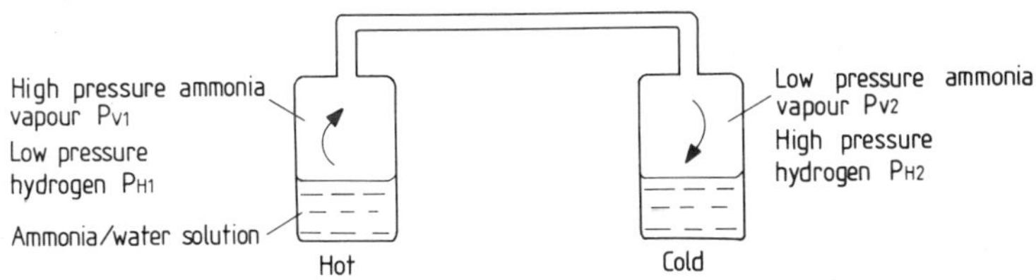

$$P_{V1} + P_{H1} = P_{V2} + P_{H2}$$

Fig. 4.36 *Equalisation of pressure using inert gas*

shown schematically in Fig. 4.37. A concentrated solution of ammonia in water is heated and the resulting mixture of ammonia bubbles and dilute ammonia–water solution rises in the vertical tube because of the reduced density of the column with respect to the dilute solution in the adjacent column. The ammonia vapour and water separate as shown. The high pressure ammonia vapour rises and enters the condenser C where it is condensed to a liquid. The liquid ammonia enters the evaporation chamber E at low pressure and meets a stream of hydrogen gas. The ammonia will vaporise and the resulting vapour will be swept away by the hydrogen gas stream. The heavy ammonia vapour will cause the mixture to descend to the water reservoir D where the ammonia will be partially absorbed. Complete absorption will occur as the remaining vapour meets the dilute concentration of ammonia and water returning through. The hydrogen, now completely separated from the ammonia vapour, will rise through G to complete the hydrogen circuit. There are seen to be three separate circuits: ammonia, water and hydrogen.

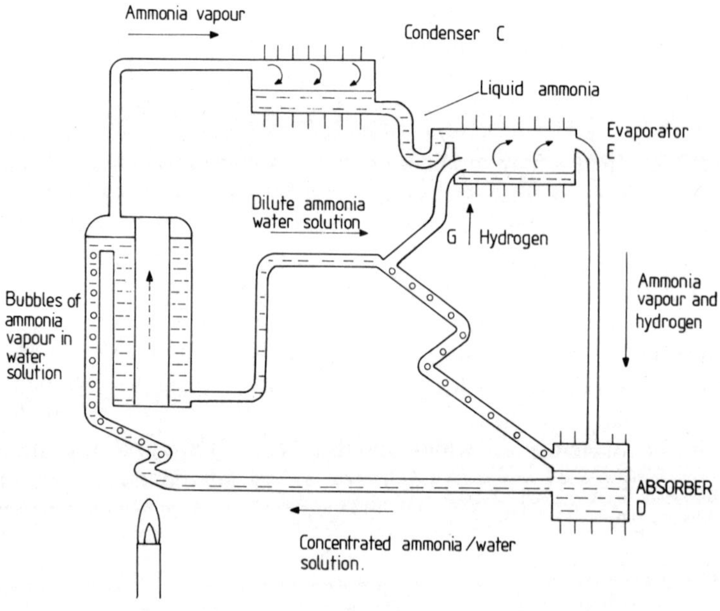

Fig. 4.37 *Electrolux refrigerator*

4.5.4 Comparison of compression and absorption cycles

As stated in Chapter 3, the compression refrigerator is a reversed heat engine and must be driven by a source of mechanical power such as another heat engine, windmill or electric motor. The absorption refrigerator is a combined heat engine–refrigerator.

The vapour compression and absorption refrigerators both operate by the evaporation of the refrigerant at low temperature and pressure and its condensation at higher temperature and pressure. They differ only in the manner in which the low pressure vapour is compressed to a higher pressure. In the case of the vapour compression machine a mechanical compressor is used, whereas in the absorption machine the low pressure vapour is absorbed into a second liquid which is then pumped to a higher pressure and the refrigerant vapour released at this higher pressure by heating the absorber liquid. In both machines the high pressure vapour is condensed and then returned to the evaporator chamber by allowing the liquid to pass through a constriction or throttle. In the vapour compression machine the energy is provided in the form of mechanical energy to drive the compressor. In the absorption machine most of the external energy is provided as heat at a temperature greater than ambient to separate the refrigerant from its carried liquid. The absorption machine is, in effect, two machines, a heat engine coupled to a refrigerator, see Fig. 4.38. A small amount of external mechanical

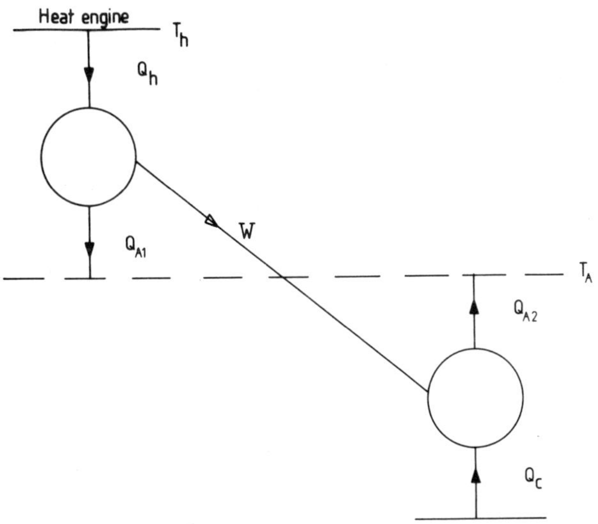

T_h	Hot source temperature
T_A	Ambient temperature
T	Cold (absorber) temperature
Q_h	Heat absorbed from hot source
Q_{A1}, Q_{A2}	Heat rejected at ambient temperature
Q_c	Heat absorbed by absorber
W	Work output of heat engine = work input to refrigerator

Fig. 4.38 *Heat engine coupled to refrigerator*

power is required to operate the pump which pumps the carrier liquid containing the dissolved refrigerant from the low to high pressure.

Tabor has compared the heat engine driven compression cycle with that of the absorption cycle and has shown that with practically achievable performance characteristics the solar operated absorption refrigerator will always be more efficient unless concentrating collectors are employed with the heat engine to give temperatures greater than 300°C.

References

1 KOLIN, I.: 'The evolution of the heat engine' (Longmans, London, 1972)

2 MAYHEW, Y. R. and ROGERS, G. F. C.: 'Thermodynamics and transport properties of fluids' (Blackwell, Oxford, 1969)

3 PATERSON, E. V.: 'The lubrication of steam engines', *The Steam Engineer*, 19, p. 29

4 NOCK, O. S.: 'The London and South Western Railway' (Ian Allen, 1965, p. 67)

5 DIAMOND, E. L.: 'The development of locomotive power at speed', *Proc. I. Mech. E.*, 1947, **156**, p. 86

6 DONALDSON, R. (ed.): 'Bicentenary of the James Watt Patent for a separate Condenser for the Steam Engine', Proceedings of Two-Day Symposium, Sept, 1969, University of Glasgow

7 KOESSLER, B.: 'Valve gear for high speed reciprocating steam engines', *The Steam Engineer*, 1946, p. 108

8 RINIA, H. and DU PRE, F. K.: 'Air engines', *Philips Tech. Rev.*, 1946, **8**, pp. 129–136

9 WALKER, G.: 'Stirling engines' (Clarendon Press, Oxford, 1980)

10 FINKELSTEIN, T.: 'Air engines', *The Engineer*, 1959, **207**, March, pp. 492–497, 3rd April, pp. 552–527, 10 April, pp. 568–571, 8th May, pp. 720–723

11 MEIJER, R. J.: 'The Philips hot gas engine with rhombic drive mechanism', *Phil. Rech. Rev.*, 1959, **20**, pp. 245–276

12 DUNN, P. D. (ed.): Proceedings of 2nd International Conference on Stirling Engines, Shanghai, June 1984

13 DUNN, P. D., RICE, G., PRESTON, C. C., LIPMAN, N., SLOWLEY, J., ALEXANDRA, W. A., RIX, D. H., BURROUGH, W. E., MARSHALL, E. L., READER, G. T., CLARKE, M. A., SLEE, R. H. and PARKER, D.A.: '20 kW UK consortium engine. Specification and manufacture', Proceedings of Stirling Engines Progress towards Reality, Reading, March 1982

14 KOHLER, J. W. L.: 'The gas refrigerating machine and its position in cryogenic technique', *Prog. Cryog.*, 1960, **2**, pp. 41–67

15 KOHLER, J. W. L.: 'The Stirling refrigeration cycle, *Scientific American*, 1965, **212**, 4, pp. 119–127

16 BEALE, W., HOLMES, W., LEWIS, R. S. and CHENG, E.: 'Free piston Stirling engines – a progress report', SAE paper No 730647, 1973

17 OKADA, G.: 'The development of free piston Stirling engines', Proceedings of 2nd International Stirling Engine Conference, Shanghai June 1984

18 WEST, C. D.: 'Liquid piston Stirling engines' (Van Nostrand Reinhold Co., New York, 1983)

19 WEST, C. D.: 'Liquid piston Stirling machines', Proceedings of 2nd International Stirling Engine Conference, Shanghai, June 1984

20 DUNN, P. D., RICE, G. and THRING, R. H.: 'Hydraulic and rotary drive Stirling engine with fluidised bed combustion/heat pipe systems', Paper No 759141, pp. 942–947, Proceedings of 10th Intersoc. Energy Conv. Conference, Newark Del, Aug 1975

21 DUNN, P. D. and REAY, D.: 'Heat pipes' 2nd edn (Pergamon, Oxford, 1982)
22 RICE, G.: 'Heat pipes for Stirling engines', Proceedings of 2nd International Stirling Engine Conference, Shanghai, June 1984
23 MEIJER, R. J.: 'Prospects of the Stirling engine for vehicular propulsion', *Philips Tech. Rev.*, 1970, **31**, pp. 168–185
24 STERNLICHT, B.: 'Stirling engine has come of age. The 21st century prime mover', Proceedings of 2nd International Stirling Engine Conference, Shanghai, June 1984
25 NAKATANK, Y., MARI, H., NARIMATSU, Y. and WATANABE, T.: 'Stirling engine development' project in Japan', *ibid.*
26 BEALE, W. T.: 'The development of Stirling engines at Sunpower Inc', *ibid.*
27 JINGWEI, Ye.: 'On the combustion and heat transfer problems of coal fired Stirling engine for use on locomotive', *ibid.*
28 WALKER, G., FAUVEL, R., Srinwasan, S., Gustafson, R. and VAN BENTHEM, J.: 'Future coal burning Stirling engines, Stirling Engines, Progress Towards Reality, Reading, March 1982

Chapter 5

The internal combustion engine: conversion of chemical to mechanical energy

5.1 Introduction

This chapter discusses methods of converting chemical energy into mechanical energy. The usual way in which this is done is to mix the fuel with air and then ignite the mixture; the hot high pressure combustion gases which are produced are allowed to expand and in doing so drive a piston or turbine. The arrangement is known as an internal combustion engine. The simplest form of internal combustion engine is a gun, and, in fact, Huyghen carried out work on such an arrangement in 1680 using a piston and cylinder with gunpowder as the fuel.

It was not until nearly 200 years later that the first successful internal combustion engine was developed by Lenoir (1860). This engine employed coal gas as the fuel and the air–gas mixture was ignited by a spark without any precompression. Shortly after, in 1876, Otto built an engine which operated on the four-stroke cycle and which can be regarded as the forerunner of the modern petrol engine. Otto used coal gas as fuel and an electric spark ignition. In the 1880s volatile liquid fuels became generally available and began to replace coal gas as the fuel. In spark ignition engines a fuel/gas mixture is compressed and then ignited by the spark. Increased compression ratio gives improved efficiency but the maximum compression ratio is limited by self-ignition of the mixture. Herbert Ackroyd Stuart overcame this difficulty by injecting the fuel after compression and ignition was achieved by contact with a 'hot bulb'. The latter was heated by a blow lamp on starting and maintained at temperature by a combination of exhaust heat and temperature rise owing to compression of the air in the engine. Diesel selected a compression ratio sufficiently high to give an air temperature which would ignite the injected fuel. The Otto (spark ignition) and Diesel (compression ignition) together account for all the modern reciprocating internal combustion engines. The possibility of improved power output by replacing the four-stroke cycle by the two-stroke cycle attracted the attention of many inventors and both four- and two-stroke cycles are in use today. The working principles and performance of modern internal combustion engines is discussed in this chapter.

We will first of all consider the reciprocating engine, and later, more briefly, the rotary engine. Recently there has been a renewal of interest in the use of fuels derived from biomass. These include biogas, alcohol, and vegetable oils such as palm oil and coconut oil and can be used either mixed with conventional fuels or alone.

5.2 Reciprocating internal combustion engine

5.2.1 Introduction

The maximum energy which can be released from a quantity of fuel will occur when sufficient oxygen is supplied for the complete combustion of that fuel.[1–3] The ratio of oxygen (or air) to fuel by weight to give complete combustion is called the stoichiometric ratio. Gasolene (Petrol) is a mixture of hydrocarbons of which hexane C_6H_{14} is a typical constituent. The ideal combustion reaction for hexane is:

$$C_6H_{14} + 9{\cdot}5O_2 \rightarrow 6CO_2 + 7H_2O + \text{energy}$$

This equation shows that each molecule of hexane requires 9·5 molecules of oxygen for complete combustion. In practical IC engines the oxygen is provided by atmospheric air. It is found that the ideal, or stoichiometric, ratio for gasolene and air is 1 to 14·7 by weight. IC engines, like humans, are air breathers. The maximum energy they can produce per cycle is limited, not by fuel, but by the oxygen required to burn it. In most engines air is sucked in directly from the atmosphere so that the initial pressure is fixed by the mechnical design of the engine and will be at some value below atmospheric pressure. Hence the nearer this pressure is to atmospheric pressure, the greater the power which can be developed by the engine. The ability of an engine to breathe in air is measured by the volumetric efficiency, which is defined as the amount of air actually in the cylinder compared to the amount of air contained by the cylinder at atmospheric pressure.

In some engines, for example large marine diesel engines, the atmospheric air is raised in pressure by a pump or turbine in order to increase the charge of air in the cylinder. This process is called supercharging. Since, presumably, the performance of athletes is also limited by oxygen rather than fuel, in principle their performance could also be improved by supercharging.

Referring to the combustion equation, it is seen that the products of the ideal combustion reaction are carbon dioxide and water (in the form of steam). In practice the exhaust will also contain oxides of nitrogen produced by the high reaction temperature which causes nitrogen in the air to combine with oxygen. For lean mixtures (air greater than the stoichiometric ratio) the exhaust will also contain oxygen and a higher level of oxides of nitrogen; and for rich mixtures (air less than the stoicheometric ratio) carbon monoxide, hydrogen, unburnt fuel and carbon (smoke) will be present. The energy released by gasolene and diesel fuel is around 45 MJ/kg.

In addition to the fuel, and the combustion efficiency, the power P from the engine will depend on the thermodynamic cycle, the volume of the cylinder and the speed.

The power P can be expressed as

$$P = \text{MEP} \times L \times A \times N \tag{5.1}$$

where MEP is mean effective pressure or the average pressure difference over the piston during the cycle, L is stroke of the piston, A is area of cylinder, and N is number of working strokes in unit time.

In a practical engine mechanical friction will reduce the power output P predicted by eqn. 5.1 to a lower value, the brake horsepower P_b where

$$P_b = \text{BMEP} \times L \times A \times N \tag{5.2}$$

where BMEP is the brake mean effective pressure and takes into account mechanical losses. A typical value for BMEP in a spark ignition engine is 10 bar. Hence the power per litre of swept cylinder volume for a four-stroke engine at a speed of 3000 rev/min would be:

$$P_b = 10 \times 10^5 \times 10^{-3} \times \frac{3000}{60} \times \tfrac{1}{2}$$

$$= 25 \text{ kW/l}$$

Note the factor ½ is necessary since in a four-stroke engine (Section 5.25) there is only one power stroke in two complete cycles.

5.2.2 Spark ignition and compression ignition

A gasolene engine inhales a combustible mixture of gasolene and air; this mixture is compressed in the cylinder and is then ignited by a spark. A diesel engine inhales air which is compressed in the cylinder and fuel injected after compression; the temperature rise due to compressing the gas is sufficiently high to cause ignition. For example, if atmospheric air is compressed by 16 : 1 the temperature rises to 600°C. Hence, in the gasolene engine, compression must be insufficient to cause self-ignition, whereas in the diesel engine it must be sufficiently high to cause self-ignition.

Classification of an engine by fuel is clearly an arbitrary business since we could, for example, use spark ignition with diesel oil, although not all fuels can be used with both methods of ignition. A more satisfactory procedure is the specification of the ignition method. We will, therefore, refer to engines as:

(i) Spark ignition (SI)

or

(ii) Compression ignition (CI)

Spark ignition is only possible for air/gasolene mixtures over a limited range of mixture strength, close to the stoichiometric ratio. For a flame to propagate through the fuel air mixture, the mixture must not be too rich or too weak. The limits of flammability lie between 7 : 1 air/fuel by weight (0·14 : 1 fuel/air) on the rich side and 20 : 1 (0·05 : 1) on the lean side; it is difficult to achieve reliable ignition with mixtures lower than 18.1. Maximum efficiency is achieved at about 16 : 1 and maximum power at about 12 : 1 and in practice the mixture strength is restricted to this range. This mixture is achieved by means of the carburettor which mixes the fuel in vapour form with the air before it enters the cylinder. It is necessary for the carburettor to operate over a pressure range in order to provide power control. In the gasolene engine the power level is varied by means of a throttle valve which controls the amount of air entering the engine. The carburettor is designed to adjust the fuel supply to maintain the required fuel/air ratio.

The situation in the diesel engine is quite different, since ignition is achieved spontaneously from the hot compressed air. Air is drawn into the cylinder, compressed, and then the fuel injected as a fine spray of droplets. Power is controlled by varying the amount of fuel injected.

5.2.3 Compression ratios: the standard air cycle

It is found that, as the compression ratio in an engine is increased, the power output and efficiency also increase, rapidly at first and then becoming rather less sensitive at high values of compression ratio.

In order to get some quantitative feel for the effect of compression ratio, it is convenient to introduce the air standard cycle. It has already been stated that internal combustion engines are not heat engines since they do not retain the same working fluid throughout the operation (fuel is injected and burnt), neither is the top temperature defined in the heat engine sense. However, in practice a peak temperature of the gas in the cylinder can be identified, and also the reject temperature will be that of the surroundings. In fact, the atmosphere is used as a cooler, and the air which is drawn in to complete the cycle can be regarded as if it were the same air that was expelled, cooled and with the oxygen renewed. Even in the spark ignition engine working with stoichiometric air fuel ratio there is 14 times the mass of air to the mass of fuel so that 80% of the working fluid (the nitrogen content of the air) is unchanged on passing through the engine. The air/fuel ratio for the diesel engine is typically 20 and in the gas turbine, with 300% excess air, as much as 94% of the working fluid is unchanged on passing through the engine. Thus for all practical purposes the performance of an internal combustion engine is the same as that of a heat engine using air as the working fluid, operating between the same temperatures and undergoing the same cyclic sequence of operations. Hence the air standard cycle is not a bad approximation to what actually happens in an internal combustion engine and provides a most convenient method of assessing the effect of varying parameters, particularly compression ratio, and also of optimising overall performance.

Even so, air standard cycles should be used with caution, since in practice both the compression ignition and the spark ignition engines operate at somewhat different conditions from the theoretical cycles.

The ideal or Otto cycle is shown in Fig. 5.1*a* and assumes an adiabatic* compression followed by a heating at constant volume (the ignition process). A working stroke of adiabatic expansion and an exhaust stroke (assumed to be at constant volume) complete the cycle.

The pressure–volume characteristic measured on an actual four-stroke engine would have a further loop as shown in Fig. 5.1*b*.

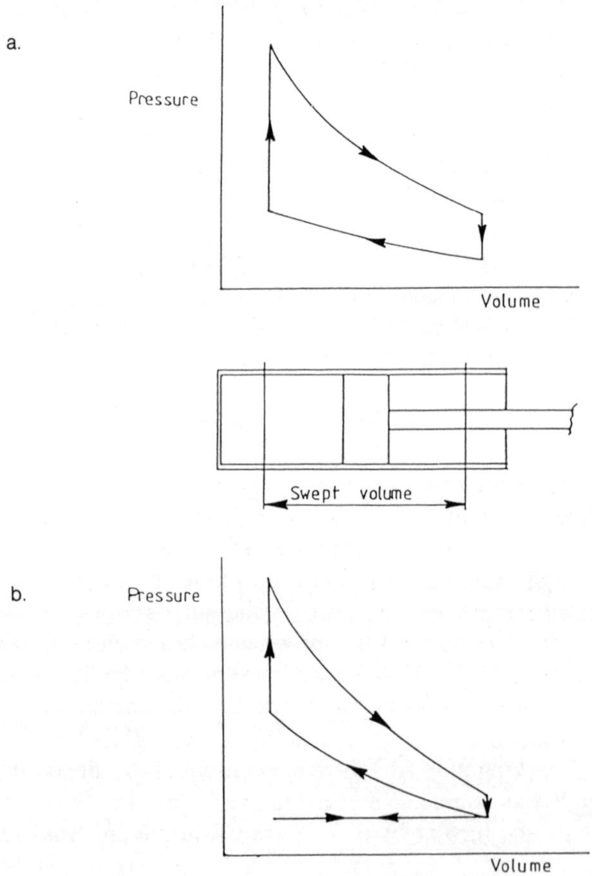

Fig. 5.1 *Engine cycles*
a Otto cycle
b Ideal four-stroke cycle

* In an adiabatic process no heat enters or leaves the gas. This is approximately true during the compression and expansion strokes of the cycle in an IC engine.

If r is the compression ratio the efficiency η is given by

$$\eta = 1 - \left(\frac{1}{r}\right)^{\gamma-1}$$

where γ is the ratio of specific heats for the gas (1·4 for air). This is plotted in Fig. 5·2 and shows how increased compression ratios above about 15 lead to little increase in efficiency. For practical spark ignition engines Tizard and Pye[3] obtained the relationship

$$\eta = 1 - \left(\frac{1}{r}\right)^{0 \cdot 258}$$

for a stoichiometric mixture.

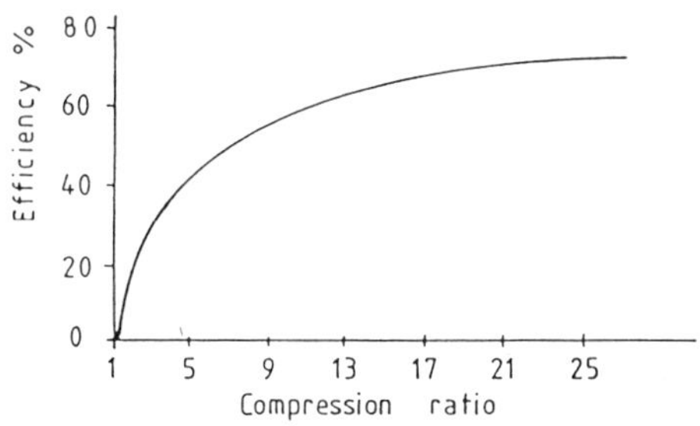

Fig. 5.2 *Efficiency of Otto cycle*

The diesel cycle is shown in Fig. 5.3. It differs from the Otto cycle in that the constant volume process corresponding to ignition in the Otto cycle is replaced by a constant pressure combustion operation on the diesel cycle. This process corresponds to the fuel injection operation.

The discontinuous processes which make up the theoretical cycles of Figs. 5.1 and 5.3 are not achieved exactly in practical engines because of the motion of the piston. The pressure–volume characteristics of both gasolene and diesel engines are similar in appearance and are as shown in Fig. 5.4. Pressure–volume characteristics of this kind are called indicator diagrams. The area of the closed curve gives the work provided by the gas in one cycle.

For spark ignition engines the compression ratio varies from about 6 : 1 to 10 : 1. In the case of diesel engines, compression ratios vary between 14 : 1 and 20 : 1.

5.2.4 fuels for internal combustion engines

5.2.4.1 Preignition and detonation: the octane number.

The efficiency of the spark ignition (gasolene) engine increases with the increase of compression of the gases before ignition. If the compression ratio exceeds a certain value the resulting rise in temperature will cause the

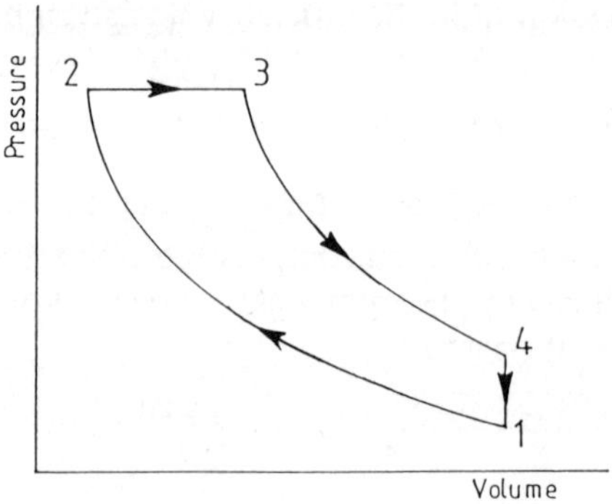

1-2	Adiabatic compression
2-3	Constant pressure expansion
3-4	Adiabatic expansion
4-1	Constant volume cooling

Fig. 5.3 *Ideal diesel cycle*

vapour–air mixture to ignite before the spark. This condition is called 'pre-ignition'. It is found that, in practice, the compression ratio is limited to a lower value than that for pre-ignition, a limit which is set by detonation or knocking caused by a portion of the charge igniting with exceptional rapidity after the onset of ignition by the spark. The resulting rapid pressure rise is accompanied by noise, loss of power and possible damage. Detonation is not completely understood but appears to occur as follows. When the fuel/air mixture is ignited by the spark a flame front travels through the volume of the cylinder.

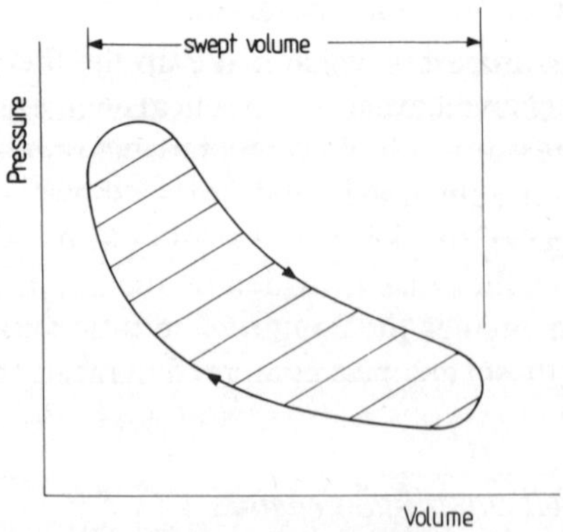

$$\left(\text{m.e.p.} = \frac{\text{Area of Indicator Diagram}}{\text{Swept Volume}}\right)$$

Fig. 5.4 *Practical indicator diagram*

The unburnt fuel is both chemically changed and raised in temperature by the shock wave from the advancing flame front and also by radiation from the flame. Under certain conditions of pressure and temperature the unburnt fuel/air will spontaneously ignite giving rise to the detonation or knock. It is found that some fuels are more prone to knock than others. Iso-octane (C_8H_{18}) has particularly good anti-knock properties whereas *n*-heptane (C_7H_{16}) is very poor in this respect. In order to classify fuels a scale has been set up: the octane number. On this scale iso-octane is given an octane number of 100 and *n*-heptane 0. A fuel is tested in a variable compression engine and the octane number is the percentage of iso-octane in a mixture of iso-octane and *n*-heptane which will cause knock at the same compression ratio. There are four standard test methods, of these the research octane number (RON) is the test normally employed for high octane gasolene.

It has been found that the octane number of hydrocarbon fuels can be raised by the addition of a small percentage of tetraethyl lead compounds which are soluble in the hydrocarbon. This 'doping' is now almost universally used for gasolene for private cars. Recently concern has been raised over the possible harmful effects of what amounts to quite significant release of lead compounds to the atmosphere. Alcohol has a high octane number and the octane volume of hydrocarbon fuels can be raised by blending with alcohol. The octane scale has been extended by the addition of a measured quantity x of tetraethyl lead to iso-octane. The relationship adopted is

$$\text{octane number} = 100 + x$$

where x is in cm^3 per US gallon, the units indicating the somewhat arbitrary nature of the scale.

In the Otto cycle, Fig. 5.1*a*, it is assumed that ignition occurs at point 2 when the piston has moved to give minimum cylinder volume a position known as top dead centre or TDC. The combustion process takes place from 2 to 3 and then the piston moves from TDC to follow the expansion stroke 3–4. The processes 1–4 can be plotted as a function of time as shown in Fig. 5.5*a*.

In practical engines a piston is driven by a crank shaft mechanism, and crank angle could have been used for the horizontal axis in Fig. 5.5*a*. TDC corresponds to a crank angle of 180° and bottom dead centre or BDC to a crank angle of 360° (0°).

The combustion process requires a finite time to complete, around 0·002s, and during this time the piston can have moved appreciably. For example, in an engine rotating at a speed of 3000 rev/min the crank shaft will have rotated through an angle

$$\frac{360 \times 0{\cdot}002 \times 3000}{60}$$

or 36°. Because of the delay time between the onset of ignition and the completion of combustion it is usual to cause ignition before the piston has reached the top dead centre position 1–2 in Fig. 5.5*b*.

Fig. 5.5*b* shows how pressure varied with time (or crank angle). if the ignition point is advanced too far, Fig. 5.5*c*, then unsteady pressure effects occur (pinking) and if the ignition is delayed too much, Fig. 5.5*c*, combustion continues through the expansion stroke with loss of power and a possibility of burning of the exhaust valves.

In a variable speed engine it is necessary to change the angle at which ignition occurs to take account of the delay time; automatic devices are fitted for this purpose in vehicle engines. The angle at which ignition occurs is changed as the speed varies by means of a centrifugal sensor. In the SI engine, power is controlled, at fixed speed, by means of cylinder air pressure. The delay time is increased at lower charge pressures and this must also be allowed

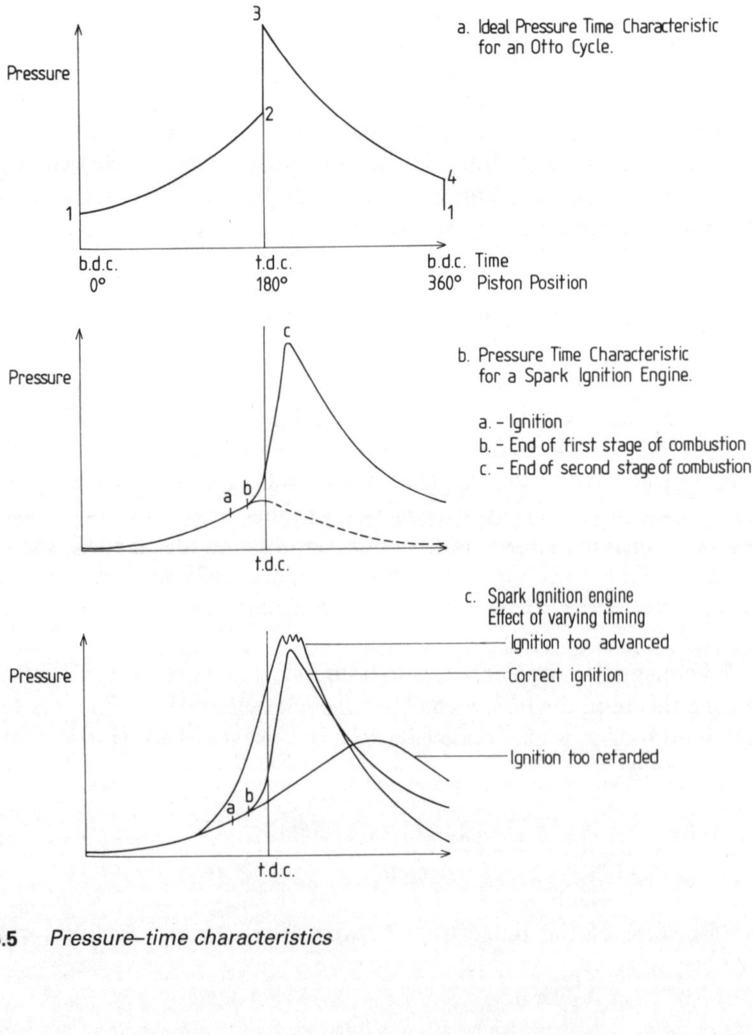

Fig. 5.5 *Pressure–time characteristics*

for. This compensation is achieved using a diaphragm which is operated by the differences in pressure between the induction manifold and the atmosphere.

5.2.4.2 Diesel knock: the cetane number. It might be thought that the criterion for selecting a fuel for a compression ignition engine is a low octane number. This to some extent is so, but since the fuel/air mixing arrangements between spark ignition and compression ignition are different, other considerations arise. In the spark ignition engine the fuel/air mixture is uniform and the fuel is in vapour form. In the compression ignition engine the fuel is injected as a jet of liquid droplets. For combustion to occur heat from the air must transfer to the surface of the droplet to first cause evaporation and then, on further heating, the creation of a vapour air layer which can then ignite. There is thus a delay between injection and ignition. If this delay is comparable to the total injection time then the whole charge will ignite together, leading to a sharp pressure rise or detonation, noise, and poor running of the engine – a condition known as diesel knock. Hence the delay time will be an important criterion in selection of fuels for compression ignition engines. The onset of knock will depend on both the fuel and the injection and engine characteristics. Fuels are compared using a standard engine and classified on an arbitrary scale, the cetane number, in a similar manner to the octane number. The comparison is made with a mixture of cetane (*n*-hexadecane $C_{16}H_{34}$) designated 100 and α-methylnapthalene ($C_{11}H_{10}$) as 0. Fuels for gasolene engines are chosen to have high octane numbers for engines of high compression ratio. Fuels for diesel engines are selected in terms of speed range; typically cetane numbers greater than 50 denote fuels suitable for high speed, greater than 35 for medium speed and greater than 25 for low speed.

5.2.4.3 Basic chemistry of fuels. The conventional fuels, gasolene and diesel, are hydrocarbons, that is they are made up from only hydrogen and carbon atoms in combination.[4] Atoms can combine with other atoms to form molecules. Some atoms, such as hydrogen, can only make one bond, but other atoms have the ability to form several separate bonds. The number of bonds of an atom is called the valency. Hydrogen is univalent whereas carbon is quadravalent, that is it can form four bonds.

Hydrocarbons are the simplest organic molecules, and they occur in two types, cyclic and non cyclic (aliphatic). In the aliphatic compounds the carbon atoms are linked together to form chains, whereas in the cyclic compounds, the carbon atoms link together to form rings.

The aliphatic type of hydrocarbons exist as alkanes (paraffins), alkenes (olefines) and alkynes (acetylenes). The paraffins are a major constituent of oil, and the simplest paraffin molecule is methane, a gas, with the chemical formula CH_4 and the structure:

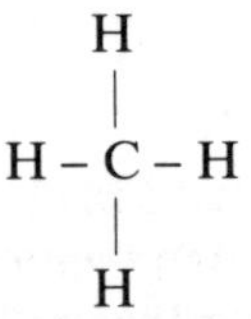

The paraffins form a series with the general formula C_nH_{n+2}. The molecules range from gases through liquids to solids (waxes). The bonds of the carbon atoms are all linked singly either to a hydrogen atom or another carbon atom and the resulting chain molecules are said to be saturated. Table 5.1 shows typical examples.

The larger molecules ($n \geqslant 4$) have the possibility of different structures for the same n. These different arrangements of the same atoms are called isomers and the different isomer molecules have different physical properties.

Another alkane group are the cycloalkanes (naphthenes) in which the molecules form saturated rings of the form C_nH_{2n}, see Table 5.2.

In alkenes not all the bonds are employed and the molecules contain double bonds, for this reason they are said to be unsaturated, see Table 5.3.

An important group, the aromatics, consist of molecules based around the six carbon (benzene) ring. They have the general form C_nH_{2n-6}, Table 5.4, and the bonds are not all saturated.

The naptheno-aromatics are a complex mixture of napthene and aromatic rings.

The alkenes and alkynes are not normally found in crude oil but alkenes can be important in petroleum produced by cracking, and alkenes and alkynes are produced as petrochemicals.

On average, petroleum contains around 85%C and 12·5% H by weight $(CH_{1\cdot8})_m$ together with small amounts of oxygen, nitrogen and sulphur.

Alcohols are formed when one of the hydrogen atoms in a paraffin is replaced by a hydroxyl group, OH. The simplest alcohol is methanol, CH_3OH, which is derived from methane CH_4. Ethanol C_2H_5OH is formed from ethane C_2H_6.

The octane number of the paraffins increases with increase in molecular weight. Alkenes and aromatics tend to have a higher octane number than that of the alkanes. The octane number of alcohols is high and they can be mixed with gasolene to raise the overall octane number. The lower calorific value of all these fuels is around 45 MJ/kg.

5.2.4.4 Conventional fuels: Conventional gasolene, diesel and gas turbine fuels are hydrocarbons. Distillate fuels are derived from the distilled components of the fractional distillation process and include gasolene, kerosene and gas oil. These are used in gasolene engines, gas turbines and high

Table 5.1 *Alkanes (Paraffins) C_nH_{2n+2}*

Methane	CH_4	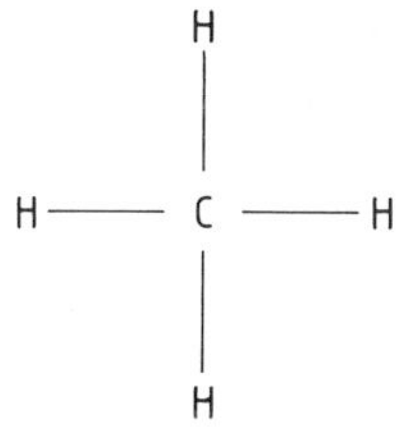
Ethane	C_2H_6	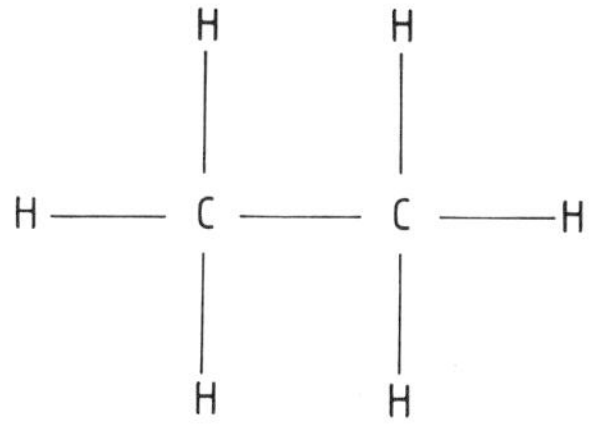
n-octane	C_8H_{18}	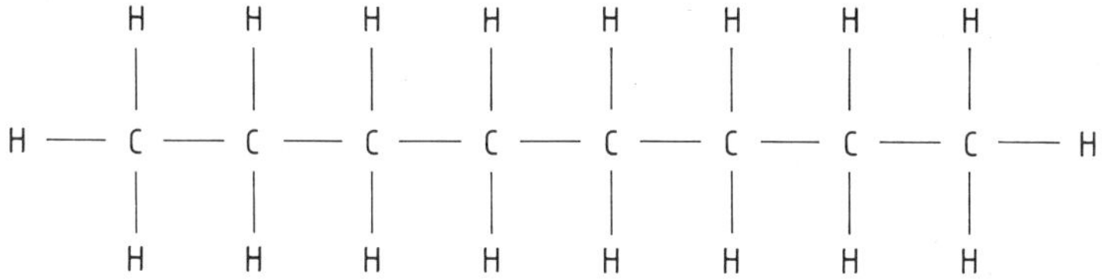
Iso-octane	C_8H_{18}	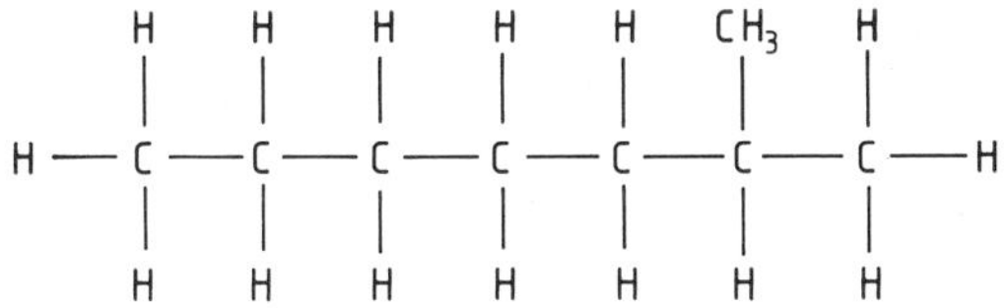
Cetane	$C_{16}H_{34}$	

n-Octane and iso-octane have the same molecular formula, C_8H_{18}, but differ in structure.

and medium speed diesel engines (> 800 rev/min and 300–800 rev/min). The viscous residual oils remaining after distillation are used in low speed (<300 rev/min) large marine and stationary engines. Residual oils require to be heated before they will flow and hence can be handled or injected into the combustion chamber.

Table 5.2 *Cycloalkanes (napthenes) C_nH_{2n}*

Cyclopropane	C_3H_6	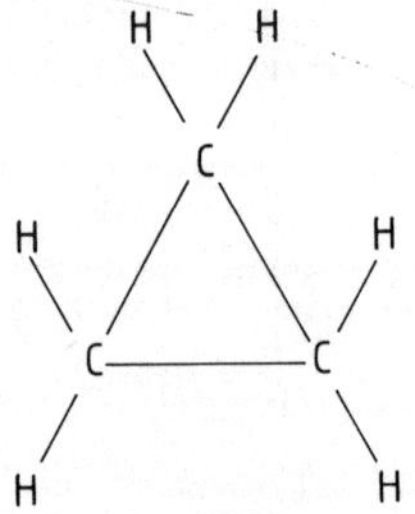
Cyclopentane	C_5H_{10}	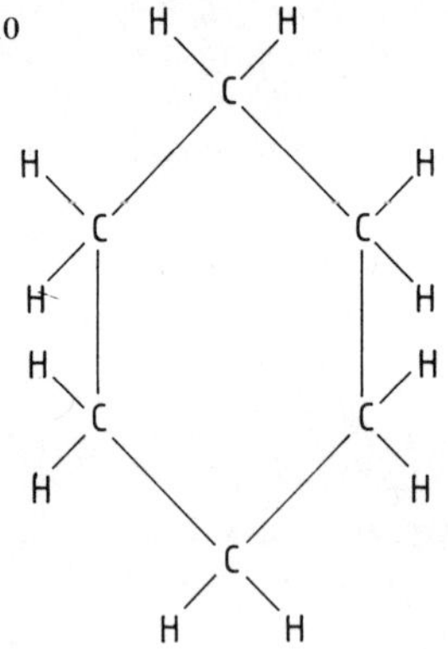
Cyclohexane	C_6H_{12}	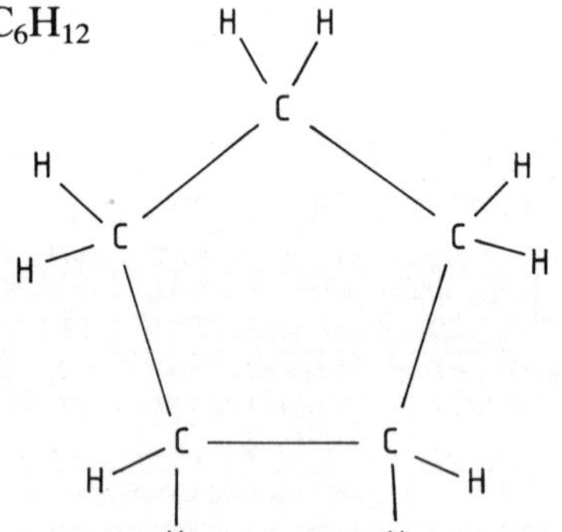

5.2.4.5 Unconventional fuels: At the present time most internal combustion engines use liquid fuels (diesel or gasolene). Internal combustion engines can, in fact, be operated on any liquid or gaseous fuel and recently there has been a renewal of interest in less conventional fuels such as biogas and alcohol.

Even solid fuels should not be ruled out; Dr Diesel carried out early

Table 5.3 *Alkenes (olefins)*

One double bond C_nH_{2n}
e.g. heptene C_7H_{14}

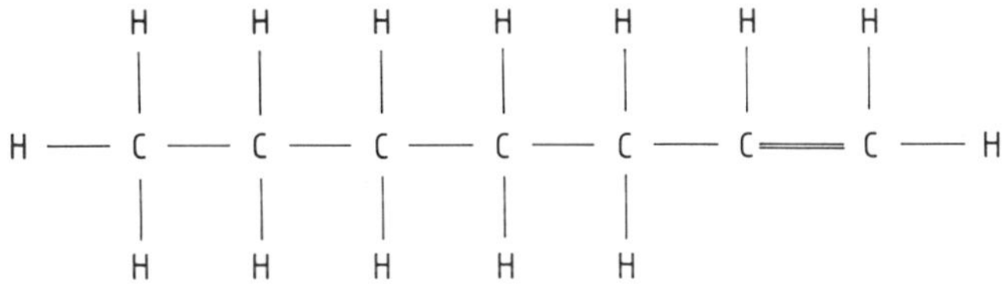

Two double bonds C_nH_{2n-2}
e.g. heptadene C_7H_{12}

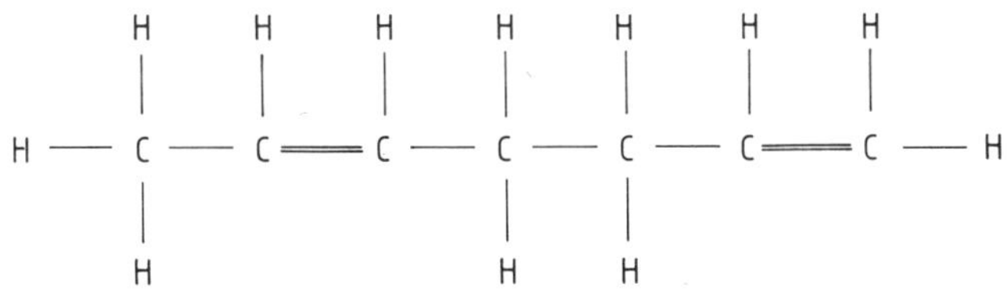

experiments on diesel engines using pulverised coal and there is current work on this fuel for use with very large engines.[5]

In the early part of this century many stationary engines were fuelled by town gas and some still exist today. The recent interest in biogas has led to the conversion of a number of stationary SI and CI engines for use with this fuel. Spark ignition can be used with biogas and only the carburettor needs to be replaced by a mixing valve to mix the air and gas. In the case of compression ignition it is necessary to add 5–10% of diesel fuel in order to obtain ignition.

Vegetable oils such as cotton seed, peanut, soya bean and sunflower can be used in diesel engines. Some difficulty may be experienced at cold conditions owing to the higher viscosity of these oils than that of conventional diesel fuel, and some authors recommend that pure diesel should be used for starting, warm up and shut down. Vegetable oils are mixable with diesel fuel in all proportions and can be used as blends, and this is sometimes referred to as 'extending' the diesel fuel.

Ethyl alcohol (ethanol) can be used in spark ignition engines either alone or in any blend with gasolene with which it is freely miscible. Unlike gasolene, which is a hydrocarbon, alcohol is a carbohydrate, that is it contains some oxygen. The energy per kilogramme is only 60% of that of gasolene. However, less oxygen is required per unit mass than for petrol and it turns out that for a given cylinder volume the heat released for both gasolene and ethanol are very similar. The stoichiometric ratio for ethanol is 9 parts of air to 1 of ethanol (by

Table 5.4 *Aromatics C_nH_{2n-6}*

Benzene	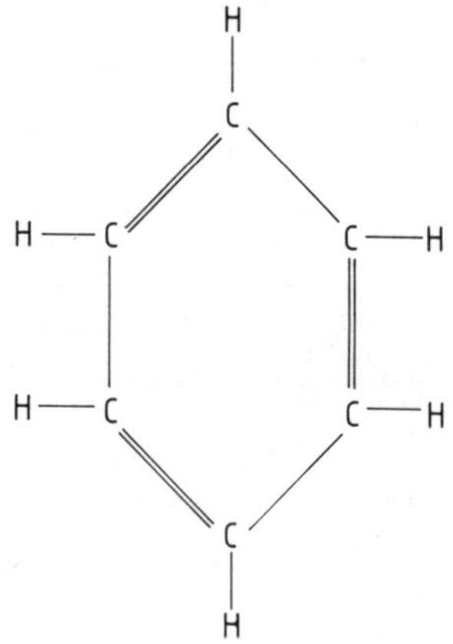
Toluene C_7H_8	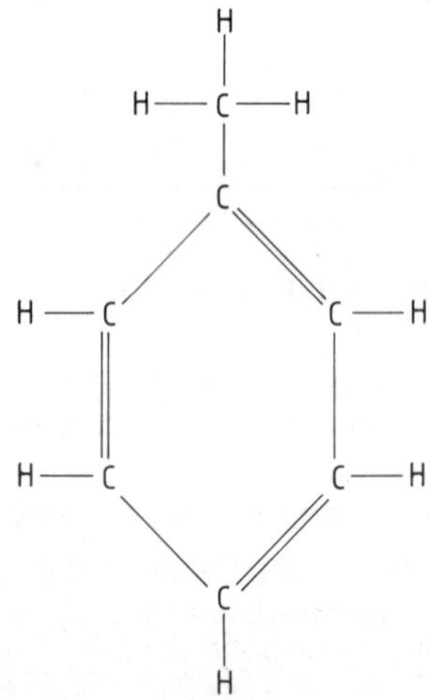

weight), hence the carburettor requires a larger diameter jet than is the case for a gasolene engine. Ethanol has a higher latent heat of evaporation than petrol, and the result of this together with the greater mass/charge is to give an improved volumetric efficiency when compared with petrol. Ethanol has a high octane number of around 110 RON which permits a higher compression ratio to be used.

One disadvantage of ethanol over gasolene for the private car is that the energy per unit volume is only 65% of that for gasolene and hence for the same volume fuel tank the range is correspondingly reduced.

In Table 5.5 the relevant properties of several fuels are given. There are difficulties in the blending of hydrous alcohol with gasolene owing to the

Table 5.5 *Properties of fuels*

		Petrol (gasolene)	Diesel	Ethanol	Methanol
Formula		C_4–C_{12} hydrocarbons	C_{14}–C_{19} hydrocarbons	C_2H_5OH	CH_3OH
Composition % (by weight)	C	85·0–88·5	86·5	52·2	37·5
	H	11·5–15	13·2	13	12·5
	O	0	0	34·8	50
Octane number					
RON		79–98	—	106–111	106–115
MON		71–90	—	89–100	82–92
Cetane number		5–10	45–55	0·5	—
Lower calorific value, MJ/kg		44	44	26·7	20

tendency of the water in the hydrous alcohol to separate out. This is discussed further in Chapter 8.

5.2.5 *Mechanical cycles: the four-stroke and the two-stroke*

In addition to specifying the method of ignition it is also necessary to specify the mechanical cycle used in a reciprocating engine. There are two common mechanical cycles; the four-stroke and the two-stroke.

The four-stroke mechanical cycle was historically the first to be used. In this cycle (Fig. 5.6) the piston compresses the gas in the cylinder, ignition takes place and the high pressure produced will depress the piston – the power stroke – the piston rises again, displacing the spent gases – the exhaust stroke. Finally the piston is depressed to draw in a fresh charge of air – the induction stroke. Suitable valving is necessary as indicated in the figure.

Clearly, if we can eliminate the two idling strokes of the piston in the four-stroke cycle, and include the induction and exhaust processes together with the compression and power in a single revolution of the crankshaft, we will have doubled the power output of the engine. Can this be done without significant loss in efficiency and without the introduction of other factors which reduce power output? This possibility early attracted the engine designer and led to the development of the two-stroke engine.

The first two-stroke internal combustion engine was designed by Sir Dugald Clark in 1881. Like the first four-stroke engine, this engine used gas as the fuel. The gas and air mixture was slightly compressed in a separate cylinder or pump and fed into the power cylinder via a poppet valve*, at a pressure slightly

* Poppet valve is the name given to the valve used in the car engine. Poppet valves are shown diagramatically in Fig. 5.6.

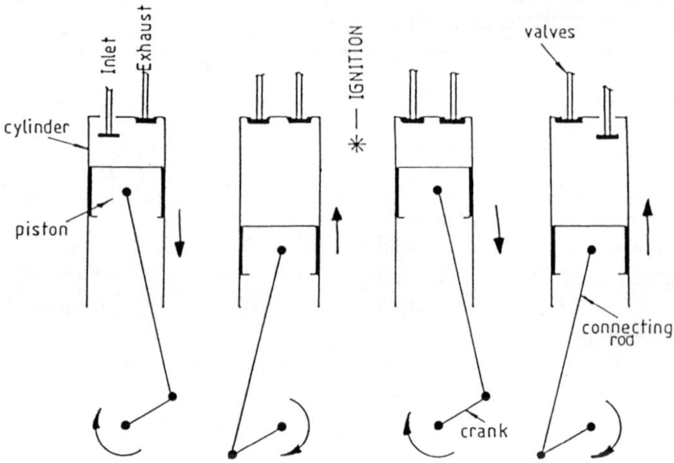

Stroke	1	2	3	4
	Induction	Compression	Expansion	Exhaust
Inlet valve	OPEN	SHUT	SHUT	SHUT
Exhaust valve	SHUT	SHUT	SHUT	OPEN

Fig. 5.6 *Four-stroke cycle*

greater than atmospheric pressure. The exhaust gases in the cylinder, due to the preceding ignition, were forced out through a port by the incoming gases. This exhaust port was uncovered by the piston at the end of its stroke. The next development was due to Day in 1891, who designed the two-port two-stroke engine. In Day's engine the crankcase was used as a pumping chamber and replaced the separate pump in the Clark engine. Gas entered the crankcase through a spring-loaded poppet valve. It was then compressed in the crankcase by the descending piston and allowed to enter the cylinder through a port uncovered by the piston at the end of its down stroke. The exhaust port was positioned opposite the inlet port and uncovered at the same time as the inlet port. Later the poppet valve on the crankcase was replaced by a third port also uncovered by the piston as it nears the top of its stroke, see Fig. 5.7.

The three-port two-stroke engine is particularly simple, having only three moving parts: the piston, connecting rod and crankshaft. We should compare this simple arrangement with the complication of a four-stroke engine with its valves, valve springs, tappets, camshafts and timing wheels. Why then is the latter almost exclusively used for the car engine? It turns out that the two main reasons for selecting the four-stroke are improved efficiency and power.

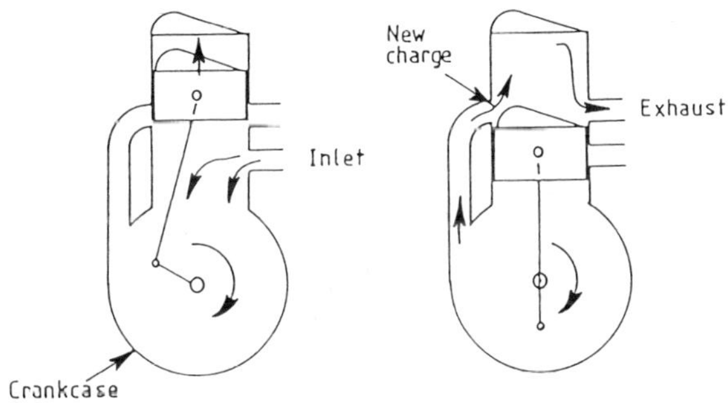

Fig. 5.7 *Two-stroke cycle*

Let us look at the reasons for the superiority of the four-stroke cycle over the two-stroke cycle. The power output from a given cylinder and thermodynamic cycle will be determined by three factors:

1 The volumetric efficiency.
2 The combustion efficiency.
3 The mechanical efficiency.

The volumetric efficiency of the two-stroke engine is low, approximately 40%. this is largely owing to the short duration of the induction period, which, for example, in an engine operating at 3600 rev/min (60 Hz) is only 1/250 s. the size of the port which can be used is limited by the need to avoid blow-back in the carburettor. In the four-stroke engine, for the same speed the injection port can remain open for more than twice as long as in the two-stroke engine. The four-stroke is also less sensitive to the other factors mentioned. In comparing typical small four-stroke engines, it is found that at low speeds the two-stroke pressure is around 2/3 of that of the four-stroke, at normal speeds this falls to 1/2 (when other things being equal the engines would give the same power output) and at high speeds it falls to around 2/5 or even less. Other factors which lower the volumetric efficiency are the temperature of the incoming gas, since the gas has been heated in the crankcase; the port area; the crankcase capacity; and the exhaust pressure. It has been said that the performance of a two-stroke engine is primarily dependent on the design of its exhaust system.

The combustion efficiency of a two-stroke engine is inferior to the four-stroke since some of the fuel charge escapes through the exhaust, Fig. 5.7 (this only applies to a spark ignition engine in which the fuel is mixed with the air before injection; in a compression ignition engine it is not necessary to inject fuel until the exhaust port has closed).

In spite of its limitations on power and efficiency, the two-stroke has the advantage of low cost, low maintenance, even torque (owing to an ignition on each compression stroke) and reliability. It is widely used for such applications as lightweight motor cycles and lawnmowers, where the first cost is important.

We have so far implied the use of spark ignition for the two-stroke engine. It is also widely used with low speed high power marine compression ignition engines; these engines do not suffer from the low combustion efficiency of the small SI engines since the fuel is injected after the exhaust port has closed.

5.2.6 Comparison of reciprocating internal combustion engines

In selecting a reciprocating internal combustion engine there are several parameters of interest. These include the size and weight of the engine necessary to provide the specified power and speed, and also the amount of fuel required to run the engine, or in other words the conversion efficiency at this power. If the engine is to be operated at variable speed and load we also require to know how fuel consumption will be affected. A further parameter of interest is the torque produced by the engine at different speeds.

The torque is proportional to a quantity brake mean effective pressure, or BMEP, usually given in units of N/m^2. The BMEP represents the effective force per unit area of piston averaged over the cycle. In the literature the BMEP and speed are plotted for constant fuel consumption. Fig. 5.8*a* and *b* shows examples for automotive gasolene and diesel engines. For smaller diesel engines, about 10 kW, the fuel consumption is around 230–240 g/kWh, which corresponds to about 34% efficiency. Small diesel engines are usually operated on the four-stroke cycle. Small gasolene engines have a higher fuel consumption of 480–360 g/kWh which gives an efficiency of 17–22%. This applies to both four-stroke and two-stroke engines, but the latter have markedly inferior part load performance. It is seen from Fig. 5.8 that the diesel has far better part load performance than that of the gasolene engine.

5.3 The Humphrey pump

The Humphrey pump[7–9] is a complete pumping engine combining both power source and pump. In an internal combustion engine, fuel and air is induced into the cylinder and when this mixture is ignited the gas pressure is increased considerably, and this pressure rise forces down the piston. In the conventional engines the piston is connected to a crankshaft by a connecting rod and the downward movement of the piston is converted to rotation of the crankshaft. This rotating shaft may in turn be connected either to the wheels of a car, or be used to drive a pump. In the latter case the pump also consists of a crankshaft which drives a piston in a cylinder, but here the cylinder contains water which is forced out by the piston to give the pumping action. Now this all seems rather round-about and one might ask if it is not possible for the expanding gases in the first cylinder to act directly on a column of water – in other words, can the water be used as a piston? In fact, of course, it can. Such a pump was invented by Humphrey at the beginning of the century. Humphrey built several very large pumps of this type, one notable example is the installation at

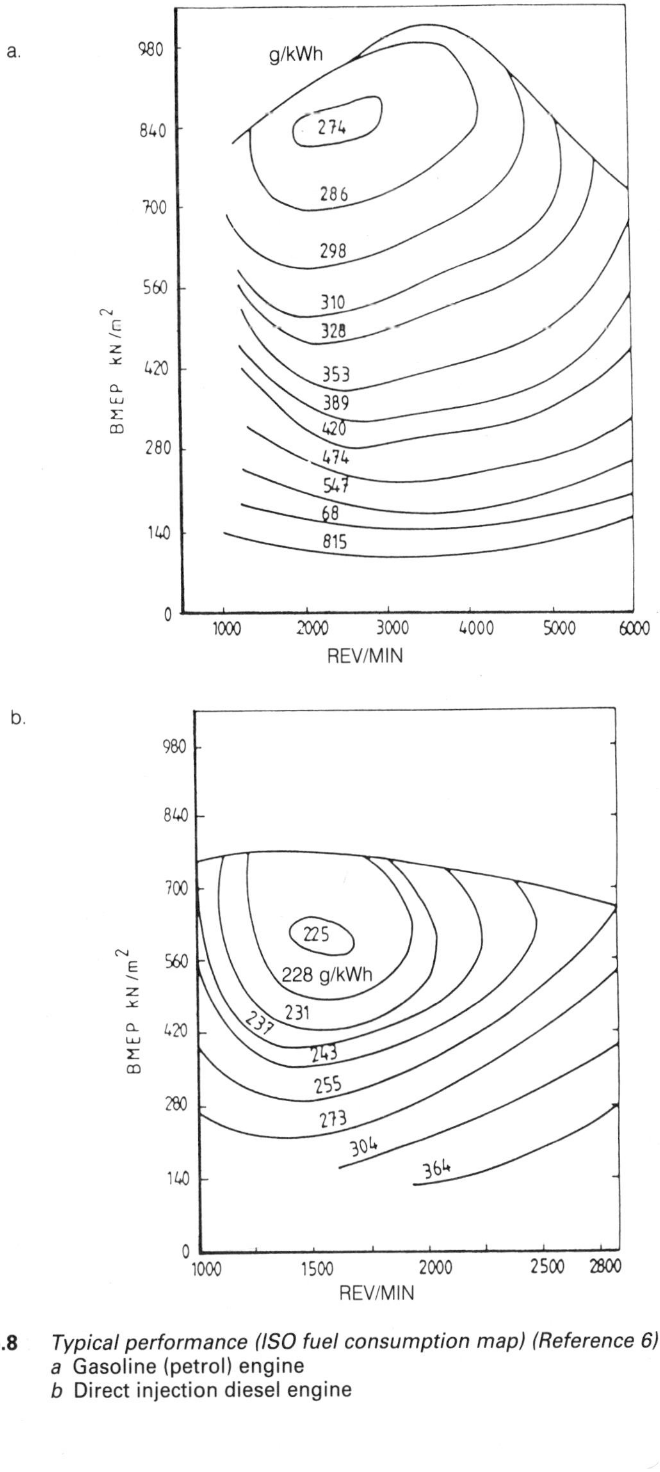

Fig. 5.8 *Typical performance (ISO fuel consumption map) (Reference 6)*
a Gasoline (petrol) engine
b Direct injection diesel engine

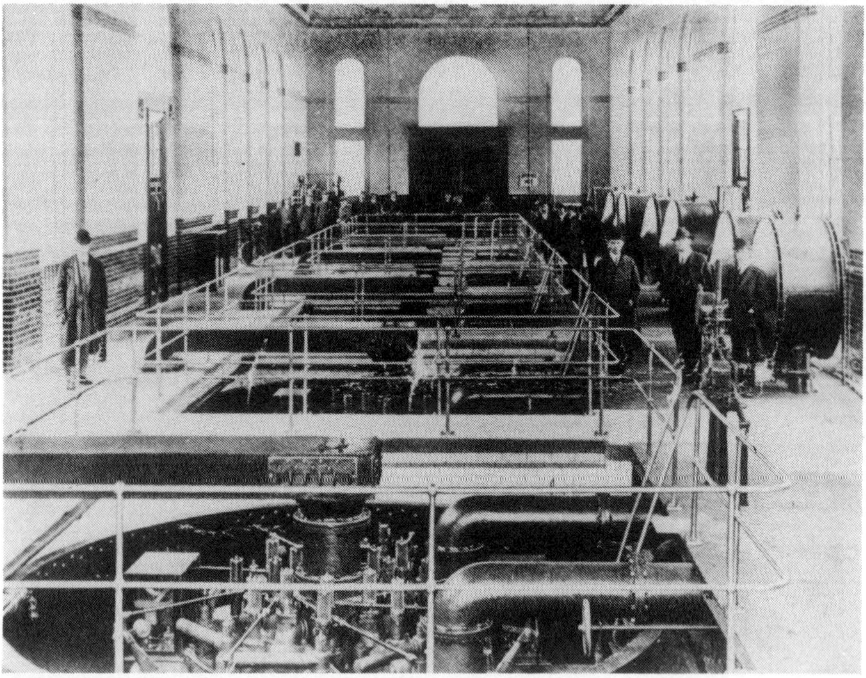

Fig. 5.9 *Pumphouse of the King George V Pumping Station*

the King George V Pumping Station, Enfield, UK, see Fig. 5.9, of four 40 million gallons per day units and one 20 million gallons per day unit. This station was opened in 1913 and used producer gas as the fuel. The pumps were operated until a few years ago when the producer gas boilers required replacement and alternative electrical pumps were installed. Interest was lost in the Humphrey pump after the First World War and much information on the development was destroyed in a fire in the inventor's office. The idea was revived at Reading University in 1969 and since this time we have designed a smaller version for use in developing countries. The main components of a Humphrey pump system are shown in Fig. 5.10. The dimensions refer to the Reading design.[8] Water is pumped from reservoir A to a higher level B.

The Humphrey pump is basically a U-tube in which water is made to oscillate by periodically exploding a charge of gas and air. The oscillating column of water acts as a flywheel. The inlet mixture valve, scavenge valve and the exhaust valve are interlocked to obtain the correct working sequence. There is no valve at the pump outlet, the pump requiring only non-return inlet valves. The products of combustion expand down to a little below atmospheric pressure, the four strokes of induction, compression, expansion and exhaust are brought about and controlled by the inertia of the water column. The sequence of operations is shown in Fig. 5.11.

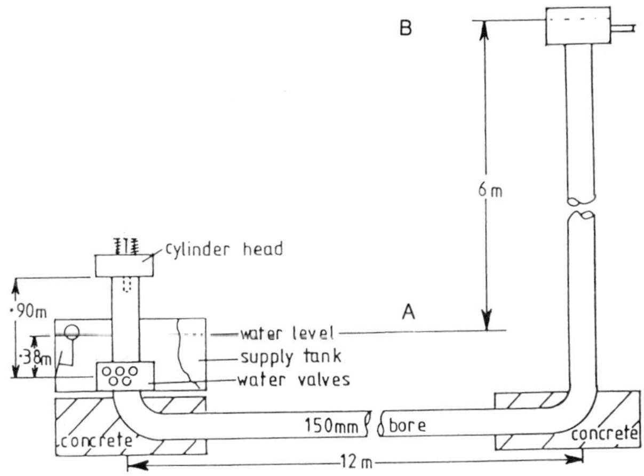

Fig. 5.10 *Simplified pump layout giving basic dimensions*

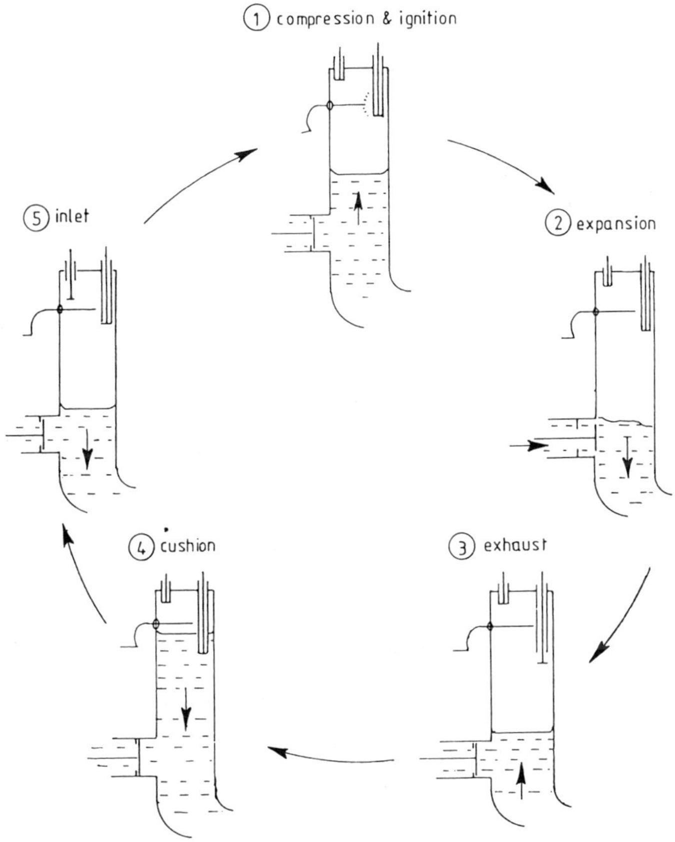

Fig. 5.11 *Operation of a four-stroke Humphrey pump*

1 The combustible mixture is contained in the cylinder with the valves closed, and ignited by a spark.
2 Expansion of the combustion products drives the water down the cylinder, and along the pipe. As the pressure in the cylinder drops below atmospheric, the water valve opens drawing in water, also the scavenge and exhaust valves open (the inlet valve being held closed).
3 The water column returns (as it oscillates in the pipework) and sweeps out the exhaust gases, until the level reaches the exhaust valve, forcing it shut.
4 The trapped volume of air acts as a cushion or 'gas spring' bringing the water column to a stop. The energy in the compressed air is returned to the water, forcing it back down the cylinder. The pressure once more drops below atmospheric.
5 The inlet valve opens (the exhaust and scavenge valves being held closed), allowing a mixture of gas and air to be induced.
6 The water returns, compressing the mixture until the pressure in the cylinder triggers another spark. The cycle repeats.

The ideal cycle is shown in Fig. 5.12*a*.

1–2	Compression of combustible mixture.
2–3	Ignition and combustion at constant volume.
3–4	Expansion to atmospheric pressure.
4–5	Pressure dropping below atmospheric as the water continues to flow under its own momentum, inducing a fresh charge of water.
5–6	Return of water column forcing out the exhaust gas.
6–7	Exhaust valve shut, compression of residual gas.
7–6	Return of water column, pressure dropping to atmospheric.
6–1	Induction of combustible mixture.

1–2–3–4–1 is the Atkinson cycle. From Fig. 5.12*a* it can be seen that the two successive piston strokes are of different length. While this is readily achieved by the liquid piston it is difficult to arrange with the metallic piston, connecting rod and crankshaft of the conventional engine. Nevertheless, the inventor of the cycle, James Atkinson, a Lancashire born engineer, constructed, in 1886, an engine which operated in this way. This 'differential' engine used opposed pistons drive by a linkage from the same crankshaft. Later Atkinson designed a single piston engine, the 'cycle' engine.

With typical figures of a compression ratio of 1·7 and a ratio of combustion temperature to compression temperature of 6·5, the ideal Atkinson cycle will give an overall efficiency of 43%, Fig. 5.12*b*. The overall pump efficiency obtained in practice is of course much lower than 43%. The large pumps at Chingford achieved efficiencies of over 20% but the small 150 mm pump developed at Reading is only 10% overall efficiency. The reduction of efficiency in the small pump is a scaling problem, the large Chingford pumps having a volume to surface area of five to ten times greater than the Reading

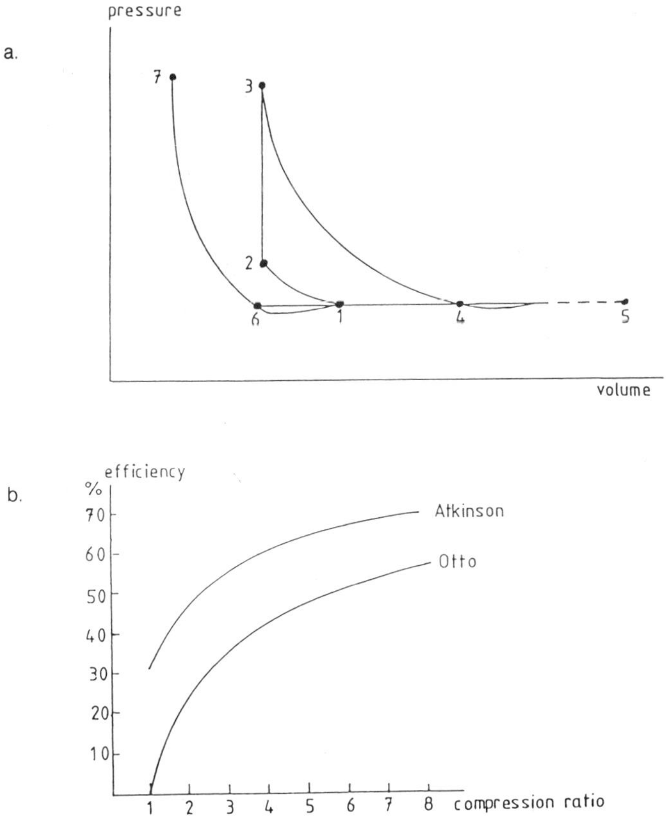

Fig. 5.12 *Atkinson cycle*
a Idealised pressure–volume diagram
b Ideal efficiency of the Otto cycle and the Atkinson cycle as a function of the volumetric compression ratio

pump and the heat loss is considerably greater, and also the presence of water droplets in the combustion space of the smaller pump absorbs large quantitities of heat. The actual cycle of the Humphrey pump does not follow the ideal Atkinson cycle, since all the burning does not take place at constant volume and there is some late burning which tends to increase the heat loss and hence lower the efficiency. The work at Reading has been directed towards making a pump that is simple in construction and can be made with a minimum of workshop machinery and technical skills. The pump is placed in a water supply tank and pumps water up a 6 m tower to a tank, from which the water is returned to the supply tank, Figs. 5.13 and 5.14, the basic dimensions are given in Fig. 5.10. The pump displaces 24 cubic metres of water per hour at an overall efficiency of 10%.

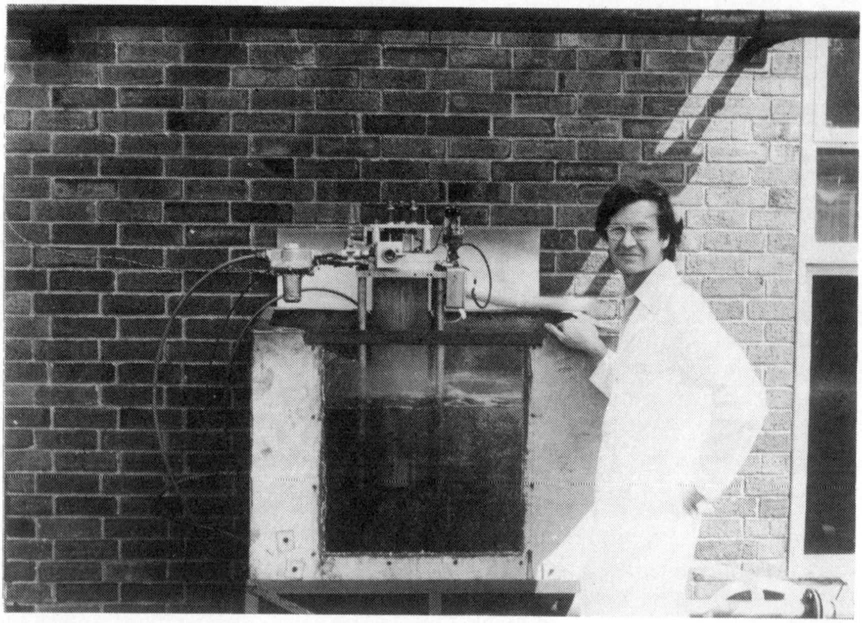

Fig. 5.13 *Humphrey pump sited in the water supply tank at Reading University*

Fig. 5.14 *Humphrey pump and tower outside the Engineering building, Reading University*

5.4 Open cycle gas turbines

The closed cycle gas turbine was described in Chapter 4 and the principle of operation of the open cycle gas turbine[10–12] shown in Fig. 4.31*d*. In the open cycle gas turbine up to 300% excess air is burnt, so that the performance characteristics discussed for the closed cycle gas turbine will also be applicable.

The two turbine types differ in that, in the closed cycle turbine, one has the freedom to choose the working gas and the operating pressure, whereas in the open cycle turbine the working gas is air, which (at the compressor inlet) is at atmospheric pressure. An important advantage of the open over the closed type is in heat transfer. In the open cycle gas turbine fuel is burnt with air and the resulting combustion gases are expanded through the power turbine and then rejected to the atmosphere so there is also no cooling requirement.

There is an efficiency benefit to be obtained from operation at high temperature; current practice is 850°C for high utilisation equipment and 870°C for low utilisation equipment.

As with the closed cycle turbine, regeneration is advantageous, particularly at part load. Compressor and expander component efficiencies in modern turbines are high (85–90%) giving an overall efficiency of around 25–30%.

Gas turbines are used for aircraft propulsion, and peak load electricity generation. In the case of aircraft propulsion the high power output and low size and weight of the turbine are of particular advantage. The rapidity with which the gas turbine can generate full load from start up is the main reason for its use in electricity generation.

References

1 ROGERS, G. E. C. and MAYHEW, Y. R.: 'Engineering thermodynamics, work and heat transfer' 3rd edn. (Longmans, 1980)

2 RICARDO, H. R. and HEMPSON, J. G. G.: 'The high speed internal combustion engine' (Blackie, 1968)

3 JUDGE, A. W.: 'High speed diesel engines' (Chapman and Hall Ltd., 1967)

4 NEUMANN, H. J., PACZYISKA-LAHME, B. and SEVERIN, D.: 'Composition and properties of petroleum' (Enke, Stuttgart, 1981)

5 TIPLER, W.: 'Prospects for the operation of diesel engines on coal or its derivatives', I. Mech. E. Conference Power Plants and Future Fuels, London 1975

6 WESTWELL, W. G. Chapter. Automotive Diesel Engines. Mechanical Prime Movers. (Macmillan, Basingstoke, 1971)

7 HUMPHREY, H. A.: 'An internal combustion pump, and other applications of a new principle', *Proc. I. Mech. E.*, 1909

8 DUNN, P. D.: 'The Humphrey pump for use in developing countries', Proceedings of a seminar on Appropriate Technology in Economic Development, University of Edinburgh, Sept. 1973

9 DUNN, P. D., JOYCE, N. G., and FULFORD, D.: 'The Humphrey pump', A Report to the Ministry of Overseas Development, Department of Engineering, University of Reading, 1977

10 COHEN, H., ROGERS, G. F. C. and SARAVANAMUTTOO, H. I. H.: 'Gas turbine theory' (Longmans, 1972)

11 RIZK, W.: 'Development in the application of gas turbines', VII World Power Conference, Moscow, 1968

12 'The jet engine' 3rd edn (Rolls Royce Ltd., Derby, 1973)

Chapter 6

Energy storage

6.1 Introduction

Sources of chemical energy, such as coal, oil and wood, can be stored for long periods with little or no loss and the energy can be released when required. This ability to store and release energy to suit the demand is of great importance, and contrasts sharply with the variable intensity and unpredictability of energy sources such as wind and direct solar power. In the case of these latter energy forms it is often necessary to convert the primary energy to another energy form for storage purposes. For example, solar energy is stored as thermal energy in a hot water tank, and the output of a small windmill may be converted to electricity and stored in a battery. The cost of storage equipment can often amount to a significant fraction of the total cost of the system.

One of the most important applications of energy storage arises in the design of self-propelled vehicles such as the car. Such vehicles require a source of energy which is compact, of low weight and which can be readily converted into controllable mechanical power. In the case of the car the energy source is gasolene which is burnt using oxygen from the air; the heat energy thus released is used to drive the engine. Gasolene (petrol) is a particularly convenient energy storage method for this application. A major weight advantage of gasolene over other energy storage methods is that the oxygen required in the combustion reaction is obtained from the air, and hence does not have to be carried by the vehicle. In fact oil in the form of gasolene or diesel fuel is so good that it has no real rival for the propulsion of small self-propelled vehicles. It is interesting to enquire why this should be so. What is so special about oil? Before attempting to answer this question we should first of all look at the various physical mechanisms which can be used to store energy and list them in order of suitability, Fig. 6.1. The ranking will depend on the criteria chosen to compare the different methods. We are concerned here with storing relatively large amounts of energy for periods of hours to weeks, and are interested in achieving this at as low a cost as possible, and since cost might be

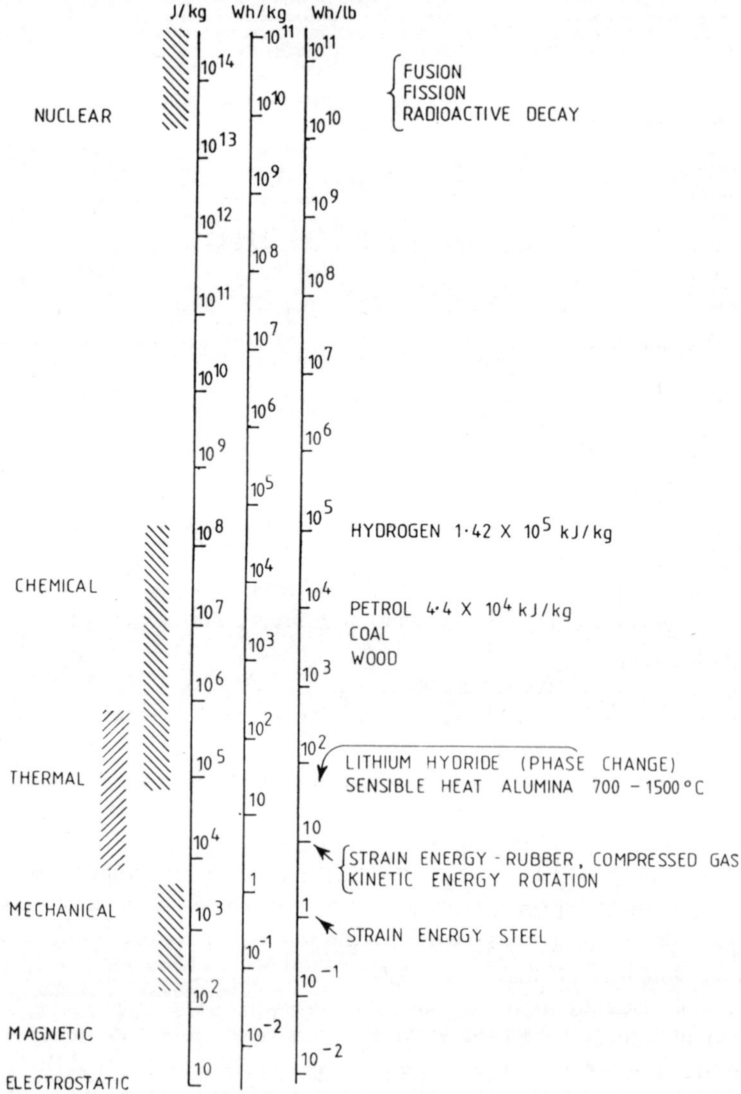

Fig. 6.1 *Comparison of different methods of energy storage*

expected to depend on the energy stored per unit mass, the mechanisms are given in terms of J/kg (electric batteries are frequently described in terms of Wh/lb or Wh/kg so these equivalents are also given in the figure). Energy stored per unit volume is also frequently of importance.

On looking at Fig. 6.1 the first thing one notices is the very large range of values between the different methods; namely, a factor of 10^{14} between nuclear stored energy and that stored by electrostatic forces. In the case of nuclear energy the particles which make up the nucleus (nucleons) are held together by very large forces and around 10 million electron volts of energy is required to remove such a nucleon. Similarly a particle released when a nuclear reaction occurs will possess this order of energy in the form of kinetic energy. One would therefore expect nuclear reactions to release energy of around 10 million electron volts per reaction. This figure is typical of nuclear reactions, including fusion reactions. Fission reactions are more energetic and release about 200 million electron volts of energy per reaction. Fission occurs in heavy nuclei so that the energy released per unit mass is comparable with that of other nuclear reactions.

Turning now to chemical reactions, here the typical energy released by each atomic reaction is of the order of a few electron volts, or less than a millionth of that for a nuclear reaction. The energy release from some chemical reactions of interest to energy storage is given in Table 6.1. Note the advantage of using atmospheric air; if only petrol is carried then the energy release is 44 MJ/kg. However, if it is necessary to carry the oxygen this figure decreases to a value of 11 MJ/kg.

Thermal energy can be stored in the form of sensible heat, which is the name given to the energy stored by raising the temperature of a body. The storage mechanism arises as follows: as the body is heated the atoms in the crystal lattice vibrate more rapidly and store energy in the form of the kinetic energy

Table 6.1 *Chemical stored energy**

	MJ/kg
Gases	
Hydrogen + air	140
Hydrogen + oxygen	15·5
Liquid	
Petrol + air	45
Petrol + oxygen	11
Solids	
Coal + air	30
Zinc + air	4·8
Sodium + sulphur	2·4
Lead + lead dioxide (lead–acid battery)	0·61

* In the table the figures given for hydrogen, petrol and coal refer to the *heat* released in the reaction. The figures for zinc, sodium and lead give the *free energy* released when the reaction occurs in an electrochemical cell. In the literature these energy differences are referred to as *enthalpy* change ΔH, and *Gibbs' free energy* change ΔG. In practice $\Delta H \simeq \Delta G$. Enthalpy and Gibbs' free energy are discussed more fully in Appendix 2. Where the fuel is combusted using air the air mass is not included in the total mass.

of these random oscillations; if the body is heated further it changes in form from solid to liquid, or in other words it melts. What has happened is that the oscillations of the atoms have become so violent that some bonds, typically one in ten, have broken and the body is no longer rigid. In the case of sensible heat, clearly, the heat capacity is less than would be required to separate the atoms completely, or less than a chemical reaction involving the material. The heat required to melt the body, involving the complete breaking of a small fraction of the bonds, might also be expected to be smaller than that for a chemical reaction. The examples in Fig. 6.1 show that this is in fact so. The energy required to melt a body is called the enthalpy of melting and replaces the older term, latent heat. This change from solid to liquid is called a phase change. In terms of energy storage it has the advantage that the thermal energy is stored, and released, at constant temperature. When the liquid is further heated its temperature will rise and it will store more sensible heat until the temperature reaches such a value that the atoms can no longer hold together and the liquid changes to a vapour, or in other words the liquid boils. Since, on boiling, all the bonds between the atoms are broken, the enthalpy, or latent heat of evaporation, is usually several times that for melting. Evaporation is inconvenient for energy storage because of the large volume change; this is not the case for the solid to liquid phase change where there is little volume change*.

A further method of using the forces between atoms to store energy is by applying an external force to strain the material. This strain energy, or mechanical potential energy, can be stored in compressed solid springs or by compressed gas in cylinders or on a larger scale in caverns in the ground.[1,2,3]. Alternatively, mechanical energy can be stored in kinetic form in a flywheel. The maximum energy stored in the flywheel will be set by the stresses in the structure, which in turn will be limited by the strength of the bonds between the atoms. Just as thermal storage might be expected to be less effective than chemical storage, the same general conclusion will apply to mechanical storage. In fact, perhaps the best way to store energy in a flywheel is to make it of wood and set fire to it! An exception to the limit set by the interatomic force arises in the case of mechanical translational kinetic energy, for example in a bullet. There is no upper limit to the energy which might be stored by this means, but the method obviously poses difficult practical problems and cannot be regarded as a serious contender. Gravitational energy, the raising of a weight, is a simple and widely used method of energy storage but requires very large masses.†

* This small change in volume has practical importance, the expansion from water to ice is responsible for water pipes bursting in winter. The water–ice transition is unusual since, in general, the solid phase will occupy a smaller volume (have a higher density) than the liquid it replaces. This characteristic of water–ice is fortunate, for had it been otherwise ice would sink and build up on the bottom of lakes etc.

† A well-known small scale practical application of gravitational energy storage is as a source of energy for church and the old-fashioned grandfather clocks.

Electrostatic and magnetic energy are included in Fig. 6.1 for completeness but have no interest for our applications. It should be said that (in this discussion) we have taken a highly restricted view of energy storage, and have limited it strictly to the terms of reference of this book. Each of the energy storage mechanisms mentioned fills a niche in the total range of energy storage requirements. For example, if energy is to be stored and released rapidly, the electrical capacitor is without rival. A microfarad capacitor charged to 50 000 volts and discharged in 1 microsecond will deliver an instantaneous power of 1250 MW, or the output of a large power station, but of course only for a very short time. In this example, although the power is impressive, the total energy stored is only 1250 J. Similarly, it is difficult to find a better solution for the power source for a small model aircraft than the rubber band, where the simplicity, low weight and high rate of power release are particularly suited to the requirement. Capacitors are not suitable for storing the relatively large amounts of eneregy required in the applications discussed here.

Returning to the problem of selection of methods of energy storage for relatively small scale applications, nuclear energy is not suitable, since it is not possible to achieve nuclear power generation on this scale at economic cost. Small reactors of a few kilowatts or so have been produced and operated successfully for military and space use, but are unlikely ever to be generally available. Small power sources of a few tens of watts output have been developed for conversion of the heat released by the decay of radioactive materials. This heat can be converted by thermocouples into electricity. Such small generators have been designed for use with unattended navigational aids, and while they have merit for this application, they are not suited either for scaling up to higher power levels nor for general use because of the high unit cost.

The energy per unit mass figures given in Table 6.1 bring out clearly the reason why petrol plus atmospheric air is so much better than alternative energy storage methods for use in the motor car. In addition to high energy per unit mass and high energy per unit volume it is relatively safe and non-toxic, can reasily be stored and transported, and as a liquid can be used more readily in the engine than a solid such as coal.

Direct solar energy can be converted by photosynthesis to chemical energy in the form of biomass and this provides a convenient method of energy storage. The characteristics of biomass are discussed in more detail in Chapter 8. We must now look at methods of storing the output from devices such as wind generators and convertors of direct solar energy. If the requirement is for heat, particularly for low grade heat (less than 80°C), then thermal storage using either sensible or latent heat can readily be achieved for short periods of a day or so. Longer term storage is possible but may prove uneconomic in many cases. The solar pond offers the possibility of low cost long term storage for medium scale plant (~100 kW). Thermal storage is described later in this chapter. Chemical storage using electrochemical cells enables energy in the form of electricity to be stored, although the currently available electrochemical batteries are considerably heavier and bulkier than a tank of oil of the same energy content. This is partly because the chemical elements chosen are so much heavier; for example, in the case of the lead–acid battery the fuel is metallic lead, whereas oil is made up from the much lighter atoms of carbon and hydrogen. Also, the lead–acid battery carries its own supply of oxygen in the form of the heavy lead dioxide. Nevertheless, secondary batteries, such as the lead–acid, can be both convenient and economic in some situations. Finally, hydrostorage in association with one of the alternative energy sources can in some circumstances be economic in spite of its large size and high capital cost.

6.2 Thermal storage

6.2.1. Introduction

Heat can be stored in one of two ways:

1 Sensible heat, in which the storage material is raised in temperature without any change in structure.

2 Phase change, in which the material changes structure by melting or evaporating and latent heat is required to cause this change.

In the case of phase change the heat exchange is effected at a constant temperature. This can be a useful practical advantage over sensible heat storage, but it does restrict the choice of materials to those in which the change occurs at the required temperature.

6.2.2 Sensible heat

Water is an obvious choice of storage medium for low temperature application such as domestic solar water heating. It has a high heat capacity (4·18 kJ/kg°C), is readily available, non-toxic and can easily be stored. Heat exchange is also simple since the water can be made to flow through both heat source and sink. Water can be used over a wide temperature range of, say, 25–95°C. For a 60°C temperature change water will store 250 kJ/kg, or $2{\cdot}5 \times 10^5$ kJ/m^3, see Fig. 6.2.

If solids (such as rock or brick) are used, then a heat exchange medium such as air or water is required. Solid thermal storage is convenient for use in buildings. For a temperature of change of 50°C rocks and concrete will store around around 36 kJ/kg, or about 10^5 kJ/m^3, see Fig. 6.3.

Table 6.2 lists the properties of a number of materials used for thermal storage by sensible heat.

6.2.3 Phase change: latent heat

Large amounts of energy can be stored by the addition and release of heat at a solid to liquid, or liquid to vapour, phase change. The energy exchange in the latter is generally several times larger than is the transition from solid to liquid.

a.

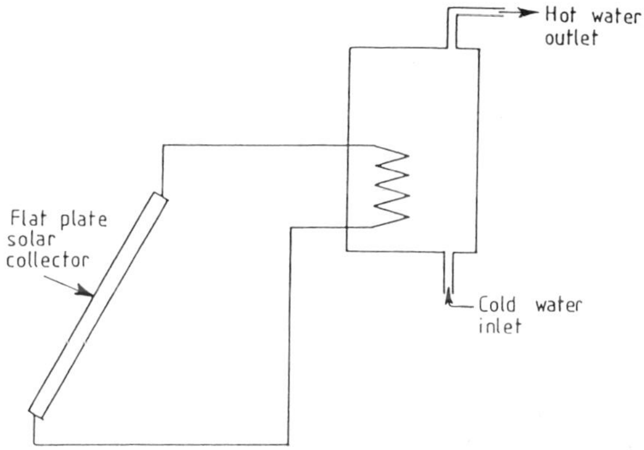

b.

Fig. 6.2 *Sensible heat storage in water tank of solar heated domestic hot water system*
a Schematic
b Flat plate solar collectors, Mauritius

For example, when ice melts 334·9 kJ/kg is released compared to 2200 kJ/kg for the conversion of water to steam at 100°C. The disadvantage of the liquid to vapour phase change is the large increase in volume which accompanies the change. For this reason energy storage is normally by solid to liquid change. The properties of some materials of possible interest are given in Table 6.3.

The energy of hydration of some salts is quite high and can be employed as an energy store. Glauber's salt ($Na_2SO_4.10H_2O$) is often mentioned as a

a.

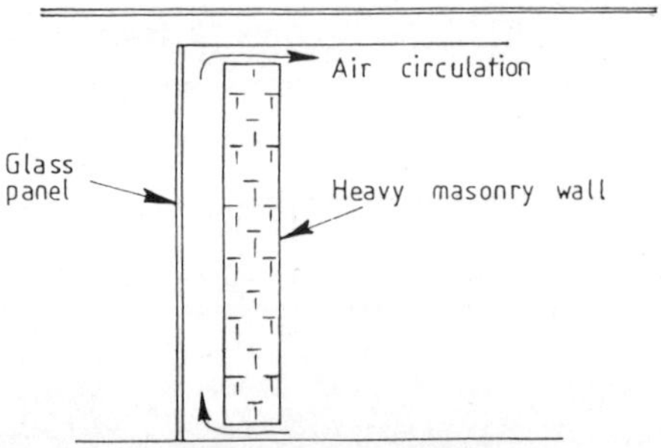

b.

Fig. 6.3 *Trombe wall*
Solar room heating using sensible heat storage in wall. Night loss can be reduced by the use of a shutter between glass and wall.
a Schematic
b Installed Trombé wall, Beijing, China

suitable candidate for domestic heat storage. It decomposes at 32°C to form a saturated water solution of Na_2SO_4 plus an anhydrous residue of Na_2SO_4. Two problems arise in the use of Glauber's salt and similar type materials for storage:

1 Stratification
2 Supercooling

It was mentioned above that, when heated, the crystalline salt decomposes, with the absorption of heat, into a saturated water solution of Na_2SO_4 plus a

Table 6.2 *Sensible heat storage (at 20°C) (various sources)*

Material	Chemical formula	Specific heat kJ/kg°C	Volumetric heat MJ/m³°C
Water	H_2O	4·18	4·2
Ethyl alcohol	C_2H_5OH	2·2	2·2
Concrete	—	1·1	2·5
Limestone	$CaCO_3$	0·91	2·5
Sand	SiO_2	0·80	1·3
Iron	Fe	0·47	3·7

Table 6.3 *Storage by phase change – solid to liquid (various sources)*

Material	Chemical formula	Melting point °C	Heat of fusion kJ/kg	MJ/m³
Ethylene glycol	$C_2H_6O_2$	−13	146	163
Water	H_2O	0	335	305
Paraffin	C_{15}–C_{16}	44–48	209	160
Sodium	Na	98	116	117
Ammonium thiocyanate	NH_4SCN	146	260	337
Sodium hydroxide	NaOH	300	226	466
Potassium perchlorate	$KClO_4$	527	1250	3150
Magnesium	Mg	651	293	499
Aluminium	Al	660	395	1650
Lithium hydride	LiH	685	3780	3100

residue of Na_2SO_4. The latter will fall to the bottom of the container as a solid layer. On cooling, the material in this layer is unable to retrieve the original 10 hydrogen atoms per molecule required for hydration. One solution to the problem is mechanical stirring; a second possibility is to prevent the sedimentary layer from forming by 'thickening' the solution with gels.

Supercooling is the second problem experienced when attempting to retrieve the heat of decomposition. Typically the mixture must be cooled to 10 or 15°C below the decomposition temperature before the reverse, hydration, reaction will occur. It is found that supercooling can be avoided by seeding with a few percent of a material such as borax.

Table 6.4 lists the relevant properties of Glauber's salt together with a number of other similar salts.

Table 6.4 *Phase change due to water of hydration (various sources)*

Material	Chemical formula	Melting point °C	Heat of fusion kJ/kg	MJ/m³
Calcium chloride hexahydrate	$CaCl_2 \cdot 6H_2O$	30	170	280
Sodium sulphate decahydrate (Glauber's salt)	$Na_2SO_4 \cdot 10H_2O$	32	251	370
Sodium carbonate decahydrate (washing soda)	$Na_2CO_3 \cdot 10H_2O$	32–35	247	360
Disodium hydrogen phosphate dodecahydrate	$Na_2HPO_4 \cdot 12H_2O$	35	265	400
Calcium nitrate tetrahydrate	$Ca(NO_3)_2 \cdot 4H_2O$	40–43	140	260

6.3 Electrochemical energy storage

6.3.1 Batteries

Energy can be released when a chemical reaction occurs. The most usual reaction is that of oxidation, that is fuel is burnt to give off heat. It is not essential that the energy release should be in the form of heat and in electrochemical reactions the energy released from the reaction appears in the form of electricity. The essential components of the electrochemical cell* are three: a source of fuel; a source of oxygen, and a means of bringing them together such that the energy released will be in the form of electricity. The simple cell is shown in Fig. 6.4. The fuel and oxidant are in the form of plates,

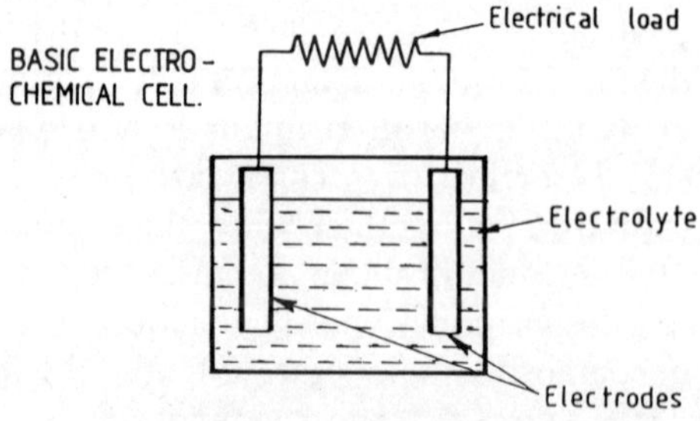

Fig. 6.4 *Basic electrochemical cell*

* The distinction between 'cell' and 'battery' is that the battery is simply an assembly of cells.

the electrodes. They are immersed in a liquid, the electrolyte, which has rather special properties. In the overall cell reaction the fuel atoms are allowed to combine with the oxygen atoms, to which they are attracted with considerable force. If the atoms are allowed to combine freely, as in combustion, then the result of this attractive collision will be the release of energy as heat. What is required is to allow the fuel atom to do work in arriving at the oxygen atom. In the electrochemical cell the work is done against an electrical field set up between the two plates. We can suppose that the an atom from the fuel electrode enters the electrolyte and moves across to the oxygen electrode. If the atom is electrically uncharged then it will be unaffected by the electric field. If, however, the process of solution into the electrolyte is such that the atom leaves behind an electron it will have a net positive charge and will interact with the electric field which will oppose its motion to the oxygen electrode. The name given to atoms which have lost or gained electrons is ion. On arrival at the oxygen electrode the fuel ion will take up a new electron when it combines with the an oxygen atom. The electrolyte must therefore have the property of allowing the passage of ions but not electrons. To the outside observer the fuel electrode will have a negative potential with respect to the oxygen donor electrode, and an electron will need to move through the external load from the fuel electrode to the oxygen electrode. It can be seen that, in addition to supplying fuel and oxygen, respectively, the electrodes must also be electrical conductors in order to allow the flow of electrons round the external circuit. This is essentially what happens in all cells, but the details vary with the particular type of cell. For example the oxygen rather than fuel may move through the electrolyte to the fuel electrode; in some cells the electrolyte takes part in the reaction, but the general principle is as outlined.

The typical open circuit voltage from a single cell is around 1·5 to 2·0 volts depending on the materials used for the electrodes. It is usual to combine cells in series to give higher values of voltage, and such combinations are called batteries.

So far we have referred only to oxidation reactions. Other electron donor materials can be used instead of oxygen and a number of materials such as chlorine and sulphur are being considered. As can be seen from Table V higher energy per unit per unit mass can be achieved in this way.

6.3.2 Classification of electrochemical cell

There are four main types of cell:

1 The primary cell. The primary cell is used once and when the fuel is consumed it is thrown away. The dry cell based on zinc as fuel, and oxygen supplied by manganese dioxide, is the best known example and is used widely for transistor radios, electric torches, etc.

2 The secondary cell. The secondary cell may be recharged, that is, the fuel may be returned to the fuel electrode by passing an electric current through the cell in the reverse direction. This charge/discharge cycle can be repeated many times. The lead–acid battery is one of the best known of the rechargeable cells; the fuel is lead, and oxygen is supplied by a lead dioxide electrode. The electrolyte is sulphuric acid, hence the name of the cell.

3 The fuel cell. The fuel cell differs from the primary and secondary cells in that its electrodes serve only as dispensers of fuel and oxidant and as conductors of electricity. The fuel and oxidant are stored externally and fed to the porous electrodes. One obvious advantage of this arrangement is that the size of the cell determines only the power output, whereas the external storage tanks determine the total energy. This leads to an overall weight and size less than that for the secondary cell for applications such as space propulsion. Apart from their use in space satellites, fuel cells are still at an early stage of commercial development.

4 Semi-fuel cells and metal–air cells. In these cells the fuel is supplied by the electrode as for a primary or secondary cell but the oxidant is supplied from the air via a non-consumable electrode. The advantage of not having to carry a separate air supply has already been referred to elsewhere.

Primary cells have very wide applications for the storage of small amounts of energy for such purposes as powering transistor radios, electric torches and hearing aids. However, they are too expensive to be used as large energy stores. The major reason for the high cost is that the cells cannot be recharged, whereas secondary cells can be recharged many hundreds of times, thus reducing considerably the capital cost/charge. The most important secondary cells are discussed in the next section. Fuel cells will not be discussed further, since, apart from rather specialised applications such as power supplies for spacecraft, they have not yet been developed to a stage when they are commercially available. It should, however, be noted that hydrogen/oxygen or hydrogen/air fuel cells are proposed from time to time for storage where they are used in association with an electrolyser which produces hydrogen gas and, if required, oxygen gas, by the electrolysis of water. The gas is stored and then used to operate a fuel cell. As an alternative the hydrogen gas can also be burnt in an internal combustion engine.

6.3.3 Secondary cells

(a) *Lead–Acid Battery*

The lead–acid battery is the work horse of the secondary cells. It employs a lead electrode as the fuel, and oxygen is stored in the form of lead dioxide on the second electrode. It is at first sight surprising that lead should have been chosen as both fuel and the base for the oxygen

Table 6.5 *Characteristics of some materials suitable for use in electrochemical cells*

Anode	Cathode	Theoretical energy density Wh/kg	Normal operating temperature °C
Pb	PbO_2 (Oxygen stored as PbO_2)	176	20–30
Zn	Air	1232	20–30
Na	S	664	300
Li	S	1500	—
Li	Cl	2200	650

store. Table 6.5 shows that because of the very high density of lead the theoretical energy density is only 170 whr/kg compared with 1232 whr/kg for zinc/air. In spite of this inauspicious start, lead, in the form of the lead–acid battery, has been developed to the stage where it is now the most important of the secondary batteries. Some reasons for the dominance of the secondary battery field by the lead acid battery are:

1 Good performance. The battery holds its charge well, can be recharged many hundreds of times, and will deliver both high and low currents.
2 The cost of the metal and also its fabrication compare well with the other alternatives, nickel, cadmium, and silver.
3 The high cell voltage of 2 V compared with 1.3 V for nickel cadmium or nickel iron enables a 12 V battery to be built up from only six cells compared to eight for nickel cadmium or nickel iron.

The overall chemical equation is:

$$Pb + PbO_2 + 2H_2SO_4 \underset{\text{charge}}{\overset{\text{discharge}}{\rightleftharpoons}} 2PbSO_4 + 2H_2O$$

with a corresponding voltage of $E = 2{\cdot}04$ V per cell.

It is seen that the electrolyte, sulphuric acid, is changed in the reaction and its concentration is reduced on discharge. Strictly speaking, therefore, the sulphuric acid is an active material as well as the electrolyte. At a temperature of 15°C a fully charged battery has an electrolyte density of 1·28 which falls to 1·22 at half charge and 1·12 or less on complete discharge. Since specific gravity depends on temperature a correction factor of around 0·04/10°C must be made for temperatures higher or lower than 65°C. It is thus possible to check the state of charge by means of a density measurement (hydrometer). Of course, the battery must be filled with acid of the right concentration, a discharged battery cannot be charged by adding concentrated sulphuric acid. The acid concentration is chosen to lie in the range of maximum electrical conductivity.

Charge and Discharge

The capacity of a battery is usually given in terms of ampere hrs at a specified time of discharge (or discharge current). Fig. 6.5 shows typical curves. The graph indicates how the capacity falls below the rated value at higher discharge currents. As an example, if the battery is operated at a current of 5 A it will give twice the Amphr than when operated at a current of 88 A. Fig. 6.5 also

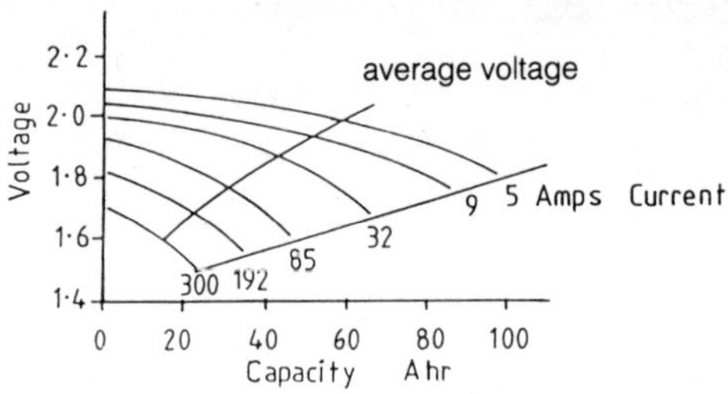

Fig. 6.5 *Effect of discharge current on capacity for lead–acid cell*

gives the average voltage during discharge. For the 5A rate the value of average voltage is approximately 2·0v or an energy of 200 whr. For a current of 88A this value falls to 87·5 whr. or only 44% of the 5A storage.

Due to internal losses not all the energy stored can be recovered. A typical efficiency is 75%. Hence if 3 kwhr is required, 4 kwhr must be stored in the battery. Further losses occur on charging, the charging efficiency is typically 80% at a slow rate of charge. Hence in order to yield 3 kwhr a battery will require 5 kwhr of energy from the source. Fig. 6.6 shows voltage characteristics for charge and discharge of a 3 cell 6V battery.

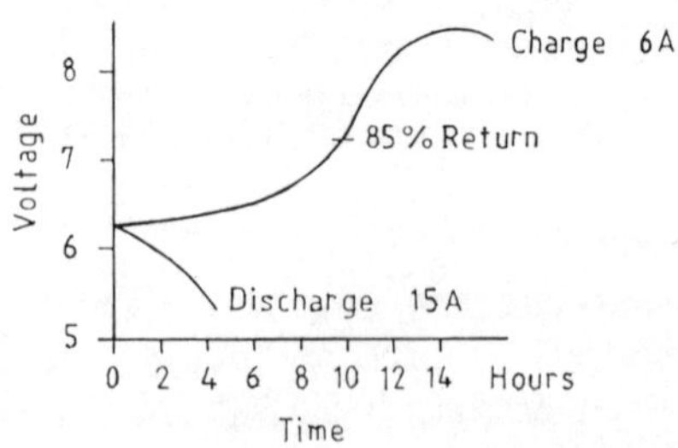

Fig. 6.6 *Charge and discharge characteristics for lead–acid cell*

Actual performance will depend on the type of use but Table 6.6 gives representative values.

Regular topping up of the electrolyte with distilled water is required.

Table 6.6 *Secondary Cells*

	Anode	Cathode	Electrolyte	Open Circuit Voltage	Theoretical Energy Density whr/kg	Practical cells whr/kg	Efficiency charge/ discharge
						5 hour rate	
Lead Acid	Pb	PbO_2	H_2SO_4	2·1	176	25–35	75
Nickel Cadmium	Cd	NiOOH	KOH	1·30	209	25–35	70
Nickel Iron	Fe	NiOOH	KOH	1·35	267	25–35	<60
Silver Zinc	Zn	Ag_2O	KOH	1·60	273	130	
Silver Cadmium	Cd	Ag_2O	KOH	1·15	171	60	

(b) *Nickel Iron Battery*
The Nickel Iron, or NiFe, battery employs an alkaline electrolyte, Potassium Hydroxide, KOH.
The batteries are robust, can be repeatedly deep discharge cycled, and have long life. they require regular topping up with distilled water due to the large amount of gas evolved at the electrodes. They have moderate energy density (on both volume and weight basis) but have poor low temperature performances. NiFe batteries are more expensive than the lead acid type.

(c) *Nickel Cadmium Battery*
This alkaline battery has become of increasing importance as a power source for small portable equipment such as electric razors. It can be sealed and is capable of high charge and discharge rates. The cost is considerably higher than that of the lead acid battery.

(d) *Summary*
The characteristics of the most important commercial secondary batteries are given in Table VI.

6.4 Hydro storage

One of the simplest methods of storing energy is to make use of the gravitational energy of water when it is pumped uphill to a reservoir, a method known as pumped storage. When the energy is required the water is allowed to flow down the hill to drive a water turbine. It has been stated elsewhere that very large quantities of water can be raised from boreholes with the expenditure of comparatively little energy; and in fact 36·5 tonnes of water can be raised to a height of 10 m for the expenditure of only 1 kwhr of energy if the process is 100% efficient. In pumped storage we have the reverse of this problem; in order to store 1 kwhr of energy it is necessary to store 36·5 tonnes of water, or in terms of volume 36·5 m^3, at an average height of 10 m. As a further illustration of the relatively small amounts of energy stored by gravitational potential we note that if a quantity of water is dropped from a height of 10 m, and assume, reasonably enough, that all the stored energy is converted into heat in the mass of water then the rise in temperature will be only about 0·024°C. This is in accordance with our common experience; we are never scalded by raindrops, and whilst if I fall off a ladder I may break a limb, the impact certainly will not raise my body temperature to a dangerous level. Thus it is seen that pumped storage will result in a need for a very large and hence costly reservoir and this is a major disadvantage of the method. A further problem is that in many cases a suitable site will be some distance from the user, necessitating a transmission system with associated losses and cost. On the credit side the energy storage is efficient; the pumping efficiency could be around 90–95% and the turbine efficiency of a similar order to give an overall energy recovery of about 80–90%. Water losses need not be significant and there will be an additional input from streams in some sites. Pumped storage is only suitable where there is plenty of water available. Rainfall catchment by the reservoir itself and evaporative losses from the reservoir are both negligible. Existing pumped storage installations are large and are used in association with the national electrical grid supply. They should not, however, be ruled out for lower power levels for isolated community use.

An example of a large modern pumped storage system is the CEGB. Dinorwig station in Wales, (4) which has recently been commissioned. Electricity generated by steam power stations is used to pump water to the head of 500 m at times of low power demand. When peak power is required both the steam power stations and the water turbo alternators generate power into the grid. The six 300 MW turbo alternator sets can be started from rest within 10 seconds and can be reversed from full pumping to full generation within a few minutes. The station has an average output of 1681 MW for up to five hours. A head of 500 m is a very high head indeed and one tonne of water at this head will have a potential energy of 4·9 MJ or 1·36 kwhr.

References

1 GLENDENNING, I.: 'Compressed air storage', IEE Proceedings International Conference. future Energy Concepts, London, 1979
2 'The world's first air storage gas turbine power station', *CME*, 17th May 1980
3 HERBST, C. M., HOEFFINS, M., STYS, Z. S. and HUNTORF: '290 MW air storage system energy transfer (ASSET) plant design, construction and commissioning', Proceedings 1978, Compressed Air Energy Storage Symposium, 1979, **1**
4 DETTMER, R.: 'Power flow from Dinorwic, *IEE News*, June 1983

Chapter 7

Applications of direct solar energy

7.1 Introduction

The amount of solar energy, or insolation, falling on a horizontal square metre of the earth's surface each year was discussed in Chapter 2. In designing a solar powered device, it is necessary to know how the power density will vary during the day, from season to season, and also the effect of tilting a collector surface at some angle to the horizontal. For most purposes solar energy can be divided into two components:

1 Direct radiation

2 Scattered or diffuse radiation

Direct radiation, as the name suggests, is the radiation received directly from the sun; it is the radiation which gives rise to sharp shadows. The other component, diffuse radiation, is the radiation which has been scattered by

a.

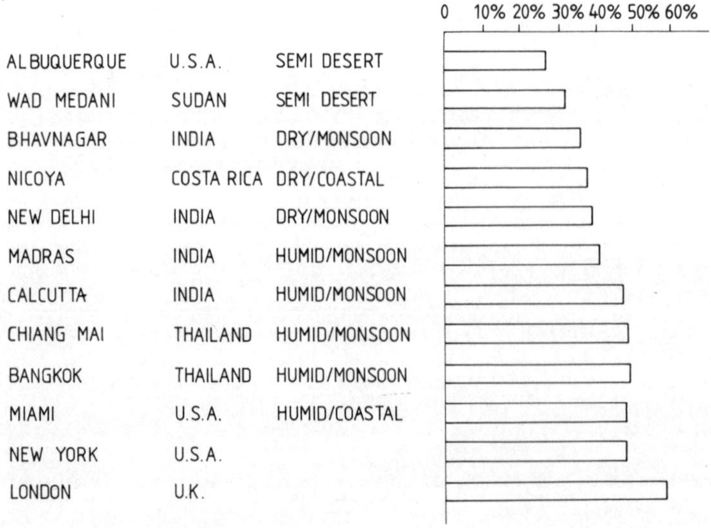

b.

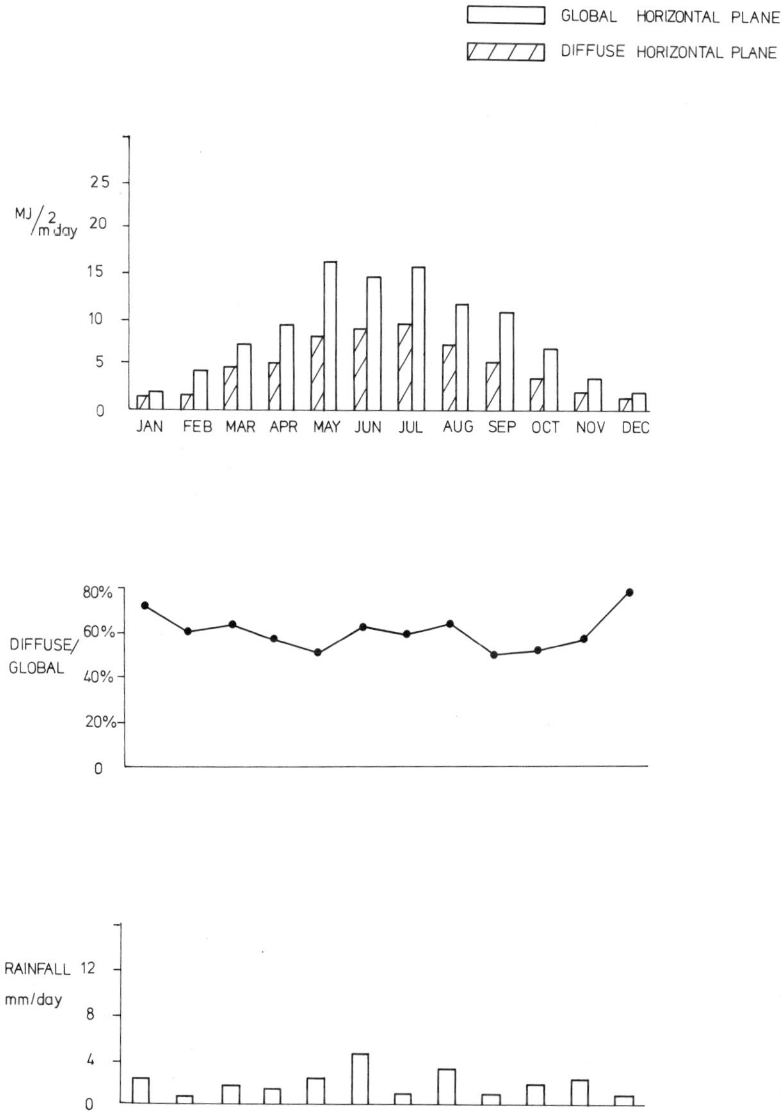

NOTE. WET JUNE AND AUGUST REDUCED GLOBAL RADIATION AND INCREASED DIFFUSE TO GLOBAL RATIO.

c.

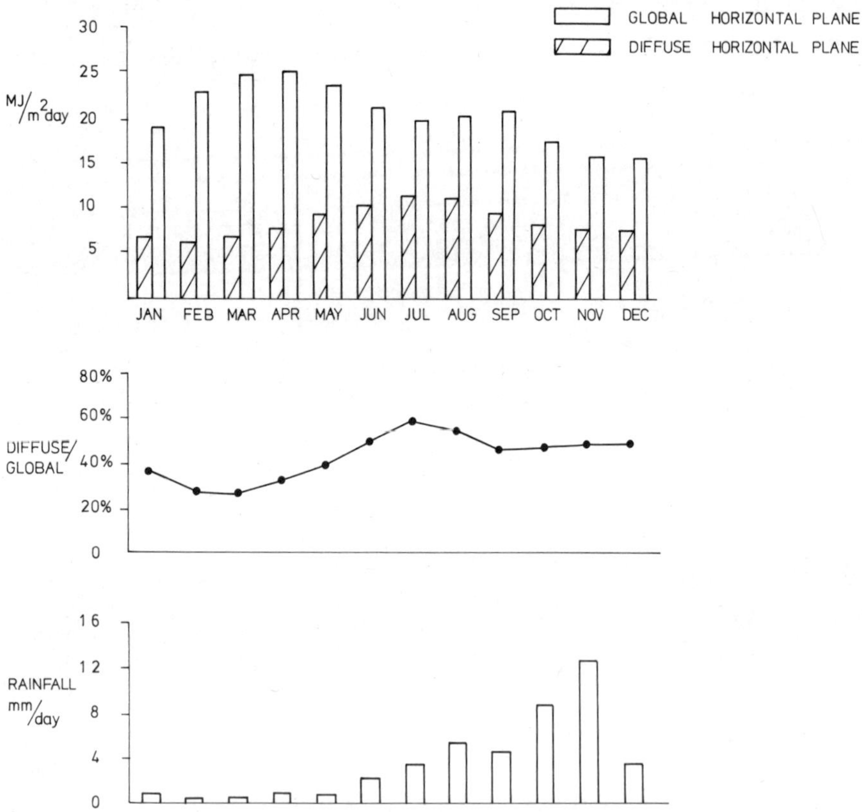

Fig. 7.1 *Diffuse radiation (Courtesy R. Bentley)*
a Annual ratio of diffuse to global radiation for a number of sites (valves are for horizontal surface)
b London, UK, 1971
c Madras, India, averaged over 21 years

clouds and dust, etc. Diffuse radiation arrives from all directions and does not give shadows. The relative proportions of direct to diffuse radiation will depend on the site, season and time of day. The diffuse component will be around 10% to 20% of the total radiation on a clear day but will rise to 100% of a much smaller total radiation on an overcast day. Fig. 7.1[1] shows the proportion of the diffuse radiation as a fraction of the total or global radiation averaged over the year for a number of sites. The instantaneous total radiation can vary considerably through the day, see Fig. 7.2.

In order to collect the maximum radiation for each unit area of collector surface, the collector should be arranged to be at right angles to the sun's direct radiation. As the direction changes owing to the sun's apparent motion, the

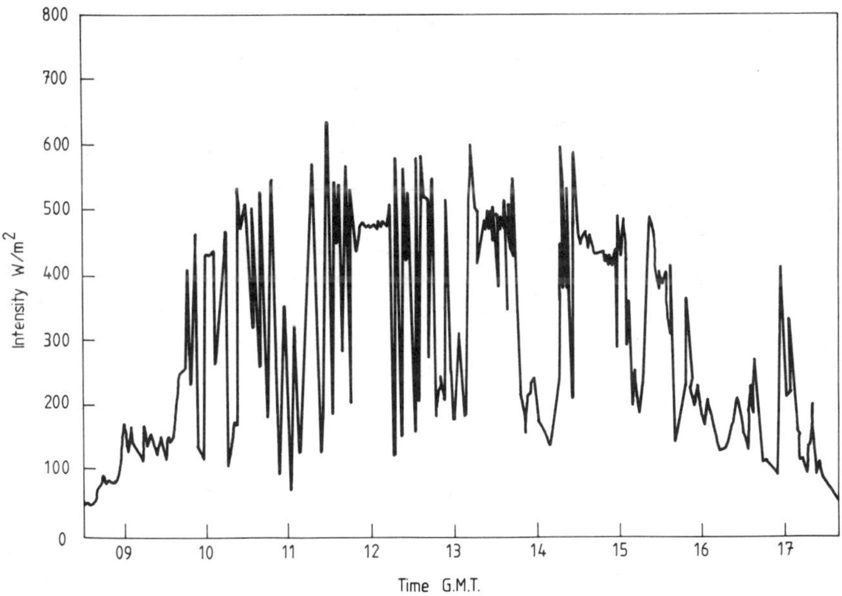

Fig. 7.2 *Variation of global solar radiation on a cloudy day in summer. Kew. U.K. (Courtesy U.K. I.S.E.S)*

collector must be moved (tracked) to maintain this maximum collection. A simple calculation shows that for direct radiation the tracking collector will receive $\pi/2$ times more energy per day than the fixed collector; in practice this figure is somewhat less, around 1·5. Much of this advantage is lost if a large fraction of the total radiation is diffuse.

Most flat plate solar collectors are fixed and should be mounted to face south in the Northern hemisphere and north in the Southern hemisphere. For a given latitude there is found to be an angle to the horizontal at which most energy will be collected over the year. This optimum angle is approximately the same as the angle of latitude for low latitudes, increasing to latitude plus 10° at 40° N or S. Where high temperatures are needed it is necessary to focus, or concentrate, the radiation by means of a mirror or lens. The effect is well known; every schoolchild will have experimented with the magnifying glass in order to start a fire. The concentrating collector focuses the incident sunlight on the collecting surface. A disadvantage of concentrating collectors is that they only focus the direct component of the incident sunlight and do not make use of the diffuse component. Their other obvious disadvantage is that they generally require some means of tracking the sun. The advantage of the concentrator over the flat plate is that by concentrating the solar radiation on to a smaller area it is possible to obtain higher temperatures than can be achieved with the flat plate collector.

Typical operating temperatures with flat plate collectors are in the range 40°C to 80°C whereas temperatures as high as 300° to 600°C can be achieved using concentrators of quite modest concentrating power and even higher temperatures are possible with high concentration ratio collectors. A concentrating collector must be aligned with sufficient accuracy to ensure that the focus coincides with the collector surface. The greater the degree of concentration the more accurate is the alignment required. Concentrators can use either mirrors or lenses. Mirrors are usually cheaper to construct than lenses, although the Fresnel lens has received some attention. The Fresnel lens differs from the normal converging lens as shown in Fig. 7.3*a* and *b*. Essentially the centre of the converging lens is removed and the lens surface commenced again. In this way a much flatter and lighter arrangement is possible. Fresnel mirrors can also be made, Fig. 7.3*c* and *d*. The advantage of the Fresnel mirror over a conventional mirror is that it requires less total depth, Fig. 7.3*d*.

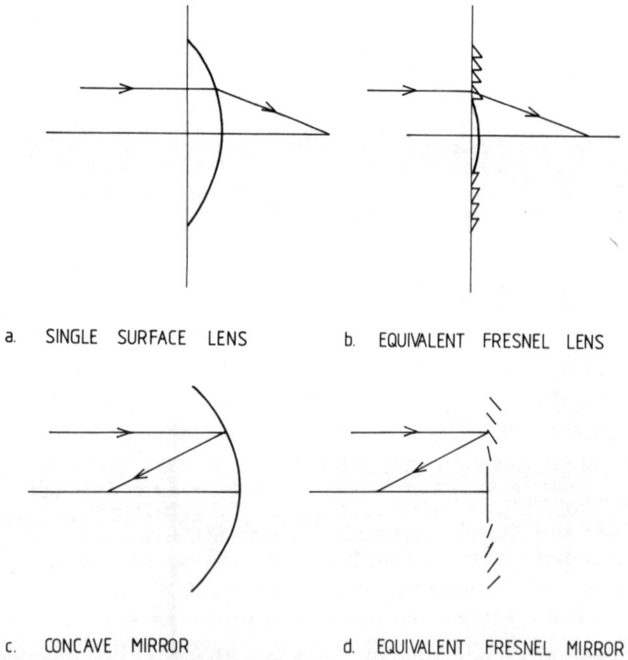

Fig. 7.3 *Comparison of lenses and mirrors*

Concentrators can be made to focus as a line, Fig. 7.4*a*, or as a point focus, Fig. 7.4*b*. The maximum theoretically possible concentrating ratio for solar concentration in two dimensions, that is a point focus, is approximately 40 000, whereas the concentration for a one-dimensional device, a line focus, is around 200. Fig. 7.5 shows some typical collectors.

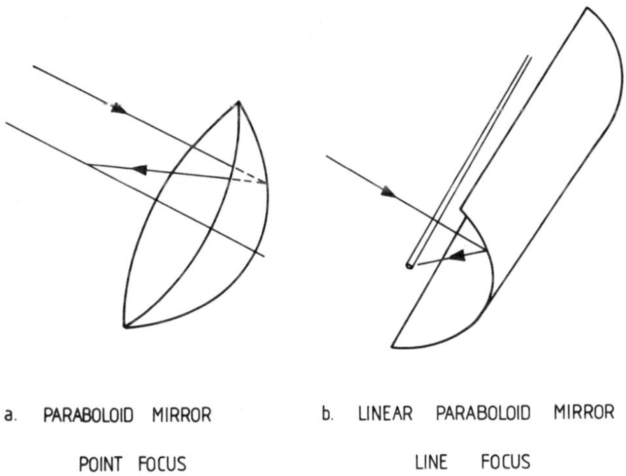

Fig. 7.4 *Paraboloid mirrors*

Solar Domestic Hot Water Heating – Essen, Germany.
(Courtesy Dornier – Friedrichshafen, Germany)

Domestic Solar Hot Water Heater Incorporated in Porch. Manilla, Phillipines

Domestic Solar Water Heaters with Integral Storage Tank. Papua New Guinea

Combined Solar Water Heater – Bathroom Beijing, China

Solar Thermal Electrical Generation using Combined Concentrators and Flat Plate Heaters. (Courtesy – Dornier Friedrichshafen, Germany)

White Cliffs Solar Thermal Power Station (140 kw Thermal, 25 kw-Electrical) N.S. Wales Australia. (Courtesy S. Kaneff)

Fig. 7.5 *Typical solar collectors*

The principal uses for solar radiation are:

1 As a source of heat, particularly for low temperature heat for such applications as domestic hot water or crop drying.
2 In the form of relatively high temperature heat to power heat engines.
3 As high temperature heat to power refrigerators and air conditioners.
4 To operate photovoltaic cells for direct electricity production.

These four applications will be discussed more fully in the remainder of this chapter but first we will summarise the main types of instrument used in solar measurement.

7.2 Solar measurement

In order to design solar devices it is necessary to know the rate at which energy arrives on a given area of the collector. Instruments which measure global radiation are named pyranometers. Fig. 7.6*a* shows an Eppley instrument which is a widely used pyranometer. The detector is made up from thermocouples to form a thermopile and is protected by a glass dome which is transparent over the wavelength range 0·285 μm to 2·8 μm, that is over the important part of the solar spectrum. For meteorological use the pyranometer

a.

b.

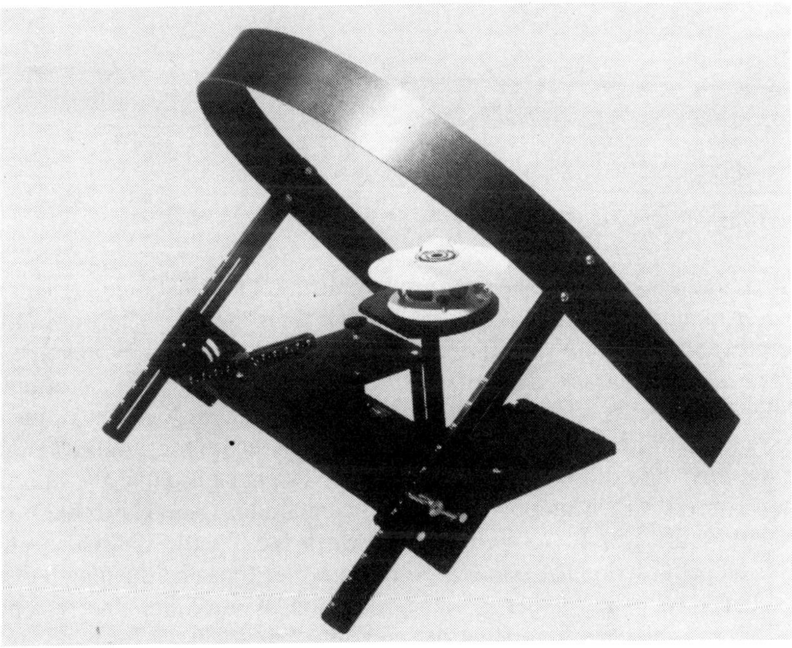

Fig. 7.6 *Eppley Pyranometer (Courtesy Eppley Laboratories Inc.)*
a Standard
b With shadow band

is mounted with its detector horizontal to give the global radiation falling on a horizontal plane (the insolation); it can also be mounted in the plane of a solar collector. The output from the pyranometer can be recorded directly or integrated to give hourly or daily radiation totals.

The direct and diffuse components of solar radiation can be measured by means of two pyranometers, the first, Fig. 7.6*a*, is of standard type and gives total radiation, the second Fig. 7.6*b*, has a shadow ring which is mounted to shield the detector from direct radiation and hence measures only the diffuse component.

Cheap pyranometers which use photovoltaic cells as detectors are now available commercially. In addition to low cost these have the advantages of rapid response time and are suited for simple electrical integration of the energy received. The response characteristics of photovoltaic cell does not match the spectral distribution of solar radiation, Fig. 7.7, and the instruments are less accurate than pyranometers having thermopile detectors.

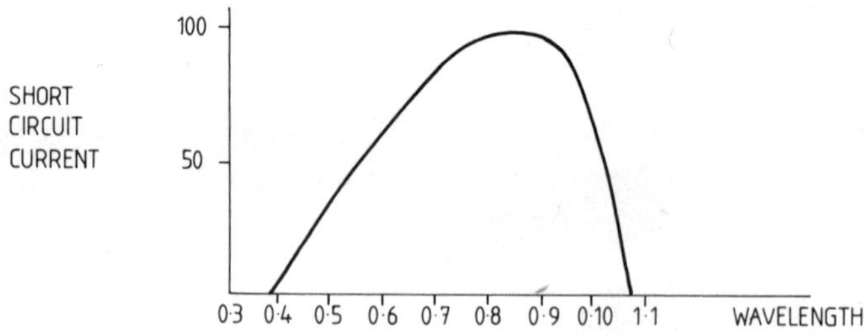

Fig. 7.7 *Spectral response of solar cell*

Direct radiation can also be measured by means of a pyrheliometer. This is a detector mounted at the end of a tube which faces the sun. The walls of the tube prevent the diffuse radiation from reaching the detector (pyrheliometers must be accurately tracked in order to follow the sun's movement accurately).

Solar instruments are expensive and the smaller meterological stations are only equipped with sunshine hour recorders which record the number of hours that the sun's disc can be seen in a given day. The most usual of the sunshine hour recorders is the Campbell–Stokes instrument, Fig. 7.8. This consists of a glass sphere of around 10 cm diameter which focuses the sunlight on to a sensitised paper strip. The paper is divided into sections and the length of the track burnt on to the paper gives the duration of sunshine. The threshold sensitivity at which a trace appears is 200 W/m^2. Empirical equations are given in the literature which enable the total energy received during the day to be estimated in terms of such data as the number of sunshine hours, the length of day, and the daily energy received at the location at the outside of the earth's atmosphere.

Fig. 7.8 *Campbell Stokes sunshine recorder (Courtesy Casella London Ltd)*

So far we have considered solar energy measurement, that is the measurement of radiation mainly in the wavelength range 0·3 μm to 3 μm which accounts for 98% of the total energy from the sun. Long wave radiation in the wavelength range 5–20 μm is important in determining heat losses from solar devices. This radiation is measured by means of radiometers, which measure the sum of long and short wave radiation. The net radiometer measures the net radiation from a surface, that is the difference between incoming* and outgoing radiation, while the pyroradiometer measures the total incoming radiation. Pyrgeometers measure only the long wave radiation. One type of pyrgeometer is based on the pyranometer shown in Fig. 7.6 but with the glass dome replaced by a silicon dome which is opaque to short wave (solar) radiation.

* Long wave radiation is received directly from the sky and also by reflection and by radiation from buildings and other surfaces. Both incoming radiation flux and the radiation flux from the collector can be large – some authors treat the two terms separately. The net loss from the collector is the difference between these two terms. The calculation of net loss is duscussed later in this chapter.

Sunlight is multicoloured and contains wavelengths over a continuous spectrum with a percentage of the energy lying between 0·3 μm and 3 μm, see Fig. 2.12. For most solar applications we are concerned only with the total amount of energy. However, where quantum processes are involved the wavelength is important. For example, in photosynthesis it is the blue and red wavelengths which are used. Similarly, with photovoltaic cells wavelengths greater than a critical value cannot be used. The spectral distribution of radiation can be measured using standard pyranometers and a series of optical filters.

7.3 Solar heat

7.3.1 *Solar collectors*

The simplest way to collect and make use of solar energy is in the form of low temperature heat. When a surface, preferably blackened so that it will absorb most of the incoming solar radiation, is placed in the sun it will get hot. If the surface is not cooled then the temperature will rise to a value such that the heat losses from the surface will equal the heat gained from the sun. If the surface is a metal plate then heat can be removed by passing a fluid through pipes soldered to it. The maximum surface temperature will be reduced to a value less than it achieved when uncooled. In this case:

$$\text{heat absorbed} = \text{heat removed by coolant} + \text{heat lost}$$

or

$$\text{heat removed by coolant} = \text{heat absorbed} - \text{heat lost}$$

The conversion efficiency η of the collector is given by:

$$\eta = \frac{\text{heat absorbed} - \text{heat lost}}{\text{incident solar energy}} \tag{7.1}$$

As the temperature of the surface increases, then the losses will increase. There are two methods by which efficiency can be raised.

1 by reducing the heat losses; or
2 by increasing the incident radiation and hence the heat absorbed per unit area.

For low temperature collectors (<100°C) the first method is adopted. Heat loss is reduced by using transparent cover plate(s), specially treated absorber surfaces and by evacuating the space between the cover plate and the absorber surface. If higher temperatures (>100°C) are required then it is necessary to increase the incident radiation by means of concentrators. In this case only the direct radiation is used.

7.3.2 *Flat plate collector*

For a flat surface collector the energy incident on the surface cannot be increased and all that can be done is to ensure that the absorber surface absorbs as much as possible of the incident radiation and that the losses from the absorber surface are reduced as far as possible. Fig. 7.9*a* shows a collector made from a single absorber surface. Some of the incident radiation will be lost by reflection; however, for a blackened surface, most (about 95%) of the radiation will be absorbed.

The heat losses are shown in Fig. 7.9*b*. The lower surface will usually have an insulating layer of material, such as a few inches of glass wool, and will lose heat by:

1 Conduction

The top surface will lose heat in two ways:

2 Convection
3 Radiation

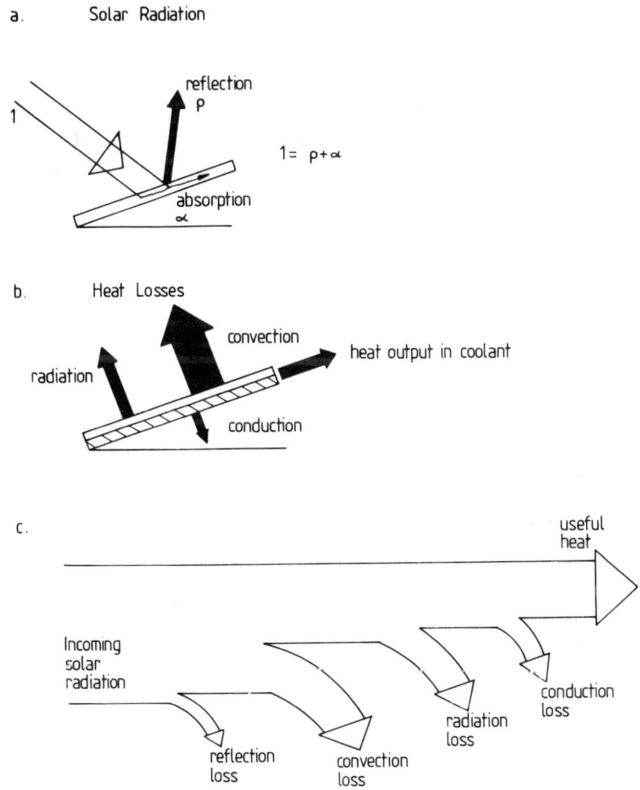

Fig. 7.9 *Single surface flat plate collector*

Fig. 7.10 *Solar heated swimming pool. Sussex U.K. (Courtesy Robinsons Developments)*

For applications such as heating swimming pools, Fig. 7.10, the temperature of the unprotected absorber plate does not need to exceed the ambient temperature by more than a few degrees and no further insulation is required. However, for such applications as domestic hot water heating, the loss from the top of the absorber plate has to be reduced. The usual way in which this is achieved is by covering the absorber with one or more sheets of glass. This is the basic flat plate collector shown in Fig. 7.11. The use of a cover plate will result in some reduction in radiation reaching the absorber plate since some of the solar radiation will be reflected from the surfaces of the cover plate and there will also be some absorption in passing through it. Nevertheless, it is found that this reduction in incident radiation is more than compensated for by the reduction in heat loss from the absorber plate at all but very small temperature differences between the absorber plate and ambient temperature, see Fig. 7.12. In the next section the main characteristics of the three heat transfer modes are summarised.

7.3.3 Heat transfer

7.3.3.1 *Conduction*. Heat is transferred through a solid body, across which there is a temperature difference, by means of vibrations of the atomic lattice which forms the body of the material. In metals heat is also carried by free

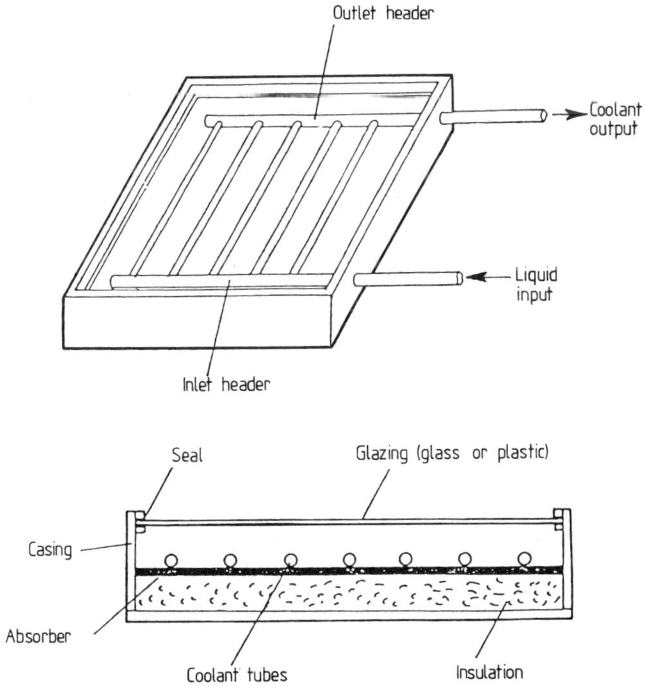

Fig. 7.11 *Basic flat plate collector*

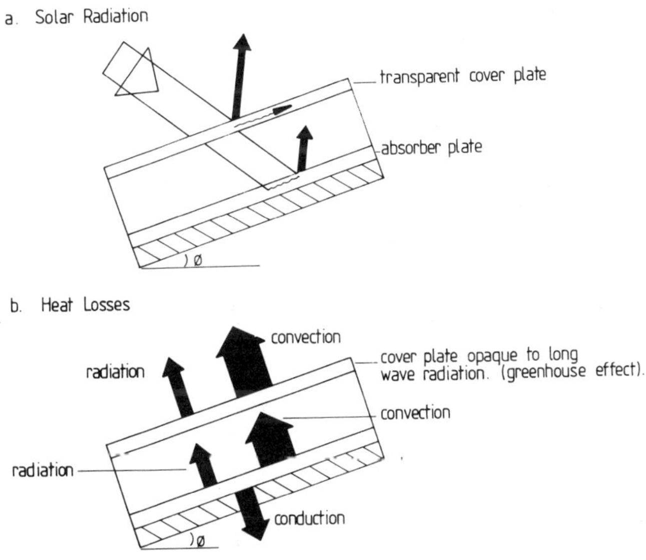

Fig. 7.12 *Flat plate solar collector with cover plate*
a Solar radiation
b Heat losses

electrons, and this contribution is much greater than that due to lattice vibration. In conduction there is no transfer of mass. The random thermal movement of the atoms in liquids and gases also lead to heat conduction. If the fluids are free to move the resulting convective heat transfer is normally much greater than the conductive transfer. When fluids are prevented from moving, for example for thin films, or when the fluids are contained in a honeycomb structure, then conduction loss may be important. Heat flow by conduction is expressed by the equation:

$$q = -k\frac{\mathrm{d}T}{\mathrm{d}x}$$

where:

q = the heat flow per unit area of cross section (W/m^2)

$\frac{\mathrm{d}T}{\mathrm{d}x}$ = the temperature gradient in the material in the x direction (°C/m)

k = the thermal conductivity of the material (W/m°C)

Table 7.1 lists the values of thermal conductivity for a number of materials of interest in solar collectors.

Table 7.1 *Thermal conductivity of commonly used materials (at 300 K)*

	k W/m °C
Metals	
Copper	385
Aluminium	205
Steel	50
Non-metals	
Glass	0·8
Concrete	0·8
Wood	0·14
Sawdust	0·06
Rock wool	0·04
Polystyrene (expanded)	0·03
Glass fibre	0·03
Liquids	
Water	0·61
Gases	
Hydrogen	0·142
Helium	0·142
Air	0·0239

7.3.3.2 Convection: Convection occurs in fluids, as follows: if a fluid such as water is in contact with a hot surface, then the fluid adjacent to the surface will become heated by conduction from the hot surface. If the fluid is flowing across the surface then the heated volume will be removed and continually replaced by colder fluid, in this way heat will be removed from the surface. This method of heat transfer is known as convective cooling (or heating). An example of such convective cooling occurs when the wind blows across the surface of a solar collector. The rate of heat removal will depend on both the temperature difference between the surface and the bulk fluid temperature and also on the velocity and characteristics of the fluid.

The method of convective cooling so far discussed is called forced convection and occurs when the fluid is made to flow by some means not connected with the heat transfer process. For example, the wind will blow across the surface of a solar collector whether it is hot or not. There is another form of convection which is important in solar collector design, this is known as natural convection. Consider a horizontal hot surface placed in still air; the air adjacent to the top surface will become hotter than the bulk air temperature. Hot air expands to a lower density, hence the hot air will be replaced by cooler air. This is natural convection; the heat transfer by natural convection will depend on the position of the surface and will be less than the case for forced convection. Both forced and natural convection may be accompanied by phase change. For example, if the fluid is water it may turn to steam at the hot surface, and when this happens the heat transfer is said to be by two-phase convection.

Convective heat transfer can be represented by the relation

$$q = h(T_s - T_f) = h \triangle T$$

where

q = the heat flow per unit area (W/m^2)
h = convective heat transfer coefficient (W/m^2 °C)
T_s = the temperature of the surface (K or °C)
T_f = the bulk temperature of the fluid (K or °C)
$\triangle T = T_s - T_f$

Since the convective heat transfer coefficient depends on both the nature of the fluid and also its velocity of flow, the actual calculation of h is somewhat complicated and formulae are given in References 2 and 3. Approximate values of h for the special case of the flat plate collector are listed in Table 7.2.

7.3.3.3 Radiation: The third way in which a hot body can lose heat is by radiation, that is by the emission of electromagnetic waves; we have already discussed the loss of energy from the sun by this means. The maximum power which can be radiated from a body at a given temperature is called the black body radiation corresponding to that temperature. The power radiated from a

Table 7.2 *Typical convective heat transfer coefficients* h_c

	h_c W/m² °C
Heat transfer between parallel plates* (separation 2·5–10 cm) with air at atmospheric pressure	3
Heat transfer from surface of cover plate where v is wind velocity at surface of plate in metres per second	$2{\cdot}3 + 3{\cdot}8v$

* In this case ΔT = temperature difference between the two plates (°C)

black body increases as the fourth power of the absolute temperature of the body and is given by Stefan's Law

$$P = \sigma T^4$$

where P = Power radiated per unit area (W/m²)

σ = Stefan's Constant = 5.67×10^{-8} W/m² K⁴

T = Temperature of the surface (K)

Fig. 7.13*a* summarises some important characteristics of black body radiation. The sun radiates approximately as a black body at a surface temperature of 5800 K. Table 7.3 gives some values for black body radiation

Table 7.3 *Radiation from a black body*

Surface temperature K	°C	Heat flux W/m²
6000		$73{\cdot}5 \times 10^6$
3000		$4{\cdot}6 \times 10^6$
2000		$9{\cdot}1 \times 10^5$
1073	800	75×10^3
873	600	$32{\cdot}9 \times 10^3$
673	400	$11{\cdot}6 \times 10^3$
473	200	$2{\cdot}84 \times 10^3$
373	100	$1{\cdot}10 \times 10^3$
353	80	880
333	60	697
300	27	459

and shows the effect of the fourth power temperature law on the power flux. Not all surfaces radiate as much energy as a black body, and, for example, it is possible to select a surface for a solar collector which will radiate as little as one-tenth of the radiation from a black body surface at the same temperature. this fraction is called the emissivity (ε) of the surface. For a surface with average emissivity ε Stefan's Law becomes:

$$P = \varepsilon\sigma\, T^4$$

The term grey body is used to define a body having an emissivity which does not depend on wavelength. Hence the power emitted will have the same variation with wavelength as that of a black body at the same temperature, Fig. 7.13*b*. Other surfaces, called selective emitters, have an emissivity which is a function of wavelength and an example is shown in Fig. 7.13*c*. It is impossible for the power radiated at any wavelength to be greater than the corresponding value for a black body at the same temperature.

a. BLACK BODY

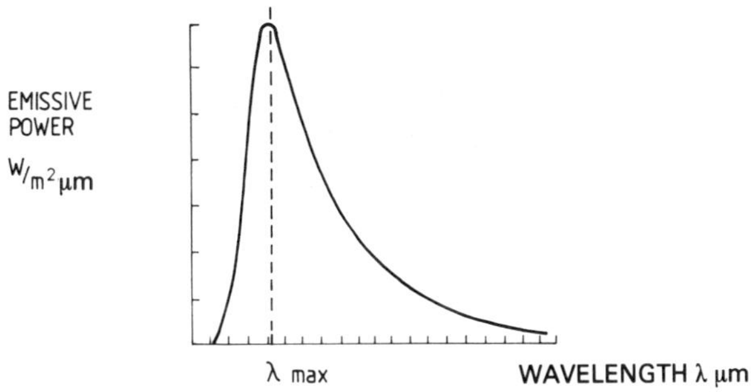

(i) WIEN'S LAW

$$\lambda_{max}\, T = 2{\cdot}885 \times 10^{-3}\,\mathrm{mK}$$

(ii) STEFAN'S LAW

$$P = \sigma T^4$$

WHERE

P = POWER RADIATED / UNIT AREA $\mathrm{W/m^2}$
T = TEMPERATURE K
σ = STEFAN'S CONSTANT = $5{\cdot}67 \times 10^{-8}\ \mathrm{W/m^2K^4}$

(iii) KIRCHOFF'S RELATION

$$\varepsilon = \alpha = 1$$
$$\rho = \tau = 0$$

WHERE

ε = EMISSIVITY
α = ABSORPTIVITY
ρ = REFLECTIVITY
τ = TRANSMISSIVITY

b. GREY BODY

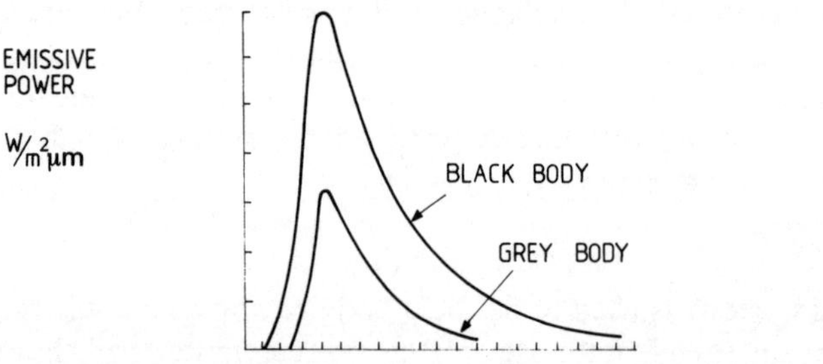

RADIATION LAWS APPLICABLE TO A GREY BODY

(i) WIEN'S LAW

$$\lambda_{max} T = 2{\cdot}885 \times 10^{-3} \text{ mK}$$

(ii) STEFAN'S LAW

$$P = \varepsilon \sigma T^4$$

ε = EMISSIVITY $\quad 0 < \varepsilon \leqslant 1$

(iii) KIRCHOFF'S RELATIONS

$$\varepsilon = \alpha$$
$$1 = \rho + \alpha + \tau$$

c. SELECTIVE SURFACE

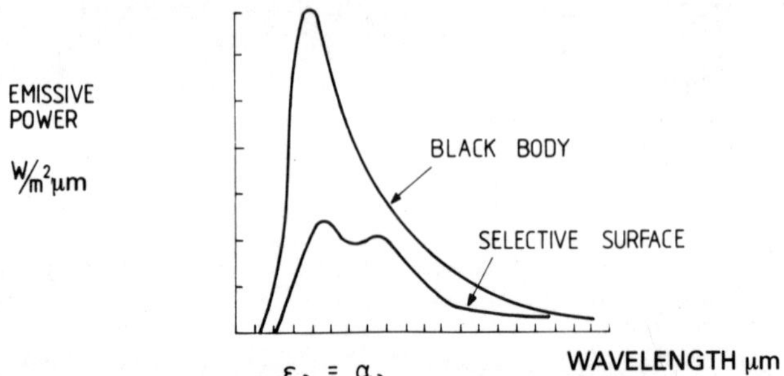

$$\varepsilon_\lambda = \alpha_\lambda$$
$$1 = \rho_\lambda + \alpha_\lambda + \tau_\lambda$$

Fig. 7.13 *Radiation relationships*

Unlike conduction and convection, no matter is required for heat transfer by radiation since the energy is in the form of electromagnetic waves. These waves can be stopped by placing opaque materials in their path; some of the incident energy will be absorbed by the material and the remainder reflected back to the source. Hence, apart from the special case of isolated bodies in space, all other surfaces will be receiving radiation from their surroundings and the net radiation loss will be the difference between the radiation from the surface and that received from the surroundings. Not all the radiation falling on a body will be absorbed, a fraction ρ will be reflected. The fraction absorbed is called the absorptivity of the surface α and that reflected is called the reflectivity ρ. If the surface is opaque

$$\alpha + \rho = 1$$

If some of the radiation is transmitted through the body the relationship becomes:

$$\alpha + \rho + \tau = 1$$

where τ is called the transmissivity. α, ρ and τ will depend on the wavelength λ of the radiation. Black body radiation covers a range of wavelengths and the values quoted normally refer to the average values of α, ρ, τ for the particular temperature.

So far we have not defined black body. A black body is defined as a surface which absorbs completely all radiation falling on it. That is, it will have an absorptivity of unity at all wavelengths. Practical surfaces will have a lower value of absorptivity α which may be different for different wavelengths. The emissivity of a black body is unity for all wavelengths but practical surfaces will have a value ε. For any surface $\alpha_\lambda = \varepsilon_\lambda$ at a given wavelength λ. If this were not so it would be possible to devise an arrangement which collected radiation from a black body and achieved a temperature higher than that of the black body, a result contrary to the Second Law of Thermodynamics. Returning to the question of the net radiation loss from a surface; if two parallel, black body surfaces are maintained at temperatures T_1, T_2, respectively, then the net loss per unit area from surface 1 will be given by:

$$P = \sigma(T_1^4 - T_2^4)$$

where $T_1 > T_2$.

If the surfaces are grey bodies of the same emissivity ε then the equation becomes:

$$P = \varepsilon_r \sigma (T_1^4 - T_2^4)$$

where

$$\varepsilon_r = \frac{1}{\dfrac{2}{\varepsilon} - 1}$$

In the general case the net loss from a surface will depend on several factors: the emissivity and temperature of the surface, and the emissivity and temperature of the surroundings, and the shape and relative position of the hot surface and the surrounding surfaces. Fig. 7.14 shows some special cases.

For solar collectors the temperature difference between surfaces is small compared to the absolute (K) temperature of the surfaces and the net loss can

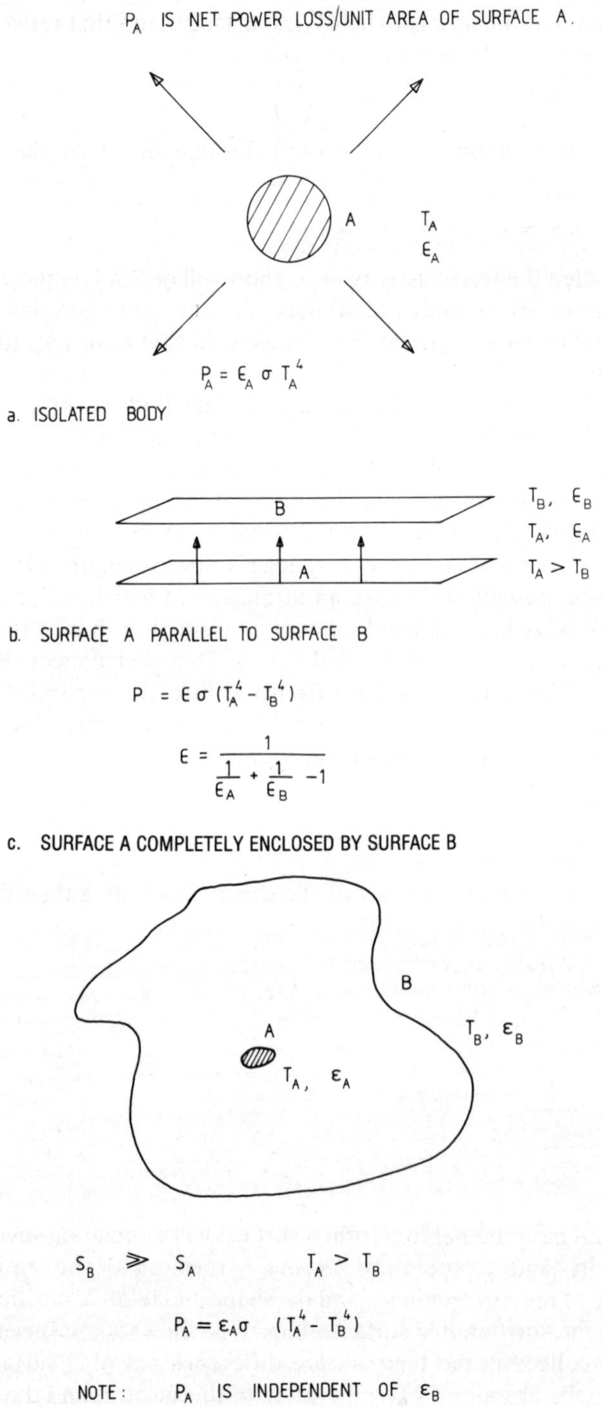

P_A IS NET POWER LOSS/UNIT AREA OF SURFACE A.
A
T_A
ϵ_A
$P_A = \epsilon_A \sigma T_A^4$
a. ISOLATED BODY
B
A
T_B, ϵ_B
T_A, ϵ_A
$T_A > T_B$
b. SURFACE A PARALLEL TO SURFACE B
$P = \epsilon \sigma (T_A^4 - T_B^4)$
$\epsilon = \frac{1}{\frac{1}{\epsilon_A} + \frac{1}{\epsilon_B} - 1}$
c. SURFACE A COMPLETELY ENCLOSED BY SURFACE B
B
T_B, ϵ_B
A
T_A, ϵ_A
$S_B \gg S_A$ $T_A > T_B$
$P_A = \epsilon_A \sigma (T_A^4 - T_B^4)$
NOTE: P_A IS INDEPENDENT OF ϵ_B

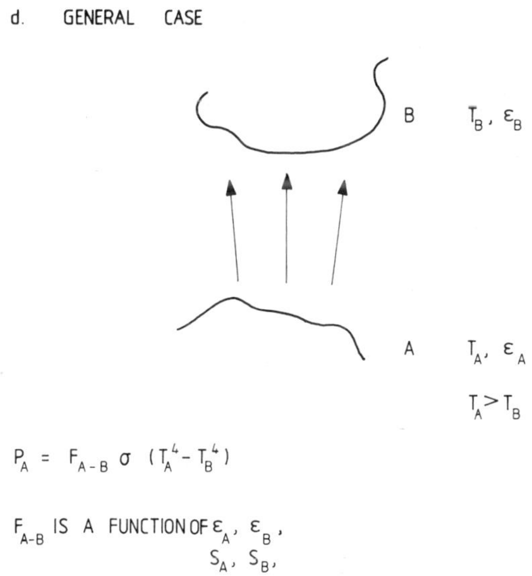

Fig. 7.14 *Radiation heat transfer special cases*

be expressed as a linear expression, using a radiation loss coefficient h_r similar to that for convection. The resulting heat flux is

$$q_r = h_r\,\Delta T$$

where ΔT is the temperature difference between the two surfaces.

For two parallel surfaces each of emissivity $\varepsilon = 0{\cdot}9$ and at temperatures of 30°C and 80°C, respectively, h_r will have the value 6·5 W/m² °C. This is typical of the case of a black-painted collector plate with a single glass cover plate, since glass is opaque to thermal radiation.

Fig. 7.14*c* shows that a body of emissivity ε_A which is completely surrounded by a second surface B will lose radiation to this surface according to the relation

$$P_A = \varepsilon_A \sigma\,(T_A^4 - T_B^4)$$

If $T_A > T_B$ we can write

$$P_A = h_r\,(T_A - T_B)$$

This case applies to the loss of radiation from the cover plate of the solar collector to the surroundings, mainly the sky. It is not in general correct to use the value of ambient temperature in this equation, particularly for clear, cloudless, sky conditions. One approximation, recommended by Whillier,[4] is

$$T_B = T_{ambient} - 6$$

Typically h_r is of the order 6 W/m²°C

7.3.4 Basic flat plate collector

When a transparent cover plate is used not all the energy incident on the surface of the flat plate collector arrives at the collector plate. About 4% is lost by reflection at normal incidence from the top surface of the glass cover plate and a further 4% from the lower surface to give a net 8% loss by reflection. The reflection is increased if the light strikes the surface of the cover plate at an angle less than 90° and the average for overall reflection is around 10%. The cover plate will not be completely transparent and some energy will be absorbed when the radiation passes through the plate. For window glass of 3 mm thickness about 5% is lost; the loss is mainly due to iron (in the form of 0·1–0·15% of iron oxide Fe_2O_3) in the glass. White glass has a low Fe_2O_3 content (<0·01%) and the absorption loss is reduced to around 2%. Hence, for a single pane of low iron glass, about 88% of the incident radiation will arrive at the collector. Of this radiation around 98% is absorbed by the collector plate or $0{\cdot}88 \times 0{\cdot}98 = 0{\cdot}84$ of the original incident radiation, Fig. 7.12*a*.

The sources of heat loss from the collector plate are shown in Fig. 7.12*b*. Heat is lost from the unprotected collector plate partly by radiation and partly by convection. Convective heat loss is particularly important when the conditions are windy. In the basic flat plate collector a glass cover plate is placed a few centimetres above the collector surface and has two effects. First, and most important, it reduces the convective loss by isolating the collector plate from the wind.

The second effect, sometimes called the greenhouse effect, is less obvious. Essentially what happens is that glass appears transparent to the short wave solar radiation but opaque to the long wave, thermal radiation, from the collector plate. Although solar radiation is made up from a wide range of different wavelengths forming a continuous band or spectrum most of the energy is concentrated in a relatively small range of wavelengths centred around the visible spectrum, Fig. 2.14. The wavelength of the radiation at which maximum radiation occurs is 0·5 μm and more than 98% of the total energy is due to wavelengths between 0·3 μm and 3 μm. The radiation from all black bodies has the same shape of power distribution as a function of wavelength but the wavelength at which maximum intensity occurs, λ_m, is related to the black body temperature by the Wien equation discussed in Chapter 2.

$$\lambda_m T = \text{constant} \simeq 3{\cdot}0 \times 10^{-3} \text{ m K}$$

The collector surface will be at a temperature around 350 K (77°C) and hence the corresponding λ_m will be 8·6 μm, or well into the infra-red range. Hence most of the radiation from the collector plate will fall in a longer wavelength range than the incident solar radiation, say between 3 μm and 30 μm. It is found that whereas glass is transparent to radiation in the visible region (we can see through glass) it strongly absorbs longer wave radiation. The

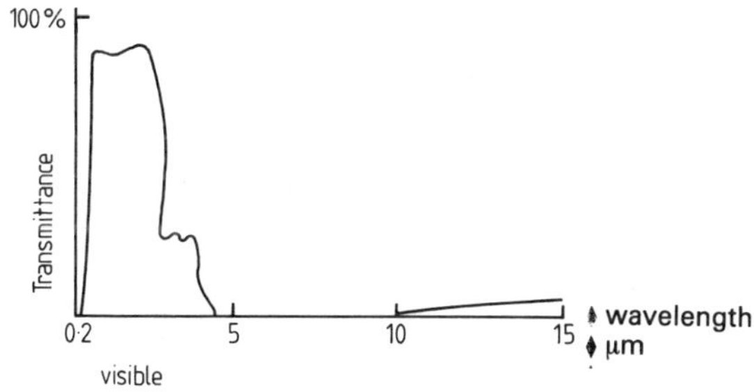

Fig. 7.15 *Transmittance of glass*

transmittance of glass as a function of wavelength is shown in Fig. 7.15. Most of the radiation from the collector plate to the glass cover plate will be absorbed (a small fraction, 5%, will be reflected back to the collector plate). The cover plate will emit radiation corresponding to its own, lower, temperature. The net effect is to reduce somewhat the radiation loss from the flat plate collector compared to the figure if the glass was transparent to long wave radiation. Not all materials which are transparent to visible radiation show the greenhouse effect, for example plastic materials such as the acrylics (e.g. Perspex) also transmit most of the long wave radiation.

The performance of a flat plate collector is often displayed as a graph such as that shown in Fig. 7.16. Eqn. 7.1 may be written in the form

$$\eta = \alpha\tau^* - \frac{h_T(T_{av} - T_{amb})}{G} \tag{2}$$

where

- α is the absorptivity of the absorber plate
- τ^* is the effective transmissivity of the cover plate $\{(1-\rho)\tau\}$
- G is the solar flux density incident on the cover plate (W/m^2)
- h_T is the total heat loss coefficient (including radiation and convection loss from the top of the flat plate collector and conduction loss from the bottom surface). (W/m^2 °C) h_T can be calculated as shown in Fig. 7.17
- T_{av} is the average absorber plate temperature °C
- T_{amb} is the ambient temperature

In practice the average absorber plate temperature is difficult to measure and it is usual to express efficiency in terms of T_i, the coolant inlet temperature, together with a correction factor, F_R, which is known as the heat removal factor.

Expressed in this way the efficiency may be written

$$\eta = F_R\left\{\alpha\tau^* - \frac{h_T(T_i - T_{amb})}{G}\right\} \tag{7.3}$$

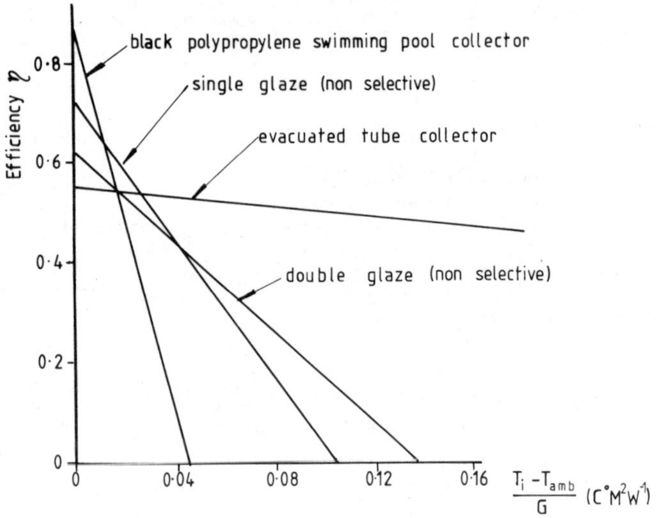

Fig. 7.16 *Instantaneous collector efficiency (based on Reference 6)*

It is seen from eqn. 7.3 that the efficiency depends on the ratio of the temperature difference between the inlet coolant temperature and the ambient temperature and the radiation per unit area falling on the plate. Qualitatively this is to be expected, since if the collector plate is operated at a high temperature with respect to the surroundings the losses might be expected to be greater and hence the efficiency to be reduced. Similarly the efficiency will increase, for a fixed temperature difference, as the solar power input increases. Fig. 7.16 shows how efficiency can be increased at higher values of $T_i - T_{amb}$ by glazing.

If higher flat plate collector efficiencies are required then means must be adopted to reduce the heat losses still further. There are a number of ways by which this can be achieved.

7.3.5 Selective surfaces

If the surface of the collector plate is painted black then it will absorb the incident short wave radiation almost completely (around 95%). With normal paints however, it will also radiate strongly in the long wave region corresponding to the plate temperature. Selective coatings[4,5,7,8] have been developed which absorb short wave radiation strongly but emit less long wave radiation. It has been stated earlier in the chapter that the absorptivity α_λ of a surface must be equal to the emissivity ε_λ for the same value of wavelength λ.

A selective surface does not violate this rule since ε is the average emissivity in the long wave range whereas α is the absorptivity averaged over the short wave range. Ideally a solar collector surface should have an absorptivity which varies with wavelength as shown in Fig. 7.18*a*.

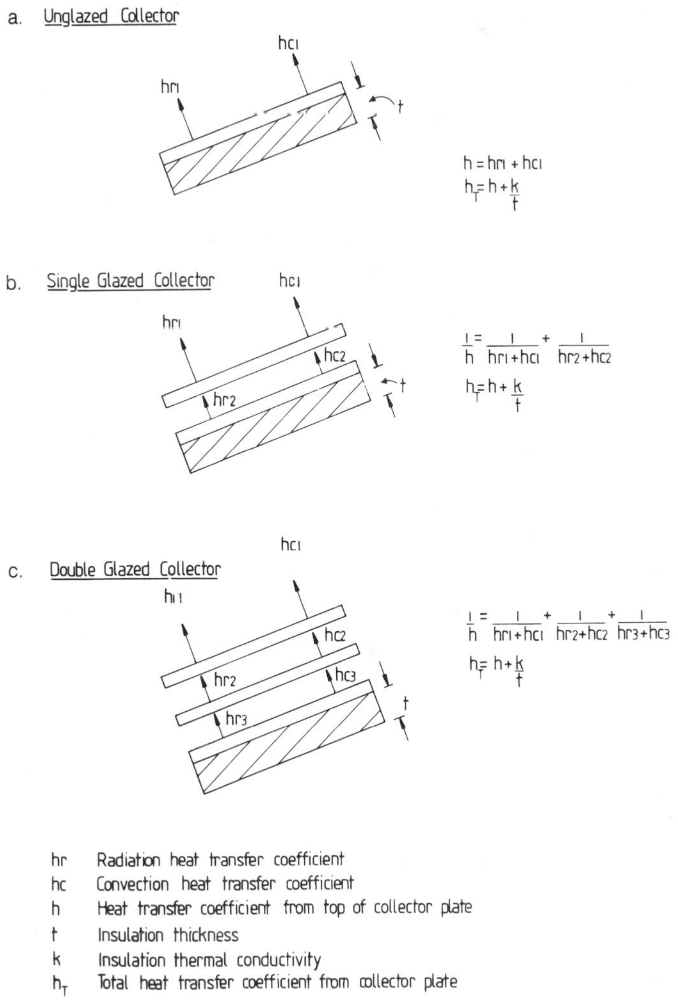

Fig. 7.17 *Evaluation of heat losses from a flat plate collector*

Fig. 7.18*b* shows the way in which α_λ and ε_λ vary with wavelength for two real surfaces and Table 7.4 lists the short wave absorptivity and long wave emissivity of a number of common selective surfaces. From Table 7.4 it is seen that some selective surfaces, such as white paint, are bad absorbers of solar radiation but good emitters of long wave radiation. Such surfaces, while unsuitable for solar collectors, are appropriate for the walls of buildings in hot countries. Most paints have a high emissivity at long wavelengths and hence the colour of domestic radiators does not affect their performance. Black paint is both a good absorber of solar radiation and a good emitter at long wavelengths. Because it is cheap it is used in low temperature solar collectors.

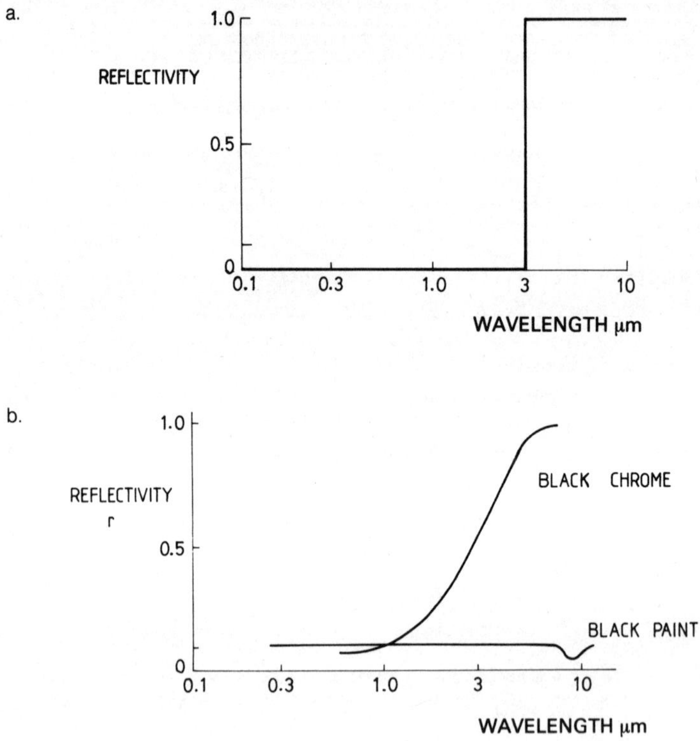

Fig. 7.18 *Selective surfaces*
a Ideal surface for absorber plate
b Reflectivity of real surfaces

Table 7.4 *Absorptivity and emissivity of surfaces*

Surface	Short wave absorptivity	Long wave emissivity	Comments
White paint	0·20	0·91	Tropical buildings
Concrete	0·60	0·97	
Black gloss paint	0·90	0·90	
Water	0·94	0·95	
Aluminium foil	0·15	0·05	Very low absorptivity and emissivity
Black sulphide on metal	0·92	0·10	solar selective surfaces
Black nickel oxide on aluminium	0·85–0·93	0·06–0·1	
Black chrome	0·87	0·09	

Aluminium foil has a low absorptivity of solar radiation and a very low emissivity of long wave radiation. Selective surfaces suitable for use in solar collectors have, typically, an absorptivity of 0·9 and an emissivity of 0·1. The effect of replacing the surface having an emissivity of 0·9 by a surface with emissivity of 0·1 is to reduce the value of h_r from about 6·5 W/m² °C to about 0·9 W/m² °C. The absorptivity of selective surfaces (0·85–0·94) tends to be somewhat less than that for the best non-selective surfaces.

In Fig. 7.19 several surfaces are shown, in each case a solar flux of 500 W/m² falls on the surface which is thermally insulated from its surroundings (for

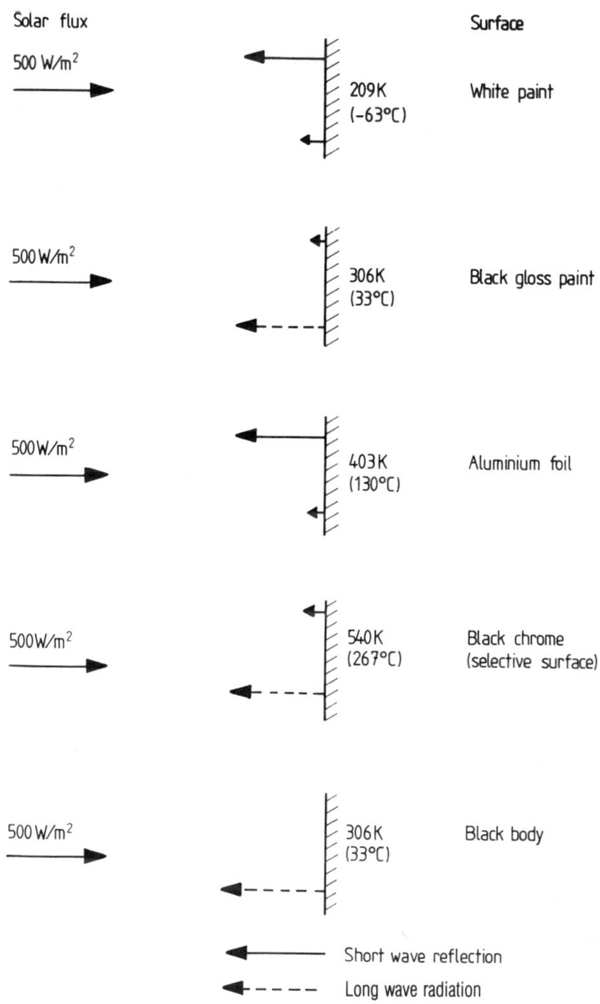

Fig. 7.19 *Comparison of the equilibrium temperature of isolated surfaces placed in a solar flux*

example, when situated in outer space). The figure shows the amount of radiation absorbed and the equilibrium temperature achieved if the heat is lost to outer space only from the single outer surface. The figure for white paint seems to imply that the wearer of a white tropical hat may be in danger of frostbite. In practice, of course, radiation from the surroundings and convective heat gain will raise the temperature markedly. Aluminium foil rises to a high temperature but this is owing to the low emissivity – the absorptivity is poor so that the fraction of incident radiation absorbed is low. These examples agree with common experience, metals placed in the sun get hot, non-metals do not to the same extent. The selective solar surface both absorbs a large fraction of the incident solar energy and has a low emissivity.

Surfaces which have selective transmission properties can be employed for the transparent cover plate. In practice selective transmission surfaces are less used than are selective absorbers.

7.3.6 Evacuated cover collector spacing

There is not too much point in reducing the long wave radiation loss from the collector to a level much less than that of the convective loss from the collector. For this reason, where high temperatures are required, attention has been given to the elimination of convective losses by evacuating the space between the cover and absorber plates. The atmospheric pressure acting on the cover plate is too great to allow the evacuation of the simple flat plate geometry and it is usual to build up an arrangement of evacuated tubes, see Fig. 7.20.

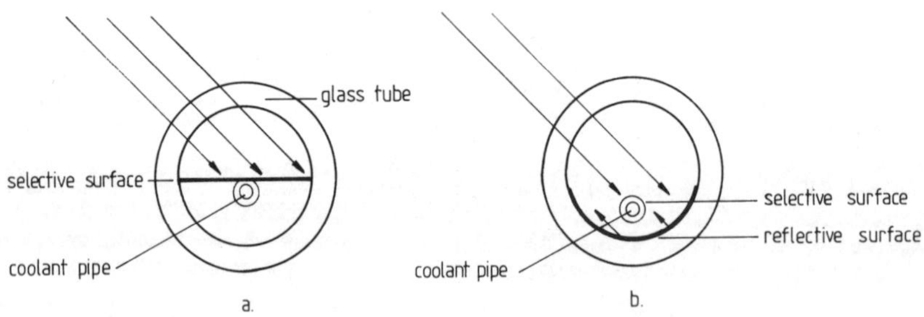

Fig. 7.20 *Evacuated tube collectors*

7.3.7 Concentrating mirrors

By silvering the rear surface of the evacuated tubes it is possible to achieve a measure of concentration. Since the absorber is now an isolated tube, the losses from the back of the absorber are eliminated. Alternatively the mirrors can be used without the evacuated space, as shown in Fig. 7.21.

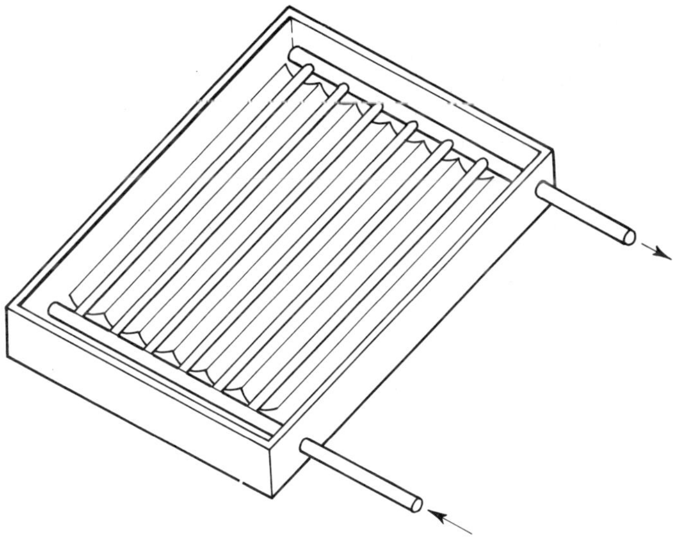

Fig. 7.21 *Collector with concentrating mirrors*

7.3.8 Heat pipes for heat removal

The extraction of heat can be improved by the use of a heat pipe. The operation of a heat pipe has been described in Chapter 4. The heat pipe has a very high effective thermal conductivity, typically 500 times that of a copper bar of similar dimensions. It therefore enables heat to be removed from the collector plate with very little drop in temperature. A further advantage of the heat pipe for this application is the fact that it can be designed to transfer heat only in one direction, hence if the solar output should fall, owing to clouds, for example, the heat pipe will prevent the flow of heat from the heat load back to the collector plate.

7.3.9 Summary of factors affecting flat plate collector performance

To summarise, the efficiency of a flat plate collector is defined as:

$$\frac{\text{Useful heat output}}{\text{Solar energy received}}$$

Since the useful heat = heat received by the absorber plate − losses, efficiency can be improved by reducing the losses. The first step is to reduce convection and radiation loss (by the greenhouse effect) by means of one or more cover plates. (Some incident radiation will be reflected or absorbed by the cover plate so that the flux received by the collector plate will be somewhat less than for the unglazed collector. Unless the absorber plate temperature differs from ambient by only a few degrees the reduction in heat losses is always much greater than the loss of incident flux.) Further reduction in losses is achieved by the use of selective surfaces to reduce radiation loss, and by evacuation of the air to eliminate convection between the absorber and cover plates. Finally, the use of heat pipes for heat extraction gives a more efficient heat extraction. The penalty to be paid for all these measures is increased cost.

Reduction in heat losses enables the coolant to be extracted at higher temperature. Eqn. 7.3 shows that a given collector performance can be defined in terms of a constant intercept $F_R\alpha\tau^*$ together with slope $-F_Rh_t$. Table 7.5 gives typical values of these constants for a number of different types of flat plate collectors. If a still higher temperature is required then the heat flux received by the surface must be increased by concentration.

Table 7.5 *Typical values of constants for various flat plate collectors adapted from Ref. 6*

Collector type	Absorber surface	$F_R\alpha\tau^*$	F_Rh_T W/m² °C
Unglazed collector (plastic)	Black paint	0·83	22
Single glazed (copper)	Black paint	0·72	7·1
Single glazed (copper)	Black chrome Selective surface	0·80	5·7
Single glazed (copper)	Copper oxide Selective surface	0·68	4·7
Double glazed (copper)	Black paint	0·62	4·4
Evacuated tube	—	0·55	0·5
Salt gradient Solar pond	—	0·40	0·41–0·8

7.3.10 Applications of flat plate collectors

7.3.10.1 Water and air heating: Flate plate collectors can be used for heating water or air to modest temperatures up to, say, 60°C at about 60% efficiency, although higher temperatures are possible if a lower collection efficiency is acceptable. The usual application for these collectors is for the provision of hot water for domestic use and also as preheaters for hot water for hotels and similar large buildings. Larger scale collectors can be used for industrial process heat.

7.3.10.2 Solar cookers: The importance of cooking and the provision of a suitable energy source has been referred to several times. Solar energy[9] offers a possible solution. Unfortunately, there are inherent problems which have so far prevented its widespread adoption. Food can be kept hot and, to some

Fig. 7.22 *Hay box cooker supplemented by direct solar heating. Manilla, Phillipines*

extent, cooked, using an insulated box – the hay box – which can be supplemented by direct solar heating, Fig. 7.22. If higher temperatures are required then concentration is necessary. The use of solar energy restricts the cooking time to those times of day when the solar radiation is available and may not be suited to communities who traditionally cook in the evening. Also, in most parts of the world solar energy may not be available at certain seasons or even on some days during the hot season; thus an alternative cooking device is necessary. As in some other conversion areas the only country in which solar cooking appears to be successful is China,[10] see Fig. 7.23.

7.3.10.3 Crop drying: Traditionally, in developing countries, crops are dried by spreading out on a flat surface in the sun, see Fig. 7.24. This method suffers from a number of disadvantages including insect and bird damage and fouling, dust, and lack of control of the drying process which affects the quality of the product and hence its value. The losses which arise during natural sun drying are significant and can exceed 10%. In many areas there is no likelihood of sudden rain but this is not the case everywhere, and it is then necessary to make provision for covering in the event of rain. This requires that someone should always be in attendance. Although modern methods using air heated by oil heaters are used to some extent, most crops are still dried in the traditional manner. Solar energy offers the possibility of an alternative which does not require a source of fuel.

Fig. 7.23 *Concentrating mirror solar cooker. Beijing, China*

Fig. 7.24 *Traditional rice drying. Beijing, China*
The rice is placed on the road and hulling achieved by allowing traffic to pass over it

All natural vegetation contains some moisture and the amount is frequently expressed in terms of a percentage as follows:

$$\frac{\text{Wet mass} - \text{dry mass}}{\text{wet mass}} \times 100$$

If there is too much moisture troubles will arise both in storage and use, and overmoist grain will spoil owing to the formation of moulds. For example, Thai rice, when harvested, has a percentage moisture content of about 24%; this must be reduced to around 14% for safe storage. Timber must be dried, or seasoned, in order to avoid cracking and distortion is use. Biomass for use as a fuel will yield greater energy if it is predried so that heat is not required to evaporate off moisture when the biomass is burnt.

In any form of air dryer water is evaporated off from the surface of the body. In order that drying should occur the drying air must contain less moisture than the saturated vapour corresponding to the surface temperature of the body. So far we have considered only the removal of water from the surface, in fact the moisture will be distributed more or less uniformly through the body and will only reach the surface by diffusion; this is a relatively slow process and is the rate determining process in the drying operation. If it is attempted to dry more rapidly the material may be damaged, for example, in the case of grain, surface hardening and cracking may occur. Diffusion is the process by which atoms in high concentration tend to drift to regions of lower concentration. The diffusion rate will be determined partly by temperature and partly by the ease with which water molecules are able to move through the body. The latter will vary with different materials. Drying will cease when the vapour pressure of water vapour in the air reaches the same value as the saturation vapour pressure at the surface of the body and the system is then said to be in equilibrium.

If unheated air is employed a large volume flow is required to dry the grain. If the air is preheated the volume required is considerably reduced. The reason why heated air is much more effective is that, on heating, the relative humidity of the air is reduced and more moisture can be absorbed before equilibrium is reached.

Solar heat can be used to raise the air temperature. Exell[11] has designed a simple drier of this type for drying rice, Fig. 7.25. The air heater is a flat plate collector of very cheap construction, made up from a plastic sheet cover plate and an absorber which is simply a layer of burnt rice husks. Atmospheric air passes over the heated layer and then through a bed of rice contained in a box constructed of bamboo matting in a bamboo frame. The air flow could be driven by a fan but Exell employs the natural convection of the hot air passing through a chimney of plastic sheet placed above the box containing the rice bed.

This design is for a mass of 1 tonne of rice to be dried from a humidity ratio of 22% to 14% in a period of three days. 36 m^2 of collector surface is required

Fig. 7.25 *Solar rice dryer. A.I.T. Bangkok. (Courtesy R. Exell)*

and the air is raised from a temperature of 30°C to a temperature of 45°C. The pressure drop across the bed is only 1/5000 of an atmosphere (2 Pa) but is sufficient to give an air flow of 5·7 complete air changes in the bed per minute.

7.3.10.4 Desalination of water: A reliable source of water suitable for drinking is an essential human need. Unfortunately much of the earth's water is in the form of saline water in the oceans and is quite unsuitable for either drinking or irrigation.[12,13] Sea water contains an average 3.5% of dissolved salts (by weight). Values vary widely, for example, the Baltic Sea is as low as 0·7% whereas the Persian Gulf is over 4%, and the Dead Sea has a value of 27·5%. Good drinking water contains up to 0·05% of dissolved salts and up to 0·15% can be tolerated. Water from deep bore holes often contains unacceptably high amounts of dissolved salts, and this 'brackish' water has a salt content of typically 0·2 to 0·5%. The possibiliity of producing potable water from sea and brackish water by some means of desalination is of considerable interest and commercial desalination plant using fossil fuel are in use in a number of countries to supplement the traditional sources of water supply. This commercial equipment is large scale and has a typical output of around $10^5 m^3$/day. Solar energy can be used as an alternative to fossil fuel to provide desalinated water. Owing to the diffuse nature of solar energy there is less advantage to be gained by scaling up to large systems than is the case for fossil fuel and small scale solar stills are likely to be used for providing water for families or small communities.

Small amounts of water can be obtained from simple solar stills of the type shown in Fig. 7.26. The saline water is situated in a shallow tray and is heated by the sun to cause evaporation. the water vapour condenses on the glass cover plate and is collected by a container as shown. It is instructive to estimate the daily output of such a device. In order to evaporate 1 kg of water at a temperature of 30°C about $2{\cdot}4 \times 10^6$ J is required. Assuming an insolation of 250 W/m^2, averaged over the 24 hours, this energy could evaporate a maximum of:

$$\frac{250 \times 24 \times 60 \times 60}{2{\cdot}4 \times 10^6}$$

$$= 9 \text{ l/m}^2\text{/day}$$

In practice heat losses will occur and the average daily yield which might be expected from a simple still is 4 to 5 l/m^2/day. This suggests an efficiency of 45–55%, which seems quite good. However, we should look at the assumptions a little more closely. What we are attempting to achieve is the separation of salt from water or vice versa. Experience tells us that when sugar is added to tea or salt to water there is no noticeable change in temperature.* While this is not quite the same thing as the energy required to add or remove salt it might be expected it to be of a similar order. It turns out that a minimum† of only 3·6 kJ/kg is required to desalinate sea water and a value somewhat less than this for the lower concentration found in brackish water. 3·6 kJ/kg is sufficient heat to raise the water temperature by less than 1°C. On this basis the simple still has an efficiency of less than 0·1%. Nevertheless, this design has uses for small isolated familes and even communities. It is, after all, litres/$ which matters, not efficiency.

In its simplest form a solar still consists of a shallow pit which is made watertight by means of a plastic or rubberised sheet, or concrete having a bitumen coating.[14] The cover can be glass or plastic; the former is preferred since, if properly cleaned, the water will condense as a film (that is it will wet the surface). Such a film will allow most of the radiation to pass through to be absorbed by the bottom of the pool. A disadvantage of some plastics is that water does not wet the surface but instead forms small droplets which reflect the incoming radiation.

One of the largest stills of this type was built about 100 years ago at Las Salinas, Chile, to provide water for a mining camp.[15] The still had an area of 4700 m^2 and gave an average of 23 m^3/day of distilled water. There are numcrous other solar stills in Australia, Spain, Tunisia, Greece and elsewhere.

* There is in fact a small change in temperature due to what is called the heat of solution. In order to remove the salt it is necessary to provide an amount of energy equal to the change in 'free energy' which is not the same as heat of solution but is of the same order of magnitude. Free energy G is discussed in Appendix 2.

† This is the energy required to recover 25% water from 3·5% NaCl water at 300 K. The minimum required for zero recovery is 3·05 kJ/kg.

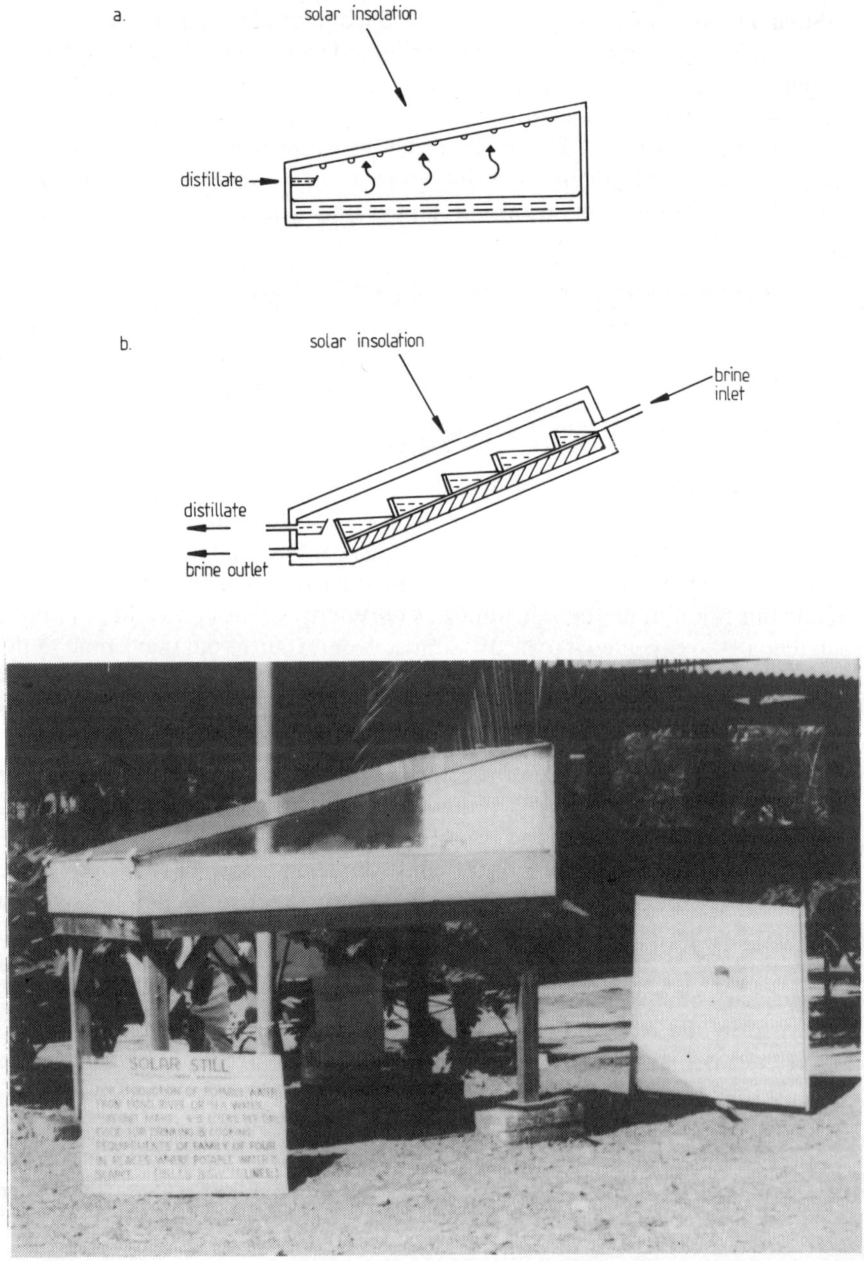

Fig. 7.26 *Simple solar still*
a Single basin
b Multi basin
c Solar still at the Institute of Alternative Energy. Manilla, Phillipines

One of the largest solar stills in use today is on the island of Patmos. This still, which was built in 1967, had a glass cover 8600 m^2 in area and a average daily output of 30 m^3.

Although solar energy is free, the low efficiency of the simple still does mean that the output per unit area is relatively low and hence the cost of the equipment per unit output is correspondingly high. While the materials used are not expensive, nevertheless it is of interest to raise the efficiency if possible. The reason why the simple solar still is so inefficient is that considerable energy must be provided to evaporate the water. Much improved efficiency is achieved in the large fossil fuelled commercial plant by carrying out the evaporation in a number of stages. Heat from the water vapour is recovered at each stage by allowing the vapour to condense and provide heating to an earlier stage of the process. In this way the overall efficiency can be raised to 1·5%, a factor of fifteen improvement. Fig. 7.27 shows the main features of a multistage flash distillation plant. Sea water is preheated by passing through condensers in each stage (C_1 and C_{20} in Fig. 7.27). The

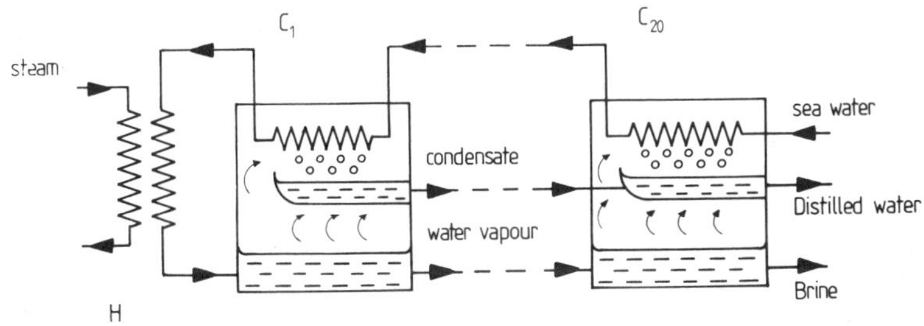

Fig. 7.27 *Multistage flash distillation*

preheated sea water is then heated to final temperature in heat exchange H which is fed by steam from a conventional boiler. The heat imput to H is the sole input to the complete system. The hot sea water is passed into an open channel in the first stage where some is evaporated (flashed). This vapour will condense on the outer surface of condenser C_1 and the condensate will fall into the tray below C_1. The process is repeated in the successive stages which operate at lower temperatures. In practice about twenty stages will be used. In order to assist the vapour flow, air is removed from the system and each stage operates at the saturation vapour pressure of the water at the stage temperature. The energy requirement is approximately 300 kJ/kg of product. Most commercial distillation plants are of this type and can have an output as high as $10^5 m^3$/day.

The multiple effect desalination system is similar in principle to multiflash distillation and is used in some commercial plant. This is also a multistage process and the major differences between the two systems is shown in Fig. 7.28*a* and *b*. Essentially the heat flow in the stages in the multiple effect system

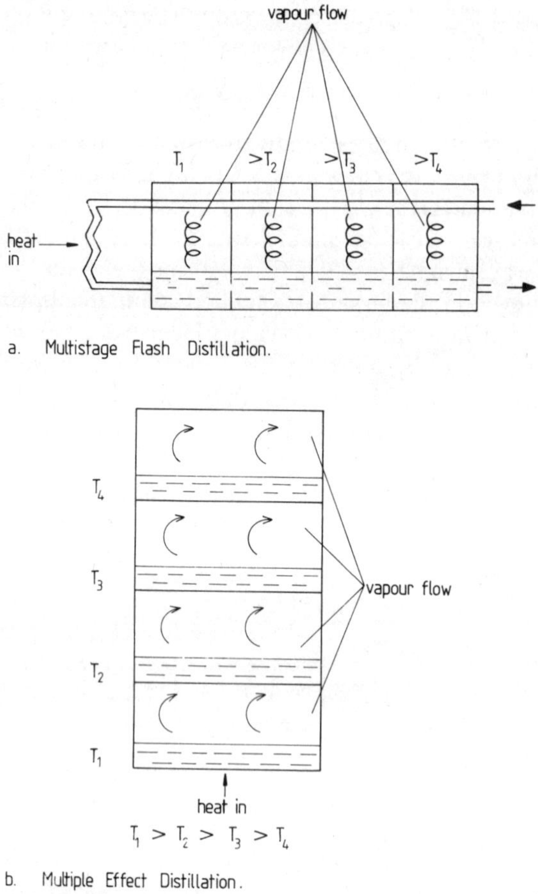

Fig. 7.28 *Comparison between multistage flash distillation and multiple effect distillation*

is in series whereas in the multiple flash system it is in parallel. The performance of the two systems is similar but the multiple flash system has the advantage for large plant.

Solar heat can be used as the heat source. Fig. 7.29 a,b shows examples of solar heated multistage desalination plants with a daily output of ~10m³ per day. Fig. 7.30 shows a design for a small solar unit which is operated with air present.[16]

Another method also used commercially is known as the vapour compression system, Fig. 7.31. Preheated saline water is pumped into vessel A, in which it evaporates, and the resulting saturated vapour is compressed in the compressor C. The effect of compression is to raise the pressure and temperature of the vapour which becomes slightly superheated so that it is

a.

b.

Fig. 7.29 *a* 10 stage vaporisation plant in La Paz, Mexico Produces up to 10m³ per day. Built by Dornier. Sponsored by B.M.F.T. (Courtesy Dornier Friedrichshafen, Germany)
b Multiple effect solar desalination plant (Courtesy C.E.A. Cadarache, France)

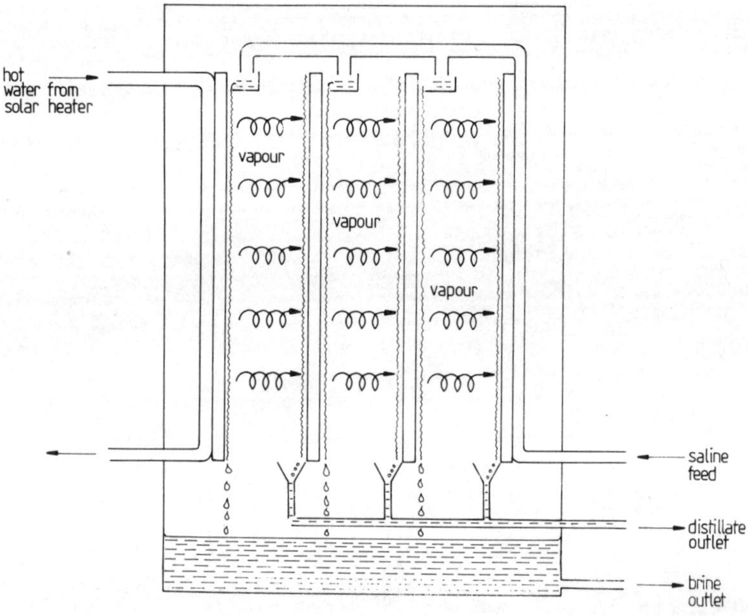

Fig. 7.30 *Multiple effect solar still*

now at a higher temperature than the liquid in A. The superheated vapour passes through the heat exchanger B in which it condenses to give pure (distilled) water; the heat given off is absorbed by the saline water in vessel A. Since the salinity of the water in A has increased, an equal quantity of hot saline water is abstracted from A as indicated. It turns out that sufficient heat is released by steam condensing in the heat exchanger to evaporate the water in

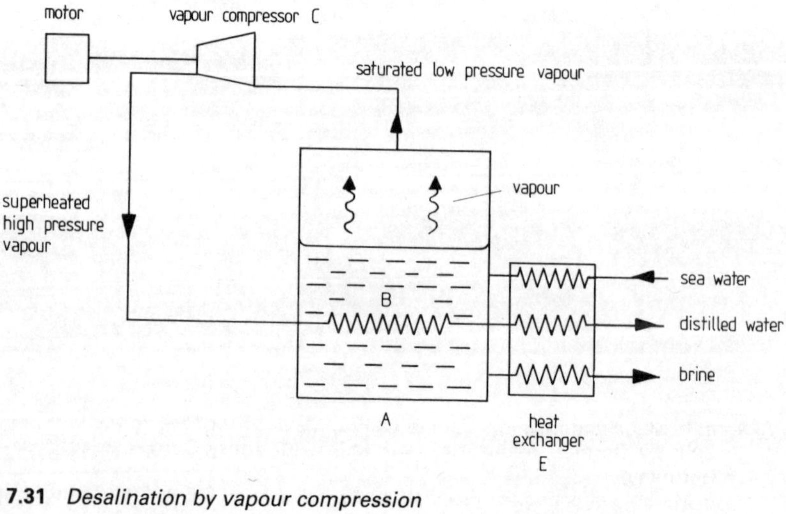

Fig. 7.31 *Desalination by vapour compression*

the container A. Also the hot distilled water and hot saline extract solution can be used to heat the incoming saline water (through a second heat exchanger E). In principle, once the process has begun there is no need to add heat and the only energy requirement is the mechanical energy to drive the compressor.

When a salt is dissolved in water the resulting ions move randomly in the water in the same manner as atoms of a gas in a container. If two vessels are joined by means of a porous wall and one vessel filled with a concentrated solution of salt and the other filled wih a dilute solution then salt ions will diffuse through in the direction of the concentrated to weaker solution and water molecules will diffuse in the opposite direction. After a time the concentration of salt in the two vessels will be the same. However, if a selective membrane is used which only allows the passage of water molecules and is non-porous to the salt ions then only water molecules will move from the lower concentration side to the higher concentration side in order to bring the concentration in the two vessels to the same value. This process is called osmosis and has an important place in biological systems. The pressure difference producing this flow of water is called the osmotic pressure and for water with 3½% NaCl at a temperature of 300 K will have a value of 30 atm. If a reverse pressure is applied then this will tend to reduce the water flow. If this reverse pressure is higher than the osmotic pressure then the flow is reversed. In effect the water is squeezed out from the salt solution. This method, called reverse osmosis, is used for desalination, See Fig. 7.32. Practical selective

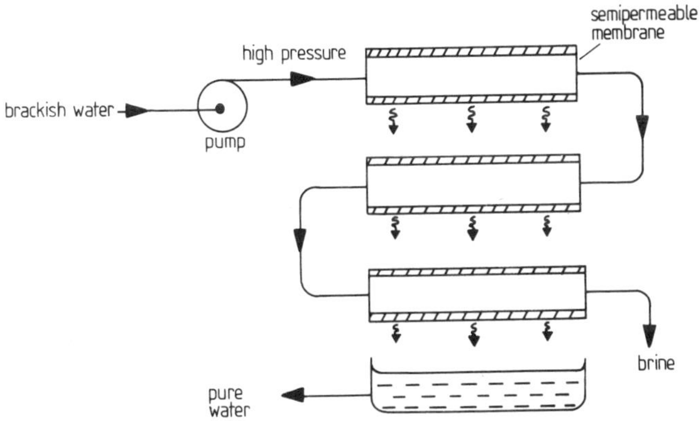

Fig. 7.32 *Desalination by reverse osmosis*

membranes, sometimes called semipermeable membranes, have been developed for this purpose. Typical values of pressure difference required by reverse osmosis desalination plant are in the range 40 to 100 atm. The process is efficient and the energy required to produce 1 kg of product is around 35 kJ.

So far we have discussed methods of desalination in which the water is removed from the salt. Since there is only a few per cent of salt it might appear to be more sensible to remove the salt from the water. There are several ways in which this can be achieved. The problem is how to cause the salt to move. Salt in solution is in the form of positive and negative charged ions which can be acted on by an electric field. This is the principle of electrodialysis, see Fig. 7.33. An electric field causes the ions to move to walls which are made of membrane selectively permeable to the particular ion.

Both electrodialysis and reverse osmosis are used in large scale plant and typical efficiencies are around 10%. Vapour compression, electrodialysis and reverse osmosis could make use of solar generated electricity as the power source. Desalination can also be achieved by freezing, since the ice is formed from fresh water and the salt is rejected during freezing, and solar ice-making could be considered for this purpose.

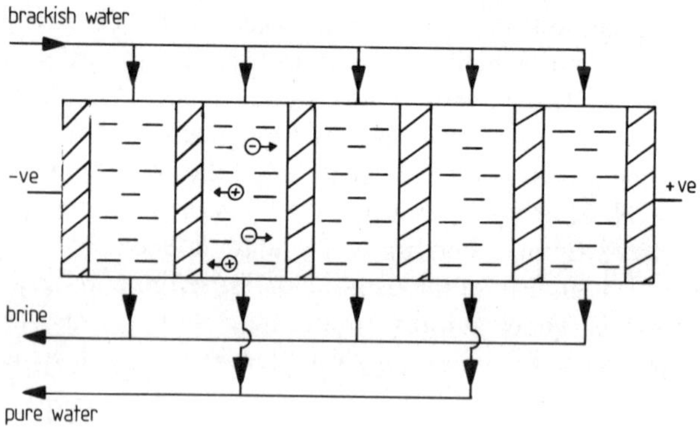

Fig. 7.33 *Desalination by electrodialysis*

7.3.11 Non-convecting solar pond

In 1902 a field geologist named A. V. Kalecscinsky reported[17] an unexpected temperature gradient effect in a natural lake in Hungary. He noted that the water at the bottom of the lake was at a higher temperature than the water at the top. A temperature as high as 70°C was recorded in July. Since cold water is denser than hot water the reverse might be expected. The explanation turned out to be as follows. Salt water is denser than fresh water and the lakes had a layer of saline water overlaid by a layer of fresh water with a non-convecting salinity gradient in between. Some of the solar energy penetrated the lower layer where it was absorbed. The warm saline water is insulated from heat loss by this non-convecting layer. Other natural solar lakes have since been observed in various parts of the world.

This effect can be used as the operating principle of a cheap solar convertor which will give an output of water at a temperature of 60–85°C. The pioneer development work was carried out by Tabor[18] in Israel and subsequently ponds of the same type have been constructed in Australia, USA, India and elsewhere.

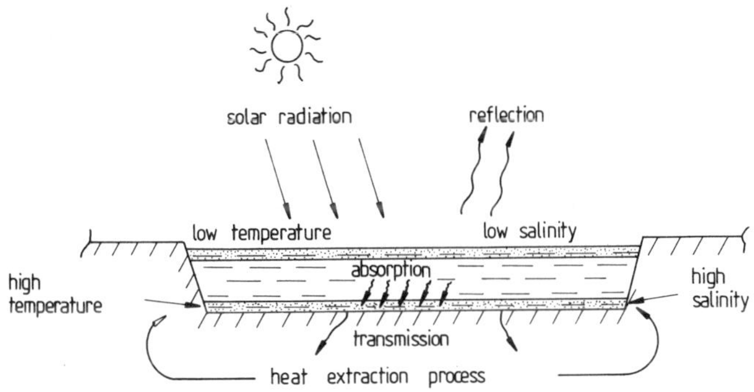

Fig. 7.34 *Solar pond*

The basic arrangement of the non-convecting solar pond is shown in Fig. 7.34. The pond is built to a depth of between 1 and 2 m. The lowest layer is arranged to have a high salinity (260 kg/m^3 NaCl) and the salinity is decreased progressively in a vertical direction until the salinity has fallen to that of fresh or slightly brackish water. The pond must be filled with care to avoid mixing between the horizontal layers. In an operating pond three distinct horizontal sectors can be identified. A bottom layer of high salinity and constant temperature, an intermediate layer which is thermally non-convecting, and a top or surface layer of about 20 cm thickness of uniform density owing to wind induced waves and other effects. Radiation from the sun falls on the upper surface. About 50% of this radiation (the long wavelength component) is absorbed in the top layer and the remainder (short wavelength component) penetrates the pond. About half of this radiation is absorbed in the intermediate, non-conducting layer and the remainder (25% of the initial incident energy) reaches the bottom of the pond where it is absorbed and its energy stored in the bottom layer of saline water.

Solar ponds with maximum temperatures greater than 100°C have been reported from Israel and the useful operating temperature has a maximum of 80–90°C. Hot water is extracted from the pond and fed through a heat exchanger before being returned. The largest electricity generating pond so far reported has an area of 7500 m^2 and is situated in the Dead Sea region. An organic liquid is passed through the secondary side of the heat exchanger and the resulting vapour used to drive a turbine and alternator. The plant produces 150 kW (peak) of electricity at an overall efficiency defined as electrical energy/solar insolation of around 1–2%. The reason for this low value of overall efficiency is discussed in the next section. The Israeli site is particularly favourable for solar ponds, being an area of high solar insolation and with readily available salt supplies. Electricity generation in the same manner may also prove economic in other areas of the Middle East and elsewhere. However, it appears likely that the more general application of the solar pond will be as a source of hot water (60°C) for process heat in industry. Such process heat ponds will have an overall efficiency of 15–40% depending on the insolation values of the particular site and the water temperature required. The cost/m^2 of the solar pond is somewhat less than the cost of the flat plate collector and the solar pond has the additional advantage of providing energy storage. The collector efficiency of the solar pond is less than that of the flat plate collector at low values of temperature difference but greater at higher temperature levels. Table 7.5. Ponds are normally lined with a membrane of butyl rubber or similar material. Quite large amounts of salt are required (approximately 1 tonne/m^2) and the cost of make-up salt and fresh water can be a significant fraction of the output energy cost.

7.4 Solar powered heat engines

7.4.1 Introduction

By the mid-1800s inventive minds were becoming attracted by the possibility of harnessing solar energy to heat the boilers of small steam engines. One of the first of these inventors was the Frenchman Mouchet who demonstrated a solar powered pump in Paris in 1866.[19] Mouchet used a conical shaped mirror which could be pivoted to track the sun and which directed the radiation on to a boiler, thus raising steam for the engine. A second Frenchman, Pifre, was interested in the experiments carried out by Mouchet and commenced a similar study; one of the best known results of his work was his demonstration of a solar powered printing press which was operated in the Tuilleries Gardens in Paris in August 1882,[20] Fig. 7.35. At this time similar experiments were being carried out in other countries, notably by Ericsson[21] in the USA and Adams[22] in India.

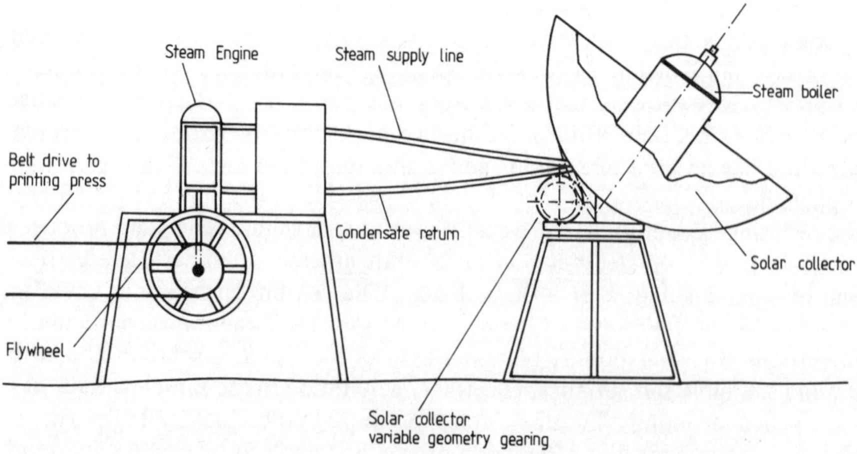

Fig. 7.35 *Pitfré's solar produced printing press. Paris. 1882*

One of the most ambitious early solar projects was the solar powered water pumping scheme by Schuman and Boys in Egypt in 1913.[23] This was said to use five solar collectors each of 250 m^2 area and to give a peak power output of 73 hp at an efficiency of 4·4%, Fig. 7.36. Following the First World War interest

Fig. 7.36 *Solar powered water pumping station. Schuman and Boys. Egypt. 1913 (Courtesy C McVeigh)*

in solar powered engines was lost until a revival of activity in the 1950s. Appropriately, in view of the early French pioneering efforts in this field, the new work was headed by the French company Sofretes. Steam is not necessarily the most suitable working fluid for solar powered engines, and other fluids, such as the freons, have considerable advantages. The early solar engine workers were well aware of these factors and experimented with alternative fluids. One of the first of these, Tellier, a Frenchman, operated an engine using ammonia as the fluid, in 1885.[24] In the early 1900s, Schuman,[25] in the USA employed ether. In addition to the reciprocating engines, small turbines are now becoming available in which freon is used instead of water. An interesting development over the last few years has been the invention of several steam pumping engines which eliminate the metal piston and replace it by the water which is to be pumped. These liquid piston engines have obvious advantages of simplicity over the traditional engines pump arrangement. Of course nothing is new and the earliest steam powered water pump invented by Savery in 1698 operated in this way.

The Savery pumping engine was invented by Captain Savery in the late Seventeenth-century and was intended for use in mines and also to pump water for domestic use in large houses. Referring to Fig. 7.37, which is taken from Savery's original patent, steam enters a vessel and displaces water up the

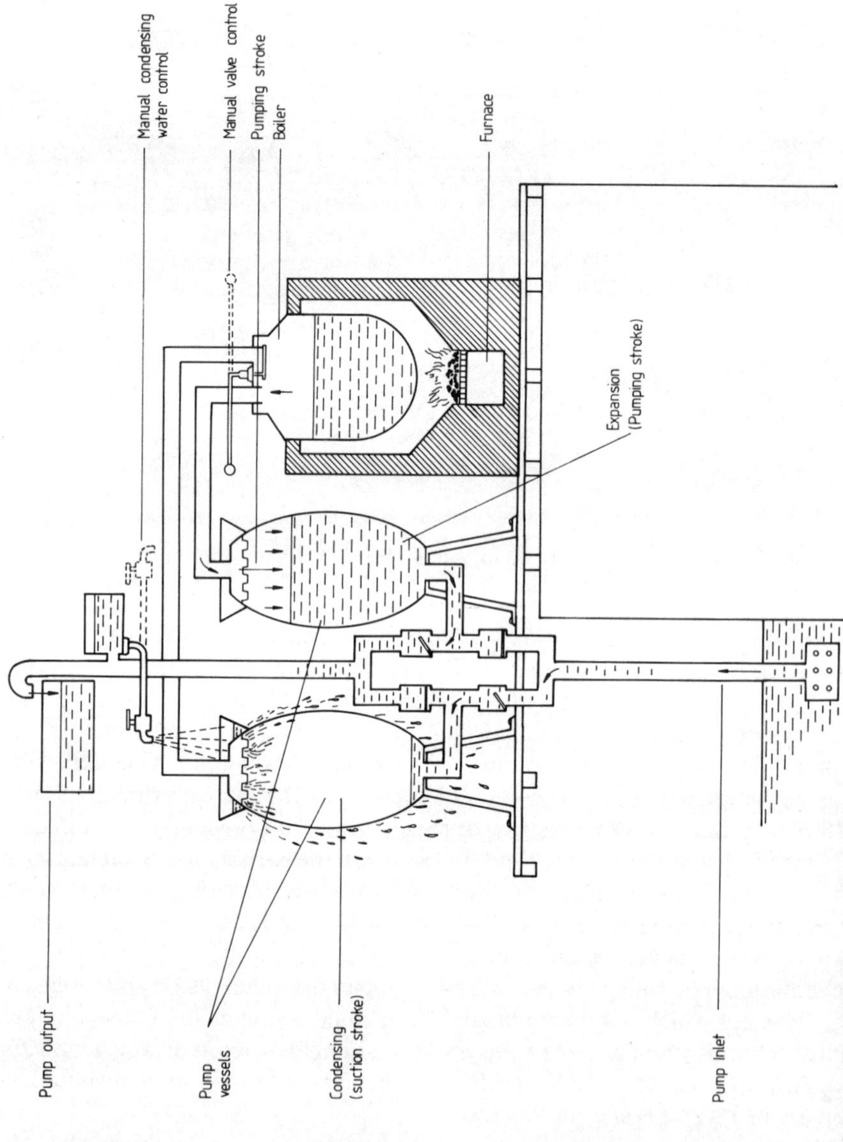

Fig. 7.37 *Captain Savery's Patent 356/1698*

vertical pipe. A valve is closed to shut off the steam supply and the steam in the vessel is condensed by means of a jet of cold water. The condensation of the steam causes a reduced pressure in the vessel and water is drawn up, through the lower pipe, to fill the vessel again. The process is then repeated. In Savery's design two vessels are used so that water is pumped alternately from each. Fig. 7.38 shows a recent proposal by Jennes[26] which adopts the Savery design for solar heating.

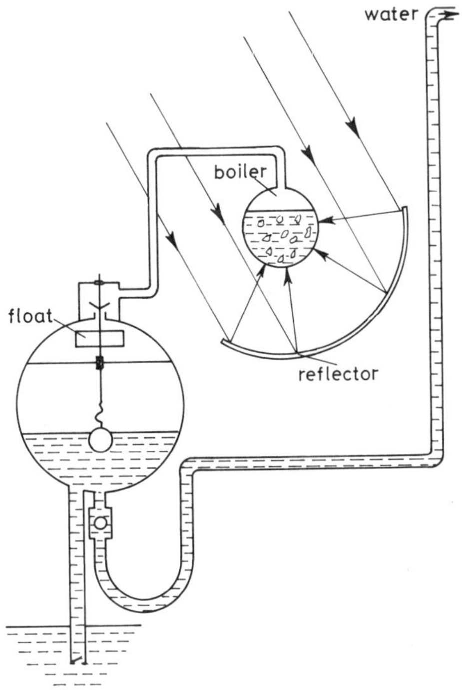

Fig. 7.38 *Proposed solar powered Savery pump (Ref. 26)*

7.4.2 Some general considerations

In order to obtain mechanical power from solar heat we need to use some kind of heat engine. We saw in Chapter 3 that the efficiency of an ideal Carnot heat engine increases as the heat source temperature is increased. In the discussion on practical heat engines in Chapter 4 the importance of operating at as high a temperature as possible was also brought out. Unfortunately, with the flat plate collector temperatures tend to be low and the efficiency of collection falls rapidly with increase in temperature. The overall efficiency of the system η_o is given by the product of the collector efficiency η_c and the engine efficiency η_e. Hence

$$\eta_o = \eta_c \cdot \eta_e$$

Fig. 7.39 shows typical collector and engine efficiency curves. For a given flat plate/engine system there will be a temperature at which the overall efficiency is a maximum.

Both flat plate collectors and concentrators can be used with Rankine engines using heavy vapour working fluids. Gas cycle engines such as the Stirling engine will require higher temperatures and are only suitable in association with concentrating collectors.

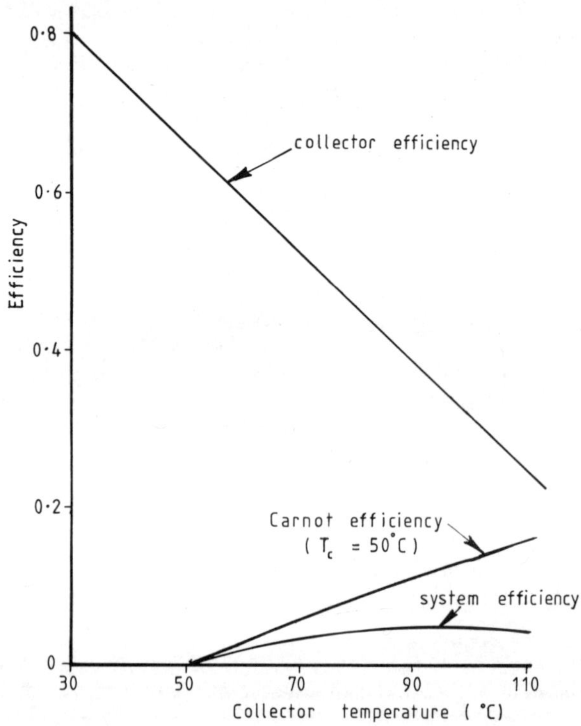

Fig. 7.39 *Overall efficiency for a single glazed flat plate collector driving a Carnot engine*

7.4.3 Heat engines using vapour as the working fluid (*Rankine cycle engines*)
The working principles of these engines have been discussed in Chapter 4. In the simplest arrangement the solar collector is also the boiler. The basic components are then:

Collector
Expander
Condenser
Feed pump

A major advantage of a water pumping application is the availability of cool water for the condenser side. Without this such engines would be far too inefficient to be of economic interest.

Some designers employ a collector system which is isolated from the engine fluid, and in this case a second circulating pump is required.

In recent years much effort has gone into the design of small solar heated pumps. One of the simplest of these is due to Payne, Fig. 7.40.[27–29] Initially the

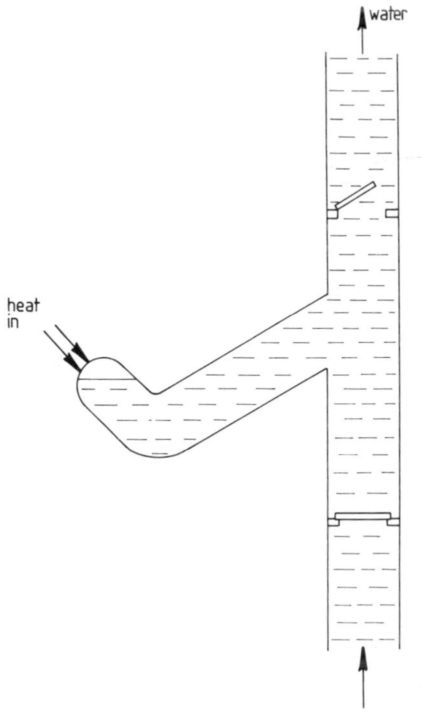

Fig. 7.40 *Payne pump (Ref. 27)*

U-tube is filled with water. On the application of heat to the sealed end of the U-tube, steam is generated and the resulting expansion forces water through the non-return valve up the vertical pipe. The expanded steam will condense on the newly exposed cold surface of the tube, causing a reduction in pressure which will allow water to be drawn up via the lower inlet valve, from the reservoir. The process will then be repeated. A further variation is the Kleen–McCracken pump,[30] Fig. 7.41. Water is heated in the boiler and the resulting steam causes water to be pumped through the vertical outlet pipe. When the steam expands sufficiently it will escape through the bend of the U-tube then will be condensed, and further water drawn in from the reservoir.

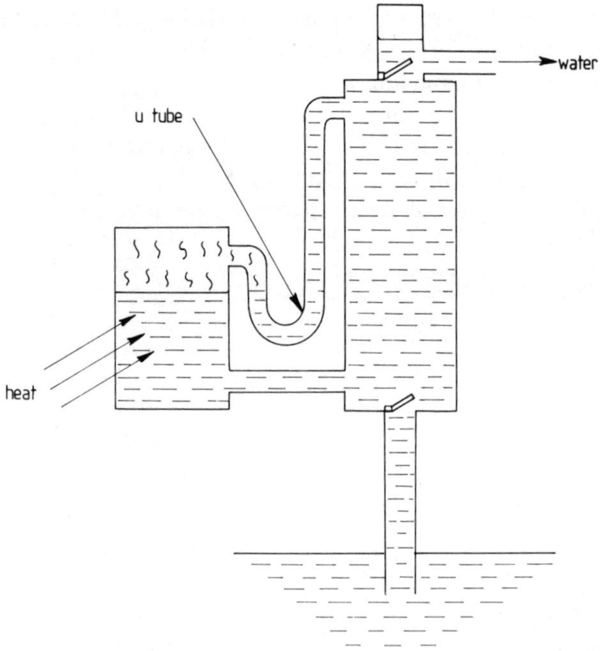

Fig. 7.41 *Kleen–McCracken pump (Ref. 30)*

One disadvantage of engines of this type is the poor efficiency owing to heat loss at the walls, and arrangements such as the use of insulating fluids or insulating floats have been suggested to separate the working fluid from the water that is to be pumped.

A further disadvantage of pumping engines which use steam as the working fluid is that a temperature in excess of 100°C is required for the pump to operate. If the working fluid can be separated from the pumped water then other substances, such as freon, can be selected. Some authors have proposed systems in which a fluid immiscible in water is used. More usually a physical separation is achieved by a metal diaphragm or a membrane. Fig. 7.42 shows a design of this type which is manufactured by Solar Pump Corporation[31] in the USA. In order to reduce thermal losses the engine is mounted in the flat plate solar collector.

The engines described so far have been intended for use as water pumps. If piston or turbine expanders are used the mechanical power can be extracted and used for electricity generation and other purposes.

A major development in water pumping was the programme started by Girardier and Masson[32] in Senegal which has led to the setting up of the Sofretes organisation in France for the manufacture of solar thermal engines. The requirement identified in Senegal was for a simple water pump which would be reliable and maintenance-free. A flat plate collector was adopted in

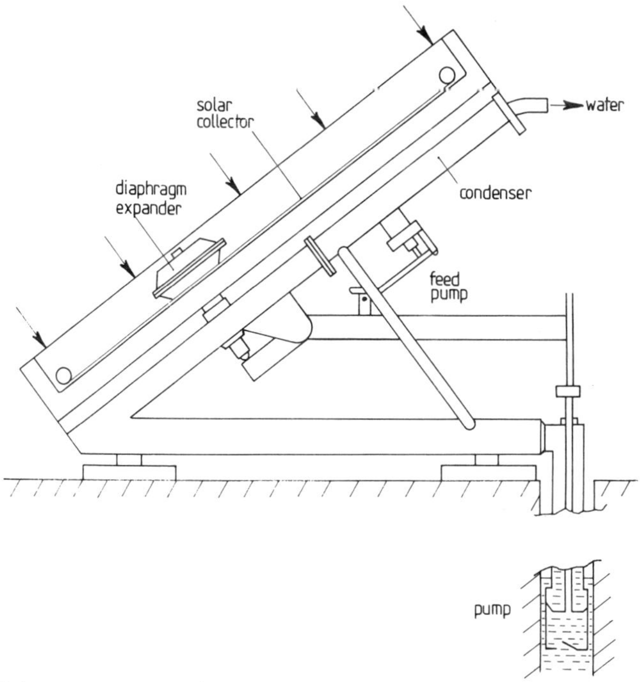

Fig. 7.42 *Solar power corporation pump*

order to avoid the complication of tracking which would be required with a concentrating collector. In the first pump methylchloride was used as the working fluid with a 6 m^2 collector surface. This machine gave a hydraulic power output of 21 W which could be used to pump 7·9–9·8 litres/min of water through a 13 m head. A reciprocating engine was used and operated at a speed of 80 rev/min. The overall efficiency was 0·4%, and problems were experienced with lubrication, thermal losses and sealing. Later engines employed turbines which drove alternators to avoid the problem of the mechanical shaft seal.[33] These engines used different working fluids, including freon 113. A similar programme has been developed in Israel by Tabor[34] and resulted in the ORMAT solar engines ranging in power from < 1 kW to the order of megawatts.

Although there has been considerable development effort in small solar pumping engines, it now seems likely that systems below 1 kW are unlikely to prove competitive with solar cell powered equipment. The reason for this is the low overall efficiency achieved (2 or 3% or less) and the resulting very large collector area that is required.

Two large irrigation plants[35] have been commissioned in the USA. Both use linear parabolic mirrors, and freon 113 as the secondary working fluid. One plant (Gila Bend)[36] produced 37 kW using a primary water temperature of 22 °C. A total collector area of 544 m^2 was employed.

Solar energy can be used to generate very large amounts of electrical power by employing an array of heliostats which focus the solar radiation onto a receiver placed in a tower where the thermal energy can be used to raise steam to operate a turbine.,[37,38] A heliostat, Fig. 7.43, is a mirror, usually around 40 m^2 in area, which can be rotated to direct the radiation on to a receiver. Other systems use a large number of concentrators which have boilers placed in their focus, the boilers are connected to a central plant where the engine is situated.

a.

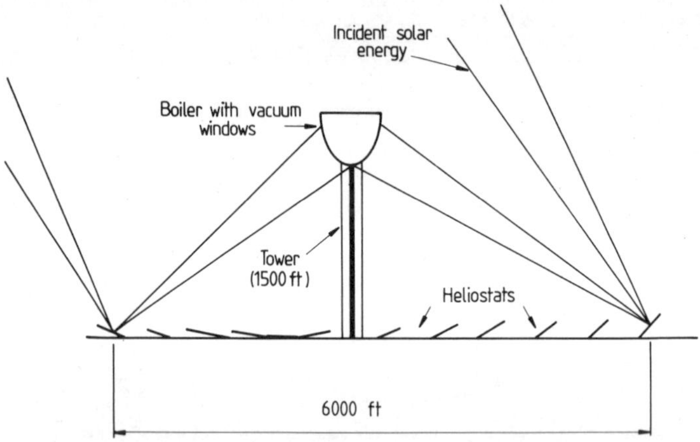

b.

Fig. 7.43 *Solar Thermal power station*

a Cross sectional view of Solar Furnace
b 1 MW(e) Eureulius. Sicily, Italy. (Courtesy Bernard McNelis)

The area of the installation can be estimated as follows. Assume an average solar power on equal horizontal surface is 250 W/m^2; if the overall conversion efficiency is 25% then 62·5 W of electrical power will be produced for each m^2 of horizontal surface. In order to generate an average power of 1000 MW an area of 10^9/62·5 or 16 × 10^6 m^2 will be required. This is the area of a square of side 4 km.

The 1000 MW is the average power and because of the variation in power density over the 24 hours a peak installed capacity of around four to six times this value will be required. Also storage will be necessary unless the generator is connected to an electricity network. The earliest project of this scale was the solar furnace constructed at Odeillo in the Southern Pyrenees in France.[39,40] An arrangement of heliostats is used to produce a peak power of 1000 kW. In this case the power is used as a furnace, as well as a source of electricity, and temperatures as high as 3800°C have been reported. Several other prototype electricity power generation plants have now been constructed. One of these is a 1 MW(e) plant commissioned by the EEC[41] and constructed in Sicily. Heliostats, of total area 7800 m^2 mirror surface, direct 4800 kW (at 1000 W/m^2 insolation) of thermal energy on to a central receiver placed on a 50 m tower. Steam is generated and used to operate a turboalternator.

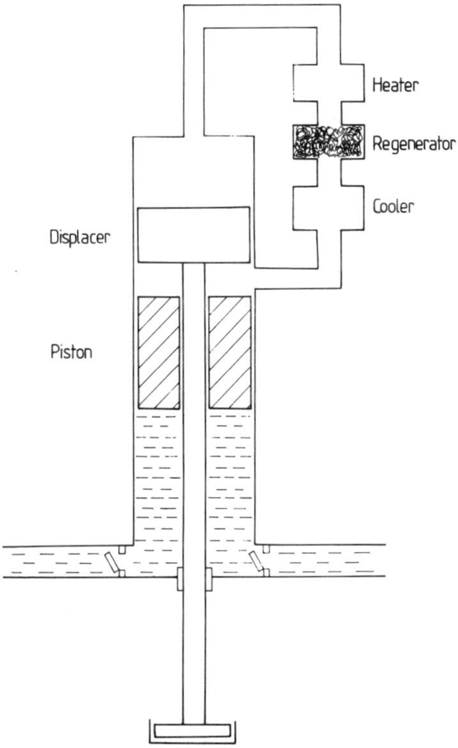

Fig. 7.44 *Beale engine*

Other stations have been constructed in France (2 MW), Japan (1 MW), Spain (1 MW) USA (10 MW) and Australia.[42]

7.4.4 Heat engines using gas as the working fluid

Following Ericsson's early work in the USA no further work on gas engines was reported until the 1950s. Since then several engines have been developed, all working on the Stirling cycle and requiring concentration.[43] Farber has reported a conventional Stirling engine producing 1/4 hp at an overall efficiency of 9%.[44]

An interesting development, due to Beale, is the free piston engine discussed in Chapter 4. The displacer and power pistons do not have a mechanical connection but are arranged to have greatly different mass, the power piston being the heavier. The resulting motion is such as to give the required phase relationship between the motions of the two pistons. In Fig. 7.44 the free piston engine is shown incorporating a water pump, the power piston acting as the water pumping piston. A version of this pump, using helium as the working fluid and using a Fresnel lens to focus the solar radiation, achieved an overall efficiency of 3·5%. The hot end temperature was 450°C.[45]

The fluidyne described in Chapter 4 has also been suggested as suitable for use with solar energy.

7.5 Solar refrigeration

7.5.1 Compression and absorption refrigerators

The principles of operation of the compression and absorption type refrigerators[46,47] have been described in Chapter 4. Since the compression type requires mechanical shaft power it can be driven using a solar thermal engine or an electric motor employing electricity from photovoltaic cells. It was shown in Chapter 4 that the overall efficiency is low when a solar thermal engine is used, hence it is preferable to use the absorption cycle if only low grade heat is available. Air conditioning systems which use this cycle with water as the working fluid and lithium bromide as the absorbent, are now commercially available, but are expensive. The typical high temperature source temperature is in the range 70–100°C and the COP around 0·8.

Exell[48] has developed an intermittant absorption solar refrigerator of the type discussed in section 4.5.3, Fig. 7.45. The source of energy is a 1·4 m^2 flat plate collector and the refrigerator, when operated near Bangkok, produced 6 kg of ice per day. The operation is as follows. The solar collector contains a solution of ammonia in water from which ammonia is distilled, by solar heating, with valve A open and valves B and C closed. The solution heats up in the morning and ammonia begins to distil appreciably by midday. The ammonia vapour is condensed in the collector which is immersed in a tank of

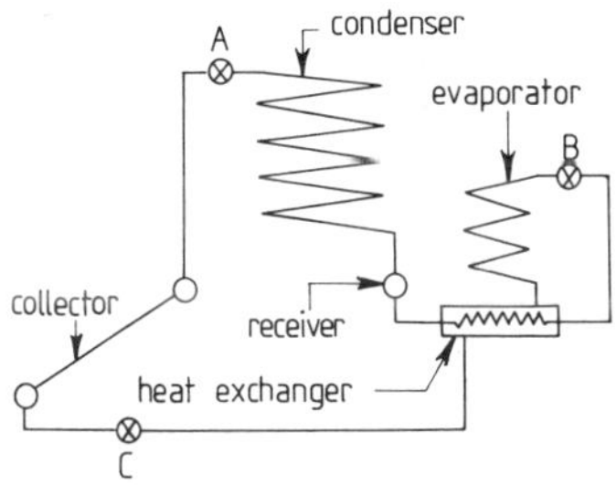

Fig. 7.45 *Exell's intermittent solar refrigerator*

cold water to assist condensation. In late afternoon valve A is closed and air allowed into the solar collector casing to cool the solution. At night valves B and C are opened and a refrigerating effect is produced by evaporation of ammonia in the evaporator. The resulting ammonia vapour is allowed into the bottom of the collector where it is absorbed into the solution and the heat of absorption escapes from the collector surface. The system can also be used to maintain a cold store at a temperature 10°C for storage of food and agricultural produce. Other similar arrangements have also been reported in the literature.

7.5.2 Dessicants

Dessicants offer an alternative for air conditioning; moist air can be dehumidified by passing through an absorbing liquid or solid and then cooled by allowing it to pick up moisture to give evaporative cooling. Triethylene glycol is a suitable liquid and silica gel, lithium chloride and calcium chloride are solid dessicants. The dessicants are reconstituted by heating.

7.5.3 Radiative cooling

There is a further method of cooling which has possible applications, for example, for use in cold rooms for fruit and vegetable storage. This method relies on the exchange of heat between the surface of the earth and the sky by long wave radiation. At night, when skies are clear, the sky temperature may be as much as 20°C less than the ambient temperature. The net heat loss under these conditions can be as much as 80–100 W/m^2. Even when the skies are cloudy, heat fluxes of 50 W/m^2 are possible. The net heat flux will obviously depend on the temperature difference between the surface and the sky. It is possible to achieve temperatures as low as 5°C when the ambient temperature is 25°C by the simple arrangement shown in Fig. 7.46. An open box is set up, having well insulated sides and botton. The top surface is covered by a thin plastic sheet (transparent to long wave radiation) and allowed to cool. During sunlight hours the top is covered by an insulator to keep out solar radiation.

7.5.4 Applications of refrigeration

Refrigeration or cooling of agricultural products is frequently desirable to prevent deterioration either for short term storage, or, in some cases, for

a.

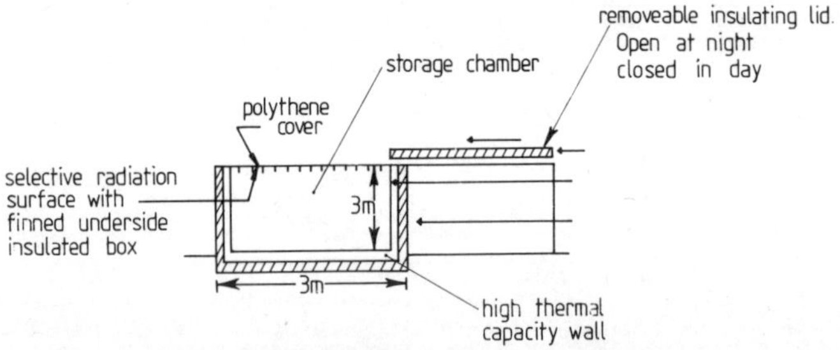

b.

Fig. 7.46 *Radiative cooling*
a Schematic
b Installation. Changmai, Thailand. (Courtesy Wasan Jompakdee)

storage over long periods. These applications can be roughly classified as follows:

1. Cooling of newly harvested crops. Crops such as berries, fruits and flowers are harvested at temperatures in the range of 25°C to 35°C and need to be cooled to a temperature of less than 12°C within 24 hours of harvesting.
2. Short term storage. Perishable crops require cooling to less than 12°C for temporary storage of a few days before local redistribution.
3. Milk chilling. Milk needs to be kept at 10°C for 24 hour storage, and 3°C for 48 hours storage.
4. Fish storage. Temperatures of −3°C to −5°C are required to store fish for periods up to one month.

5 Long term storage can be divided into cooled storage of crops, such as potatoes, when temperatures should be maintained at around 2°C or 3°C, and meat, which requires refrigeration to temperatures in the range −10°C to −30°C.

In developing countries there is a requirement for the refrigeration of vaccine and other medical supplies in clinics. For short term storage a temperature of 4°C may be adequate but for long term storage vaccines require around −20°C.

7.7 Direct generation of electricity – photovoltaic conversion

7.6.1 Metals, semiconductors and insulators

In metals, the atoms forming the material readily lose one or more of their electrons. We can regard a metal as a rigid lattice of positively charged ions immersed in a sea of free electrons. If a voltage is applied to the metal these electrons will move freely in the electric field due to the voltage, thus giving rise to an electric current. Since the resistance to electric current flow in metals is very small, they are good conductors of electricity.

In insulators the atoms share their outer electrons with adjacent atoms, and these electrons are fixed and are unable to move in an electric field. The effect of temperature increase is to dislodge a few electrons so that in practice a very small current will flow in the presence of an electric field. Table 7.6 shows the very large difference in electrical conductivity between metals and insulators.

There are some materials having a structure which is the same as that on insulators but where the outer electrons are much more easily dislodged than

Table 7.6 *Electrical conductivity (at 300°K)*

Material	Conductivity S*	
Silver	$6{\cdot}3 \times 10^7$	Metals
Copper	$5{\cdot}85 \times 10^7$	
Aluminium	$3{\cdot}5 \times 10^7$	
Steel	$1{\cdot}07 \times 10^7$	
Germanium	$2{\cdot}2$	Semiconductors
Silicon	5×10^{-4}	
Window glass	10^{-12}–10^{-13}	
Mica	10^{-11}–10^{-15}	Insulators
Polythene	10^{-15}–10^{-17}	

* The SI unit of electrical conductivity is the sieman written S. The sieman is $ohm^{-1}\ m^{-1}$.

are the electrons in the former. At normal temperatures the vibrations of the ions in the lattice are sufficiently strong to dislodge some of these electrons leading to an electrical conductivity which lies between that of the metals and the insulators, Table 7.6. Such materials are called semiconductors. Unlike metals whose electrical conductivity decreases slightly with increase in temperature, the electrical conductivity of semiconductors increases sharply with increase in temperature. Because of this strong temperature dependence, semiconductor devices must not be allowed to rise to a higher temperature than their rated value.

Semiconductor behaviour can be explained in terms of the energy gap model shown in Fig. 2.15. As the temperature of the device is increased more electrons have sufficient energy to jump the energy gap and become free to move in an electric field.

7.6.2 Photovoltaic effect

A photon of energy greater than the energy gap can raise an electron from the bound to the free state away from its parent atom. The increase in electrical conductivity of a semiconductor when produced in this way is called the photoconductive effect. If the effect of the incident photons is simply to free the electrons it will only increase the electrical conductivity of the material and the electrons will eventually fall back to the ground state, giving up their energy in the form of heat to the lattice. How then can this energy be extracted in the form of electrical energy in an external load?

It is perhaps helpful to consider a mechanical analogy. If a bucket of water is filled in a pond and then raised to some height above the surface of the pond, the water in the bucket will have potential energy with respect to the pond , in rather the same way as the electron when raised through the energy gap. If now the water is simply poured back into the pond this potential energy will be converted to heat in the pond. If, however, we pour the water into a pipe which contains a water wheel, then the potential energy of the water will be extracted as the water flows back to the pond through the pipe, Fig. 7.47. Thus what we need to achieve in the semiconductor is some way of removing the electron before it falls down through the energy gap and allowing it to return to the semiconductor only after it has lost its extra energy by performing work in an external load. In order to achieve this removal of the electron it is necessary to modify the pure 'intrinsic' semiconductor described in the previous paragraph. In silicon crystals one atom shares four electrons with its four neighbours, one with each neighbour, Fig. 7.48. If one of these electrons is freed, for example by an incident photon, it will move freely through the lattice. In addition the vacancy left by this electron will be filled by an electron from an adjacent atom, in turn leaving a new vacancy. This process will continue with the vacancy moving, if there is an electric field applied to the material, in the opposite direction to the electron which originally filled it. This vacancy behaves rather like a positively charged electron and is called a hole. Thus if an electron in

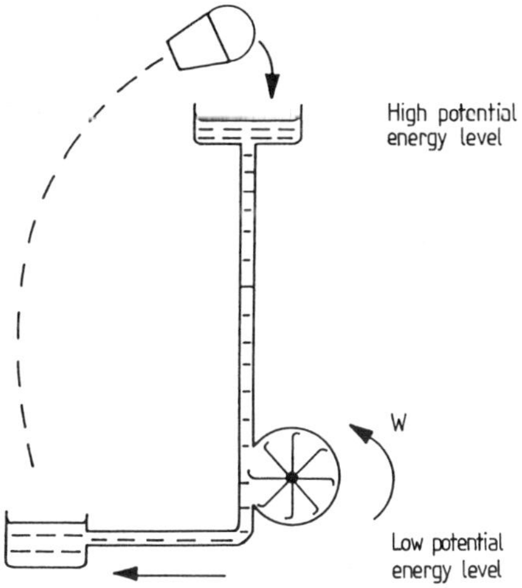

Fig. 7.47 *Mechanical analogy to photovoltaic cell*

silicon is broken away from its bond an electron-hole pair is produced in the material. If a small number of the silicon atoms are replaced by atoms of phosphorous then the phosphorous atom will take up a position in the lattice replacing a silicon atom. It will share four electrons with its neighbouring silicon atoms but in addition will possess an additional electron which will be readily detached and free to move. Such a process is called 'doping' and the material so formed is termed an extrinsic semiconductor. If instead of using phosphorous we had chosen a material such as boron as the doping material, the boron atoms would have taken the place of some of the silicon atoms but in this case the boron atoms can supply only three electrons to their neighbours so that one bond adjacent to each boron atom would not be filled. The effect of this is to create a hole which will move freely in the material and will behave in the same way as a positively charged electron. The two types of extrinsic semiconductor are referred to as *n*-type and *p*-type respectively. If we join an *n*-type and a *p*-type semiconductor to form a *p–n* junction, initially electrons will flow from the *n*-type material to fill up the empty bonds in the *p*-type material. The *n*-type material in the junction region will be left with a net

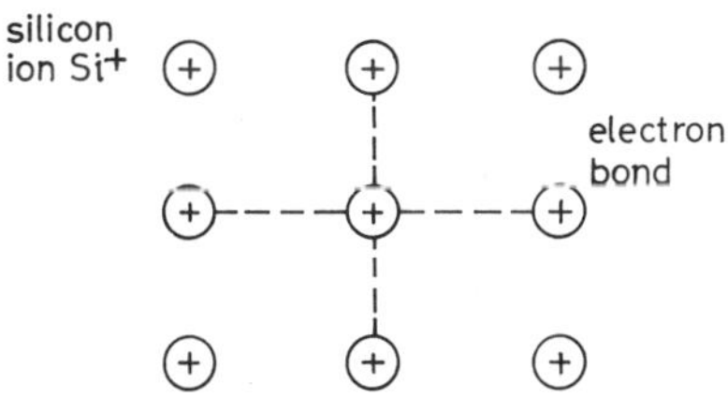

Fig. 7.48 *Silicon crystal, 2-dimensional diagram*

positive charge and no free electrons and the region of the *p*-type material will similarly have a negative charge and no free holes. This is called the depletion layer. The flow of electrons will cease as the retarding potential builds up and a state of dynamic equilibrium will exist. If now the junction region is irradiated with light having a frequency sufficiently high to create electron-hole pairs near the junction, the electrons will be swept to the *n*-type region by the electric field in the junction and similarly the holes will be swept to the *p*-type region. If an external connection is made to the two materials the electrons will pass through it, returning to the *p*-type material to recombine with the holes. This is the photovoltaic generator, or solar cell, and the effect of such a generator is to convert the photon energy into electrical energy. It is important that as far as possible the electron-hole pairs are generated in or near the region of the junction; if not they may recombine in the material, in the same way as if in our mechanical analogy we failed to pour the water down the pipe, but missed and threw it back into the pond.

Silicon photovoltaic cells are usually made by taking a flat wafer of *n*-type silicon and then diffusing boron into the top surface layer to produce a *p*-type material, Fig. 7.49.

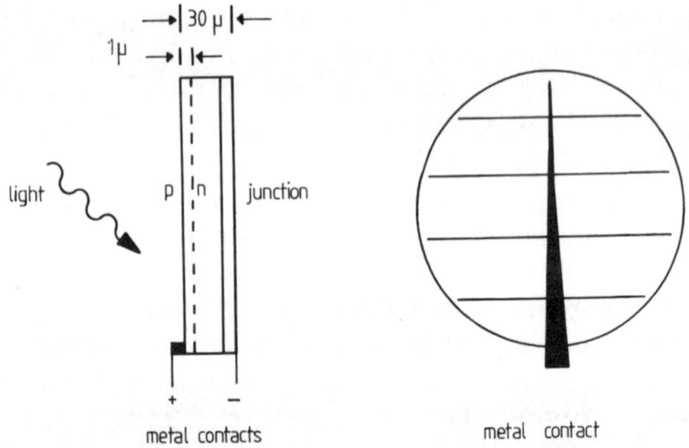

Fig. 7.49 *Silicon photovoltaic cell*

A practical difficulty in the design of solar cells is in the provision of the top electrical contact, since this contact will absorb the incoming light and prevent it reaching the junction. For this reason it is usual to provide a fine grid of conductors which leave as much area exposed to the light as possible. The wires cannot be too widely spaced, however, otherwise surface recombination effects, and also high internal resistance, limit the performance of the cell. The top surface of the cell is treated to reduce reflection losses and the cells are encapsulated behind a transparent window to protect the cell from the weather and from mechanical damage. To date most development work has gone into the single crystal silicon cell and the current state of development is described below. Work is also taking place on other materials and these are also discussed.

7.6.3 Characteristics of photovoltaic generators (solar cells)
The number of hole pairs generated in a photovoltaic generator will depend on the intensity of radiation falling on the cell. For this reason it is usual to specify the performance of a cell by giving the peak power output when irradiated by a standard of radiation, the standard 'sun', which is defined as 1 kW/m² at 1·5 air mass, and the cell is maintained at a temperature of 25°C. The output of the cell under these conditions is called the peak watt output. It is important to note that the radiation in the tropics, average over the day, is about 250 W/m² or about 1/4 of the standard sun. This reduction is more in higher latitudes, hence it is necessary to install at least four times the mean power required in terms of the peak power rating. A further increase in installed capacity of around 10% will be required to take account of higher operating temperature and reduction in performance owing to dust etc.

Typical current voltage characteristics for a commercially available cell are shown in Fig. 7.50. Commercial solar cells are made in the form of discs of

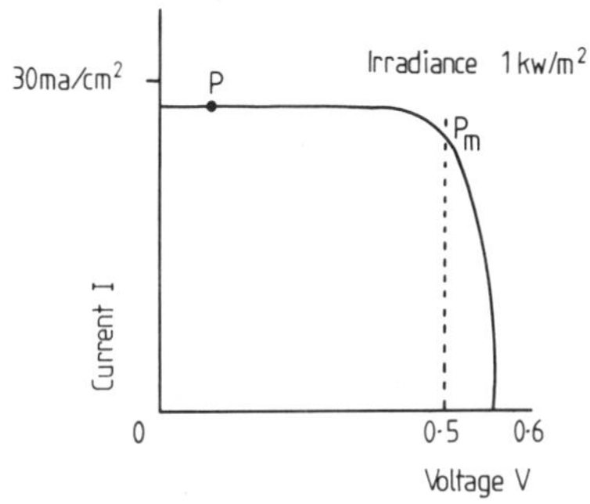

Fig. 7.50 *Current–voltage characteristic of a photovoltaic cell*

about 10 cm (4 inches) diameter and connected electrically to similar cells in a series/parallel arrangement to give the required voltage and current. The operating point P can be chosen to give maximum power output at a given intensity of radiation. The peak power is somewhat lower than the product I_{sc} V_{oc} owing to the curvature of the current voltage characteristics. At maximum power the conversion efficiency will be around 12%. Fig. 7.50 refers to standard irradiance of 1000 W/m² temperature of 25°C. If we assume that a load of variable resistance is connected across the cell output terminals then when the load is zero, that is when the cell is short circuited, it will deliver a current of rather less than 30 mA for each square centimetre of active cell area. As the load resistance is increased then the current will stay constant and the output voltage will rise. The operating point P will move along the horizontal portion of the characteristic as shown in Fig. 7.50. As the operating point moves, the power delivered to the load (Volts × current) will increase until it reaches a maximum value at point P_m. further increase in load resistance will result in only a small increase on voltage but a sharp fall in current and the power output will fall. Eventually the current will fall to zero as the open circuit condition is reached and the open-circuit voltage will have increased to around $V_{oc} = 0{\cdot}6$ V, Fig. 7.51. At the maximum power output condition P_m for

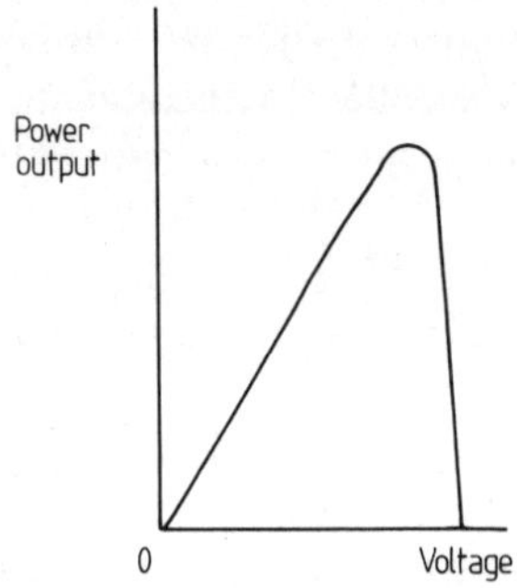

Fig. 7.51 *Variation of power output with choice of operating point*

the cell characteristic shown in Fig. 7.51 the voltage will be 0·48 V, the power output per unit area 12·7 mW/cm² and the efficiency

$$= \frac{(12{\cdot}7 \times 100)}{100}$$

$$= 12{\cdot}7\%$$

Since for most of its operating time the cell will be working at values of irradiance much less than the rated value, it is important to see what performance can be expected at lower irradiance levels, Fig. 7.52. Fig. 7.53 shows, as might be expected, that the short circuit current increases more or less linearly with level of irradiance. The open-circuit voltage is less sensitive to irradience and increases only logarithmically with irradiance. The irradiance can be increased above that level for one sun by means of concentrators. It is found that, under these conditions, the short-circuit current continues to increase linearly to very high values of intensity (500–1000 suns). In order to take advantage of this it is necessary to cool the cell to maintain normal operating temperature and also to design the cell to have a low value of the series resistance of the top contact.

The dotted lines in Fig. 7.52 refer to constant power output and the cell will deliver this power level at operating points where the power curves intersect with the current voltage characteristic of the cell.

It is of interest to see how the maximum power will vary with irradiance. Since the current is proportional to irradiance and the voltage remains almost

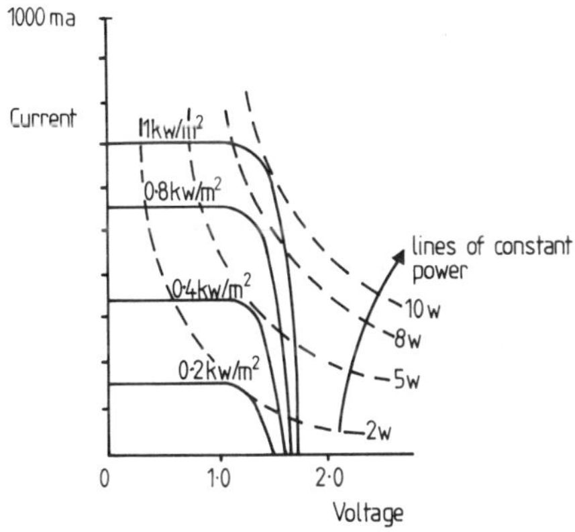

Fig. 7.52 *Current–voltage characteristics of typical photovoltaic module for different values of irradiance*

constant, the power output will be approximately proportional to irradiance. The efficiency will remain approximately constant, Fig. 7.53*b*. This will, of course, only apply provided that the load is continuously adjusted to ensure that the operating point remains at the optimum position. This will not be the case if the load is of constant resistance, and Fig. 7.54 shows how the operating point will move with change in irradiance for this case. Fig. 7.55 shows how the load must be continuously matched to the changing characteristic in order to give maximum power output.

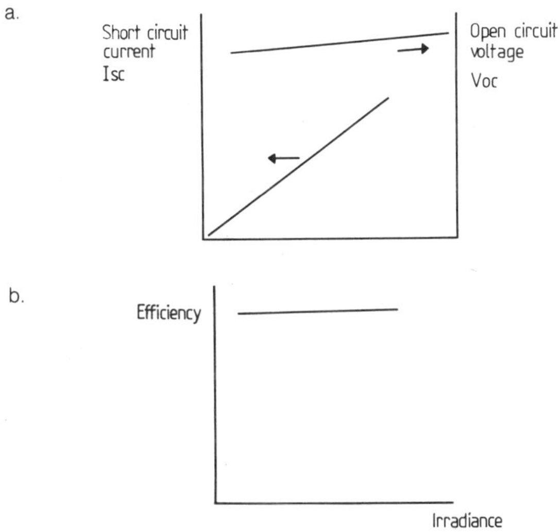

Fig. 7.53 *a* Short circuit current I_{sc} and open circuit voltage V_{oc} versus irradiance
b Efficiency versus irradiance

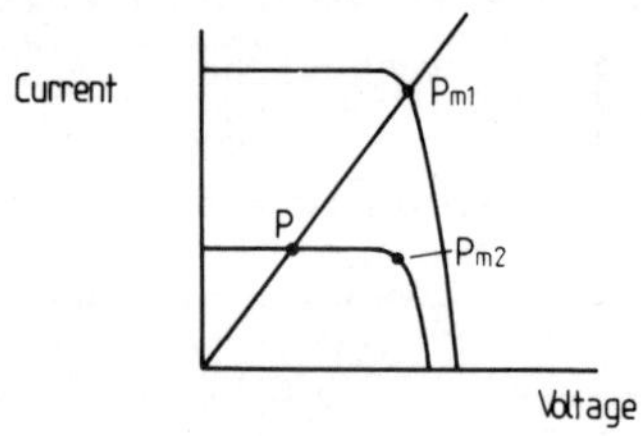

Fig. 7.54 *Load line for resistive load*

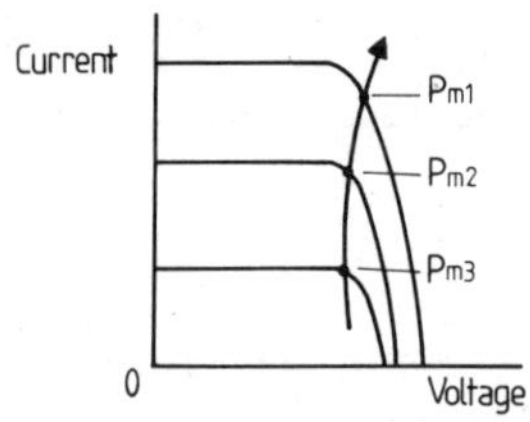

Fig. 7.55 *Load line for matched load as irradiance is varied*

The effect of small temperature changes on cell performance are shown in Fig. 7.56. it is seen that the short-circuit current rises and the open-circuit voltage falls as the temperature is raised. The overall effect is to reduce the maximum power output and the corresponding efficiency by about 0.5% for each degree rise in temperature. The variation of I_{sc} and V_{oc} with temperature is shown in Fig. 7.57.

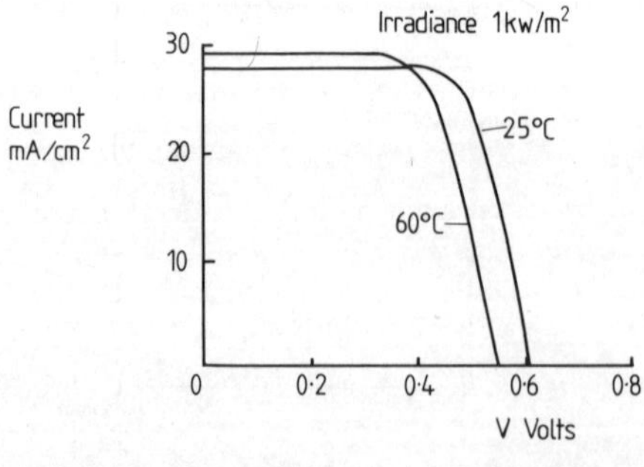

Fig. 7.56 *Effect of temperature on voltage–current characteristic*

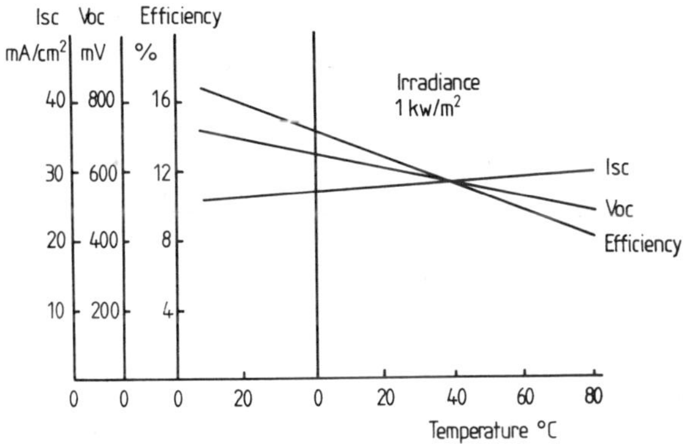

Fig. 7.57 *Temperature dependence of efficiency, open-circuit voltage and short-circuit current*

7.6.4 Developments in solar photovoltaic cells

Silicon cells have now been developed to a high standard of reliability. They have been used extensively as power sources for space vehicles and are beginning to find terrestrial applications. The goals for development are:

1 Higher conversion efficiency
2 Reduced cost

The highest efficiency reported[49] is 19% and the upper theoretical limit is about 25%. The achievement of reduced cost is the major objective in silicon photovoltaic cell development.

There is also some concern at the high energy costs of the present generation of single crystal silicon cells which are so energy intensive that they may require several years of operation before they produce as much energy as was required for their manufacture. This is not important in special small scale applications but would be of great importance if solar arrays were to be used for large scale power production. Currently there is considerable development on new routes for the manufacture of silicon cells including the use of polycrystalline material. Preliminary work on very thin (2 μm) films of amorphous silicon has shown promise of greatly reduced costs, the efficiency to date of practical sized cells has not exceeded 5–6%. Such cells could significantly reduce the cost/watt of photoelectric conversion.

Cadmium sulphide cells have also been developed and the maximum conversion efficiency achieved has been 9% and the average efficiency of 6%, there are, however, problems in achieving long life.

The cell that has received most development attention, apart from silicon, is gallium arsenide. The cost of these cells is very high but efficiencies as high as 25% have been reported, and since the cell can be operated at high concentration ratios (500 to 2000) the cost of the cell per watt can be reduced considerably. This reduction must, however, be set against the increased complexity of the tracking support structure.

Schottky barrier cells represent an alternative to the semiconductor *p–n* junction cells described previously. These cells consist of a semiconductor–metal junction. It has been found necessary to interpose a thin oxide layer between the metal and the semiconductor and for this reason Schottky cells are sometimes referred to as MIS or metal–insulator–semiconductor cells. Preliminary work with single silicon crystal as the semiconductor has yielded an efficiency of 12% and somewhat higher efficiencies have been measured with gallium arsenide.

7.6.5 Costs and cost projection

Currently (1985) the cost per module of the silicon cell is around $8 per peak watt. The support structure and storage adds a similar amount to give an overall figure of about $16 per peak watt. Experts seem to agree that there is little likelihood of the cost of single crystal silicon falling below $2 or $3 per peak watt. Further reduction will require the success of one of the newer developments.

7.6.6 Electrical storage

Most applications of photovoltaic cells will required some form of storage. The usual method is to make use of storage batteries. Either lead–acid or cadmium nickel can be used but the charging characteristics of the former are more suited for use with solar cells.

7.6.7 Applications

Solar cells are claimed to have long life, need little or no maintenance, and have high reliability. While this could well be true, further field trials under various tropical conditions are required to substantiate this claim. Because of the high cost only small scale applications of special character, where cost is not of prime importance, are likely at present. Such applications include:

1 Communications equipment
2 Navigational aids
3 Educational television and radio
4 Domestic lighting

There are also a number of field trials at the 1 kW level (peak) of water pumping equipment, Fig. 7.58 a,b.

An unusual and controversial proposal for the large scale generation of electrical power from solar energy was made by Dr Peter Glaser of Arthur D. Little Inc. in 1968.[50,51] It was proposed that large arrays of solar cells should be placed in synchronous orbits around the earth (22 600 miles) and the electrical energy generated should be converted to microwave frequencies (3×10^9 Hz) and beamed down to receivers on the earth's surface where the frequency would be changed to power frequency (60 Hz) and fed into the electrical power

a.

b.

Fig. 7.58 *Photovoltaic Water Pump*
a Babouge. Mali. (Courtesy P. Fraenkel)
b Port Moresby. Papua New Guinea

network, Fig. 7.59. Since this first proposal a number of design studies have been carried out. One of these[52] considers a 5000 MW power station having a satellite of area of 75 km^2 and a mass of 35×10^6 kg. The terrestrial receiver

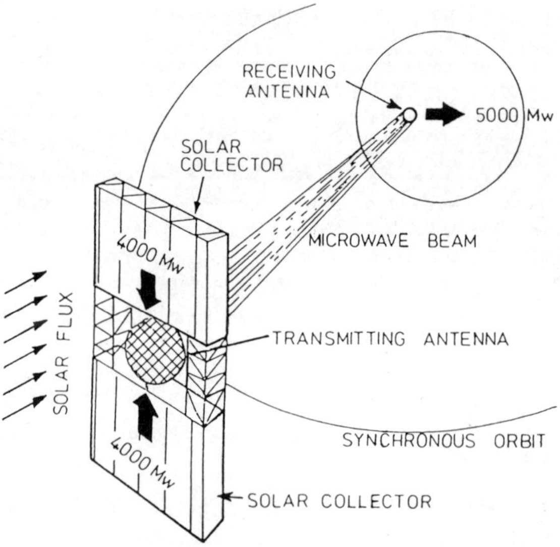

Fig. 7.59 *Schematic for a satellite solar power station*

would have an area of 13 km × 10 km and the power density at the earth's surface is 23 mW/cm^2 (about one-quarter of the maximum solar radiation level). An estimate for the overall efficiency of such a system gives a figure of 7%.[53] Since the station is situated in space it will receive the full extra terrestrial power of 1.4 kW/m^2. Since it will generate electricity continuously, storage will not be required. Twice, yearly, at the equinoxes, the earth's shadow will shield the satellite from the sun's radiation for a period of about one hour only.

A number of problems need to be considered, and these include cost, the problem of installation and maintenance and possible safety risks from the levels of microwave radiation in the beam. It has been estimated that the considerable energy costs of the project would be recovered within four years.[54] Other estimates range from 1 to 20 years.

Less spectacular, but probably more technically realistic and economically viable, are terrestrial-based power plant in the megawatt and multimegawatt range.[55] The largest prototype plant of this type to date has been built by ARCO Solar at Carissa Plain in California. The electrical output is 5 MW and is generated by photovoltaic modules that track the sun in two axes and incorporate mirror boosters.

References

1 BENTLEY, R. W.: Private communication
2 DARLEKAR, B. V. and DESMOND, R. M.: 'Engineering heat transfer' (West Pub. Co., St. Paul, 1977)
3 MCADAMS, W. H.: 'Heat transmission' 2nd edn (McGraw-Hill, New York, 1954)
4 WHILLIER, A.: Design factors influencing solar collector performance. Low temperature applications of solar energy' (A-SHRAE, New York, 1967) Chapter 3, pp. 27–40
5 TABOR, H.: 'Selective surfaces for solar collectors. Low temperature applications of solar energy' (ASHRAE, New York, 1967)
6 BAKER, E., FLORA, C. T., GASTELOW, J. P. and MCCAFFREY, J.: 'Solar heating and cooling systems design for Australian conditions' (Pergamon Press, Sydney, 1984)
7 WATSON-MUNRO, C. N. and HORWITZ, C. M.: 'Selective surfaces solar energy' *in* MESSEL, H. and BUTLER, S. T. (eds) (Pergamon, Oxford, 1975)
8 SERAPHIN, B. O.: 'Solar energy conversion solid state physics aspects' (Springer-Verlag, Berlin, 1978) Chapter 2
9 'Solar cooker manual' (Brace Research Institute, McGill University, Canada. Aug. 1982)
10 FANG, P. H. and SWANNICK-NUNN, S.: 'Chinese solar cookers', *Appropriate Technology*, **5**, 4
11 EXELL, R. B.: 'Basic design for a simple solar rice dryer', *Renewable Energy Review Journal*, 1980, **1**, 2
12 SPIEGLER, K. S. (ed): 'Principles of desalination' (Academic Press, New York, 1966)
13 HOWE, E. D.: 'Fundamentals of water desalination' (Dekker M., New York, 1974)
14 HEADLEY, O. and SPRINGER, B. G. F.: 'A natural convection solar crop drier', Proc. UNESCO Congress, The Sun in the Service of Mankind, Paris, July 1973
15 FRICK, G. and HIRSCHMANN, J.: 'Theory and experience with solar stills in Chile', *Solar Energy*, 1973, **14**, 4 pp. 405–413
16 DUNKLE, R. V.: 'Solar water distillation. The roof type still and a multiple effect diffusion still', *ASME Quarterly Trans.*, 1961, pp. 859–902
17 KALECSCINSKY, A. V.: 'Ungarische Warme und Heise Kocksalzeen', *Ann. D. Physik*, 1902, (4) **7**, 4, pp. 408–417
18 TABOR, H.: 'Solar ponds (non-convecting)' *in* DIXON, A. E. and LESLIE, J. D. (eds) 'Solar energy conversion' (Pergamon, Toronto, 1979)
19 MOUCHET, A.: 'La Chaleur Solaire' (Gautheir-Villiers, Paris, 1879)
20 PIFRE, A.: 'Nouveaux Resultats d'Utilization de la Chaleur Solaire Obtenus a Paris', *Comptes Rendus de l'Academie des Sciences*, 1880, **91**
21 ERICSSON, J.: 'The sun motor', *Nature*, 1884, **29**, Jan 3 and 1888, **38**, August 2
22 ADAM, S. W. *Scientific American* 38–39 376. 1878.
23 HALLY, G.: 'Sunpower—its commercial utilization', *Inst. of Engineers and Shipbuilders in Scotland*, 1914, **57**, April 21
24 TELLIER, C.: 'The utilization of solar heat for elevation of water', *Scientific American*, 1885, **52**, Oct. 3
25 ABBOT, C. G.: 'The sun and the welfare of man' (Smithsonian Institute, New York, 1929)
26 JENNES, J. R.: 'Some considerations relative to a solar powered Savery water pump', *Solar Energy*, 1961, **5**, pp. 58–60
27 PAYNE, P. R.: 'A new steam engine cycle', Proceedings ICE CE Conference, San Francisco, 1974
28 PAYNE, P. R., GREENWOOD, S. W. and BROWN, R. G.: 'Recent developments with the water pulse jet', Proceedings, 13 Inter Society Energy Conversion Engineering Conference, San Diego, August 20–25 1978
29 MCHUGH, C. J.: 'Power propelled boat', US Pat. No 1200960, Oct 1916
30 MCCRACKEN, C. D.: 'The solar powered thermopump', Conference on Solar Energy, Tucson, Arizona, 1955

31 'Specification manual for solar pump' (Solar Pump Corporation, Nevada, 1979)
32 GIRARDIER, J. P. and MASSON, H.: 'Solar motors with flat plate collectors', *Solar Energy*, 1966, **10**, 4
33 GIRARDIER, J. P. and ALEXANDROFF, M.: 'Les Moteurs Solaires et l'habitat pour les sones arides. Realisations actuelles et perspectives', ISES Conference Paper E82, Paris, 1973
34 TABOR, H. and BRANICKI, L.: 'Small turbine for solar energy power package', Paper 35, UN Conference on new sources of energy, Rome, 1961
35 TALBERT, S. G., FISCHER, R. D., ALEXANDER, G., FRIELING, D. H. and EBBING, T. A.: 'The development of a 37 kW solar powered irrigation system', Proceedings ISES Conference New Delhi, 1978
36 BRANDVOLT, G. E.: 'Solar irrigation program status', Proceedings ISES Conference 1930 1 1934, New Delhi, 1978
37 HILDEBRANDT, A F., HAAS, G. M., JENKINS, W. R. and COLACO, J. P.: 'Large scale concentration and conversion of solar energy', *Trans. American Geo Phys. Union*, 1972, **53**, 7, 684–692
38 VANT-HULL, LL.: 'The solar tower as a source of thermal energy', Proceedings Solar Conference, Toulouse, March 1976
39 TROMBE, F. and LE PHAT VINH, A.: 'Thousand kW solar furnace, built by the National Center of Scientific Research in Odoillo (France)', *Solar Energy*, 1973, **15**, 57
40 TROMBE, F.: 'Solar furnaces and their application', *Solar Energy*, 1957, **1** 2, pp. 9–15
41 STEPHENS, W. H.: 'The European Community 1 MW(e)', Proceedings of the Conference on Solar Thermal Power Generation, UK, ISIS, July 1978
42 KANEFF, S.: 'The White Cliffs Solar Power Station', Proceedings Solar World Progress, Perth WA, August 1983
43 EIBLING, J. A. and TRAYSER, D. A.: 'A 50W portable generator employing a solar powered Stirling engine', *Solar Energy*, 1967, **11**, 324
44 FARBER, E. and PRESCOTT, F. L.: 'Closed cycle, solar, hot air engines. Part 1: A ¼ h.p. engine', *Solar Energy*, 1965, **9**, 4, pp. 170–174
46 GROSSMAN, G. and HOHANNSEN, A.: 'Solar cooling and air conditioning', *Prog. Energy Combust. Scir.*, 1981, **7**, pp. 185–228
47 FARBER, E. A.: 'Design and performance of a compact solar refrigerator system', *Engineering Programme at the University of Florida*, 1970, **24**, 2, pp. 17–38
48 EXELL, R. H. B., KORNSAKOO, S., JARIMAPAS, B. and SOPANRANNAVIT, S.: 'Experiments of a prototype solar powered refrigerator for remote areas', Paper 0371, International Solar Congress, New Delhi, 1978
49 GREEN, M. A., BLAKERS, A. W., SHI, Jiqun, KELLER, E. M. and WENHAM, S. R.: 19.1% efficient silicon solar cells', *Applied Physics Letters*, June 1984
50 GLASER, P.: 'Power from the sun. Its future', *Science*, 1968, **162**, 3856
51 GLASER., P.: 'The feasibility of a satellite solar power station', *Mechanical Engineering*, 1973, **95**, 4, 64
52 HANLEY, G. M.: 'Evolution of satellite power system concepts', Proceedings of the 13th Intersociety Energy Conversion Conference 20–25 Aug, San Diego, 1978
53 DENMAN, O. S.: 'From Sunlight in space to 60 Hz on earth. The losses on the way', *ibid.*
54 WOLF, M.: 'Cost goals for silicon solar arrays for large scale terrestial applications', Ninth EEE Photovoltaic Specialists Conference, Silver Spring, Maryland, 1972

Chapter 8

Biomass conversion

8.1 Introduction

Biomass is important in world energy terms: in Chapter 2 estimates were quoted which suggest that it may provide as much as 15% of the present world energy supply. Most of this biomass is used in the developing countries, where, in many cases, it is the major component of the national energy supply. Some general characteristics are listed in Table 8.1. The simplest way in which to use biomass as an energy source is to burn it to provide heat, for example for cooking and comfort. This traditional method still accounts for the greater fraction of the material used today. We should not overlook the role of biomass in the form of food which provides human and animal muscle power. Muscle power is an important source of mechanical power in the rural areas of developing countries. In addition to its use as food and by direct combustion, biomass has other energy uses which require some preprocessing. These

Table 8.1 *Biomass characteristics*

Advantages	Disadvantages
Widely available	Low conversion efficiency giving low yield per hectare
Technology of production and conversion understood	Competing demands for land use
Suitable for small or large scale	Water content high
Needs only low light intensity	
Needs only low temperatures (5–35°C)	
Incorporates storage	
No pollution	

Table 8.2 *Biomass conversion for energy use*

Process	Energy form
Food	Muscle power (human, animal)
Combustion	Heat
Pyrolysis	Oil, char, gas
Gasification	Gas
Fermentation	Gas (methane) Liquid (ethanol)

methods of conversion are summarised in Table 8.2. One of the oldest methods of conversion is that of charcoal production. Charcoal is widely used as a fuel for cooking. Pyrolysis or destructive distillation is the process by which wood is heated in the absence of air, the gases and vapours which are given off are collected and can be used as fuels or chemicals. Charcoal remains as a solid residue. Hence pyrolysis is a more efficient method of conversion than the simple charcoal plant. If some air is allowed to enter the convertor, partial combustion occurs and the solids and liquids can be converted to gaseous fuels. The process is known as gasification. Fermentation in the absence of air (a process known as anaerobic) can be used to produce alcohol or methane (biogas*). Biomass can be burnt either with or without pre-drying, but in the pyrolytic process water must be driven off either before or during the process. Fermentation processes are 'wet' processes. In the case of alcohol production a dilute mixture of alcohol in water (approximately 9% alcohol by volume) is formed and the alcohol must be separated from the water by a distillation process.

Biogas appears as a gas so that no separation process is necessary. Biogas, which is a mixture mainly of methane (~60%) and carbon dioxide (~40%) by volume appears in the form of a gas which can be burnt directly. For some applications it may be necessary to remove the carbon dioxide by a process known as 'scrubbing'. Hydrogenation of biomass to give liquid fuels is under investigation in a number of countries but will not be discussed here since it is unlikely to be developed on a small scale in the foreseeable future.

8.2 Energy from biomass

8.2.1 Basic chemistry of the reactions

In the photosynthetic process energy from the sun is used to break down the stable molecules carbon dioxide and water, to remove some of their oxygen

* Biogas is the name given to the mixture of methane and carbon dioxide from the fermentation 'digester'.

and build up complex organic molecules. The result is the various forms of biomass, all of which are largely made up from cellulose and similar molecules. The overall chemical reaction for the formation of cellulose* can be written:

$$6CO_2 + 5H_2O = C_6H_{10}O_5 + 6O_2$$

The biomass produced can then be stored until the energy is required. In order to release energy from the biomass the chemical reaction must be reversed by providing oxygen in an environment such as a fire which will allow the reaction to proceed at a suitable rate. Hence:

$$C_6H_{10}O_5 + 6O_2 = 6CO_2 + 5H_2O + \text{energy}$$

For each kilogram of cellulose oxidised, 17·6 MJ of energy is released. In practical terms the energy can be released from the biomass as follows:

1 eat it and make use of the resulting muscle power
2 burn it

or convert to a secondary fuel which can be

3 solid, e.g. charcoal
4 liquid, e.g. by fermentation to alcohol or by simple extraction by mechanical pressing as for coconut oil
5 gas, e.g. by digestion to form methane.

All the conversion routes attempt to retain the basic constituents carbon and hydrogen in the original biomass but to arrange them into a form more suitable for storage, transport or use. Inevitably energy will be required to cause this conversion, there will also be losses, thus the overall energy available from the resulting fuel will be less than that of the original biomass. In conversion by pyrolysis the biomass is broken down into its various components, solids, tars, liquids and gases by heating in the absence of air and the components separated and collected.

8.2.2 Combustion

Organic matter is a somewhat complex material, however, for the purpose of considering the energy stored it can be regarded as a mixture of carbon, hydrogen and oxygen. For example, wood has the composition by weight of around 50% carbon, 43% oxygen and 6% hydrogen, and the remainder is mainly inorganic ash. The simplest way to release the energy stored is to combust the material using oxygen in the air. It is useful to work out the

* Cellulose is a member of the chemical family carbohydrates. As the name suggests, carbohydrates are made up from carbon, hydrogen and oxygen and have the general form $C_x(H_2O)_y$. Carbohydrates are divided into three groups:
Monosaccharides, containing six carbon atoms, e.g. glucose $C_6H_{12}O_6$.
Disaccharrides, made up of two monosaccharides, e.g. sucrose $C_{12}H_{22}O_{11}$.
Polysaccharides, made up of large numbers (500 or 1000) mono- or disaccharides linked together. Cellulose and starch are examples of polysaccharides and have the general formula: $(C_6H_{10}O_5)_n$ where n is the number of units in the chain.

volume of air theoretically required to burn a 1 kg of wood. The wood will contain 500 g of carbon which will burn according to the reaction:

$$C + O_2 = CO_2$$

This means that since the mass of one carbon atom is 12 atomic units and that of an oxygen molecule 32 units a total mass of oxygen equal to

$$500 \times \frac{32}{12} = 1333 \text{ g}$$

will be needed. Similarly, in order to combust the 60 g of hydrogen by the reaction

$$H_2 + \frac{1}{2}O_2 = H_2O$$

a further amount of oxygen equal to

$$60 \times \frac{16}{2} = 480 \text{ g}$$

will be required. However, there is already 430 g of oxygen in the wood which has to be subtracted from this total to give a net oxygen requirement of

$$1333 + 480 - 430 = 1383 \text{ g}$$

Now the composition of air by weight is 23·3% oxygen and 76·7% nitrogen so that the total mass of air required to combust 1 kg of wood is 6·0 kg. Since the density of air at 27°C is 1·18 kg/m³ this mass is equivalent to a volume of air of about 5 m³. In a practical woodburning stove, rather more, say twice, this quantity would be needed because of poor mixing between the inflammable gases given off by the wood and the air. The combustion of wood is a very complicated process indeed – as wood burns it decomposes into charcoal and a complex mix of tars, resins, vapours and gases. It is these latter, volatiles, which contribute most of the energy released. The rate of release of the volatiles depends on the size of the wood and temperature of the fire. It is factors such as these which make the technical design of wood stoves so difficult, difficulties further compounded by social and economic restrictions.

8.2.3 Fermentation to alcohol

The chemical equation for the feedstock, glucose, can be written $C_6H_{12}O_6$. Hence the total relative weight of the glucose molecules will be the sum of the weights of 6 carbon atoms, 12 hydrogen atoms and 6 oxygen atoms or a total of 180. The fermentation reaction will convert the glucose into two molecules of ethyl alcohol,* each of molecular weight 46. The overall reaction is given by:

$$C_6H_{12}O_6 \rightarrow 2C_2H_5OH + 2CO_2$$

* Ethyl alcohol C_2H_5OH is also known as ethanol.

The ethyl alcohol will remain in solution but the carbon dioxide will be released in the form of a gas.

The reaction is slightly exothermic, that is heat is released; however, the amount of heat is small and most of the energy stored in the glucose is converted into stored energy in the ethanol. The energy released when one gram molecule of glucose is burnt (the gram molecule is the molecular weight in grams or 180 g) is 2·82 MJ (15·6 MJ/kg). Similarly, the energy released from two gram molecules of ethanol or 92 g, is 2 × 1·37 MJ or 2·74MJ. The latter figure is equivalent to 21 MJ/l.

To summarise, this reaction predicts that 1 kg of glucose will yield 0·51 kg of ethanol, or 0·64 l, or an energy equivalent of 13·6 MJ.

8.2.4 Anaerobic fermentation

Biogas is produced by the digestion of vegetation or dung. Dung, like vegetation, is largely cellulose. The process of digestion, although it goes through complex stages, can be represented by the overall reaction

$$C_6H_{10}O_5 + H_2O = 3CH_4 + 3CO_2$$

Substituting the atomic weights as before it can be seen that 1 kg of cellulose should yield, theoretically, 0·29 kg of methane or 415 l at STP. From Table 8.3 it is seen that this is equivalent to 13·6 MJ or about 1/3 kg of petrol. Since, in practice, the process will not be completely efficient it is found that dry dung, when converted to methane, will yield the equivalent of around 1/10 to 1/5 of its weight in petrol.

8.2.5 Discussion

In considering the energy potential of biomass it is helpful to keep these simple chemical relationships in mind. In the case of charcoal, combustion is simply the oxidisation of the carbon, whereas with methane, in addition to carbon, hydrogen is also present. In the case of wood and alcohol some of the combustible material can be regarded as already oxidised by the oxygen contained in the material and the energy released on combustion will be due to the oxidation of the remaining carbon and hydrogen atoms. It was shown that for each kilogram of wood burnt a minimum of 1·4 kg of oxygen is required, and since this is in the form of air this means around 6 kg of air. Thus if air came in solid lumps, like coal, some six times the effort would be required to shovel in the air as is necessary for the wood.

The energy released when a unit mass of material is burnt is usually expressed in terms of calorific value and given in units such as MJ/kg. Biomass contains both hydrogen and carbon and the resulting combustion products are water and carbon dioxide. The water in the exhaust is normally in the form of steam and if this is condensed then about 2·3 MJ/kg of energy will be given up as latent heat. If, however, the steam is released without condensing then this heat is lost. Normally the figures in the literature refer to the former case and

Table 8.3 *Approximate calorific values of various fuels*[4]

	Gross MJ/kg	Net MJ/kg	Gross MJ/m³	Net MJ/m³
Solids				
Anthracite	34	34	45 000	45 000
Peat (30% moisture d.b.)	16	15	15 000	14 000
Charcoal	30	30	11 000	11 000
Wood: (oven dry)	20	19		
(air dry) 15% moisture d.b.	17	16		
30% moisture d.b.	15	14		
Sunflower stalks	20		10 000	
Wheat straw	18			
Beef cattle manure (dry)	14			
Bagasse	21·5			
Rice hulls	16·7			
Liquids				
Gasolene (petrol)	45	42	33 000	31 000
Fuel oil	43·5	41·1	41 300	39 000
Paraffin (kerosene)	46·5	43·5	36 300	33 400
Diesel	45·9	42·9	38 400	36 000
Methanol	22·7	20·0	18 000	15 900
Ethanol	30·1	27·2	23 900	21 600
Vegetable oils:				
Sunflower oil	39·3		36 000	
Cotton seed oil	39			
Soya oil	39·5		36 000	
Safflower oil	39·4		36 000	
Maize oil	39·5		36 000	
Rapeseed oil	39·4		36 000	
Olive oil	39·3		32 000	
Peanut oil	40		32 000	
Gases				
Methane	56	50	37	33
Coal gas	40		20	
Propane	48	44	92	85
Biogas (65% methane by volume)	20		25	
Wood gas	15–20		11	

d.b. refers to dry weight biomass. In practice there may be up to 30% moisture content for air dried biomass. This will reduce the heat released in two ways:
(*a*) The volatiles may not ignite and hence their energy value will be lost.
(*b*) Heat will be required to evaporate the water content of the biomass.

The figures for gases refer to a temperature of 15°C (288 K) and a pressure of 101·3 KPa (one atmosphere).

are known as the gross or higher calorific value. If the latent heat of the steam is deducted the resulting value is called the net or lower calorific value. In practice there will be a further reduction below the HCV since biomass will have a moisture content which will require heat to evaporate off water and a further reduction in useful energy release is necessary. The figures given in Table 8.3 are for oven dry material unless otherwise stated. Fig. 8.1 indicates the energy yield from wood having various moisture contents.

The basic data for this section is summarised in Tables 8.3 and 8.4. Table 8.3 lists the calorific values for various fuels and Table 8.4 summarises the basic chemical relations.

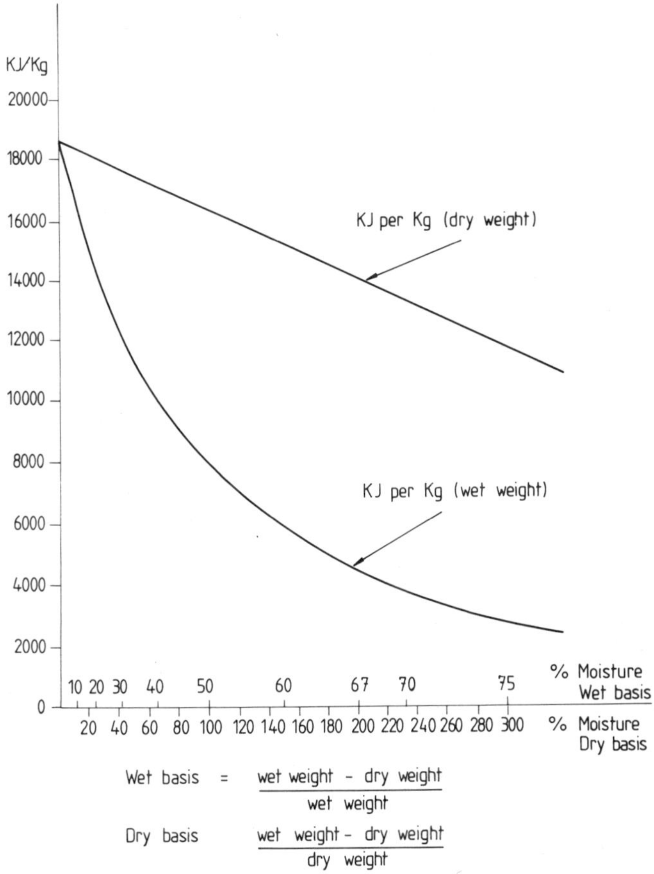

Fig. 8.1 *Calorific value of wood (based in Reference 1)*

$$\text{Wet basis} = \frac{\text{wet weight–dry weight}}{\text{wet weight}}$$

$$\text{Dry basis} = \frac{\text{wet weight–dry weight}}{\text{dry weight}}$$

Table 8.4

Element	Chemical symbol	Atomic weight		Oxidation (combustion)	Reaction energy MJ/kg
Hydrogen	H	1		$H_2 + \frac{1}{2}O_2 = H_2O$	142
Carbon	C	12		$C + O_2 = CO_2$	32·7
Oxygen	O	16		—	—
Compound		**molecular weight**	**% composition by weight**		
Methane	CH_4	16	C 75 H 25	$CH_4 + 2O_2 = CO_2 + 2H_2O$	56
Glucose (sugar)	$C_6H_{12}O_6$	180	C 40 H 6·6 O 53·3	$C_6H_{12}O_6 + 6O_2 = 6CO_2 + 6H_2O$	15·6
Cellulose	$C_6H_{10}O_5$	chain molecule	C 44·4 H 6·2 O 49·4	$C_6H_{10}O_5 + 6O_2 = 6CO_2 + 5H_2O$	17·6
Wood		mixture of cellulose lignins ash	C 50 H 6 O 43		18–20

8.3 Muscle power

Muscle power is still an important source of mechanical energy in the rural areas of developing countries, see Fig. 8.2. The average useful power output from a man working in an active occupation is around 1/20 hp averaged over an 8 hour day. It is, of course, possible to work at a much higher power output for short periods of time, for example, in running upstairs about 3/4 hp is required, but this level of power could not be sustained for any length of time. 1/20 hp is a very small amount of power; on this basis if a dynamo were used to generate electricity to power a 100 W electricity bulb three men would be required to turn the dynamo (assuming an efficient dynamo). It is interesting to estimate the efficiency of human muscle power. If we assume 1/20 hp for an 8 hour day and 250 days/year this is equivalent to 100 hp h/year or 270 MJ/year. If the

a.

a Man powered irrigation (S. Vietnam)

b.

b Firewood Collecting (Thailand. Courtesy Wasan Jompakdee)

c.

c Ox-cart carrying firewood (Thailand. Courtesy Wasan Jompakdee)

d.

d Water buffalo ploughing (Thailand)

Fig. 8.2 *Muscle Power*

average daily food consumption is 2000 kcal or 3051 MJ/year the efficiency of conversion of food to mechanical power is 7%. This of course is the net efficiency, and is what remains after deducting the energy required for other activities and for internal processes in the body. If we assume that in a farming community only half the people do useful work (that is, we add in the food requirements of children and the elderly) then the efficiency figure falls to 3·5%.

In any population individuals differ greatly in the extent of their physical activity. The food requirement will vary with weight, age, sex, climate (temperature) and occupation. The FAO[3] have produced tables which enable the average food requirements to be predicted. As an example consider a Thai rice farmer of 25 years of age and 50 kg weight. The tables predict a daily food energy requirement of 11·3 MJ made up as follows:

Sleep	8 hours	1·6 MJ
Work	8 hours	6·1 MJ
Non-occupational activities	8 hours	3·6 MJ

The total food requirement for a given animal can be divided into two components. The first is the energy required for the metabolism of the animal at rest. This depends on mass and is shown in line A in Fig. 8·3 and is produced as heat. Further energy is needed to carry out external work. Line B shows the

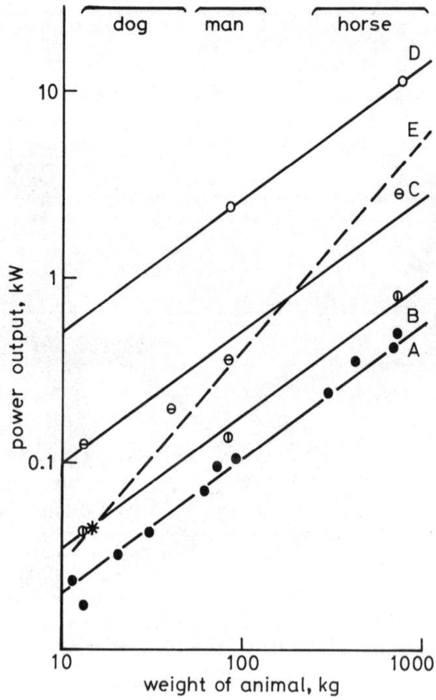

Fig. 8.3 *Power output of animals as a function of their weight (Ref. 2)*

Courtesy Pergamon Press

power which can be maintained all day, line C the maximum steady power (5–30 minutes) and D a brief effort of a few seconds.

Animal power is still widely used in developing countries. Table 8.5 shows an estimate of the number of draught animals in India. If we assume that each of these 80 million animals is capable of a power output of 1/2 hp this gives a total figure of 40 × 10^6 hp or 30 000 MW. The total installed capacity of the Indian electricity grid is 26 000 MW or rather less than the figure for animal power.

The food requirement of a typical draught animal (e.g. 550 kg buffalo) working 6 hours per day is 70–100 MJ. In some situations animal feed can be

Table 8.5 *Estimate of work animals in India (Reference 5)*

Bullocks	70 million
Buffaloes	8 million
Horses	1 million
Camels	1 million

provided from crop waste and forage and can be regarded as 'free'; if, however, a food crop is required then this can constitute a significant demand on land utilisation.

There are estimated to be 13 million animal drawn carts in India and most of the rural transport is by this means.[5] Over half the farms are less than two hectares and are too small to support mechanical power use. Currently much animal power is used inefficiently – and often the animals are treated with great cruelty. Animals are badly fed and cared for, lack veterinary attention and the equipment they operate is not well designed. Many yokes are particularly unsuited and lead to various problems including a cancerous condition, neck gall, which appreciably shortens the animal's life. The output of oxen can be improved by the improved design of harness and implements.[6]

Human muscle can provide power for light industrial machinery. The lathe drive shown in Fig. 8.4 has been used since the Middle Ages.

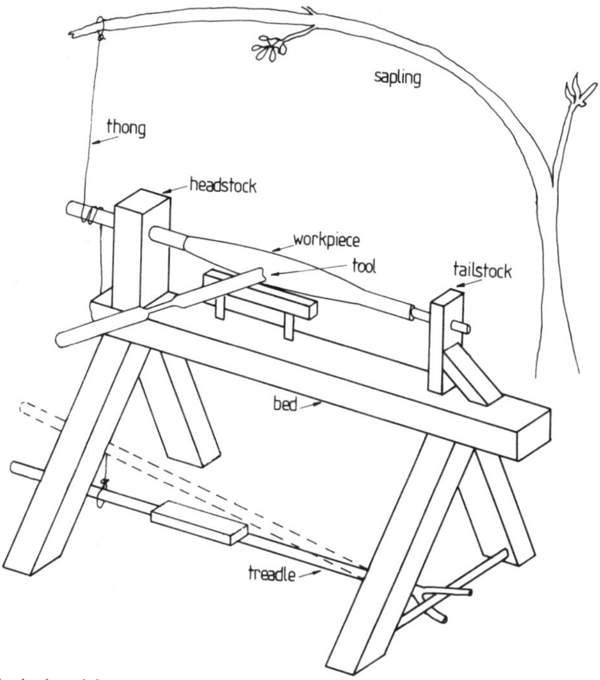

Fig. 8.4 *Early lathe drive*

The bicycle provides a very effective way of matching the leg muscles to the load, Fig. 8.5. Fig. 8.6 shows a plot of various forms of transport and brings out clearly the advantage of the bicycle for this purpose. Papanek refers to a multipurpose transport based on the bicycle and used in the Vietnam conflict and Wilson has published information on a number of similar developments.[7–11]

Fig. 8.5 *Bicycle park, Hangchou, China*

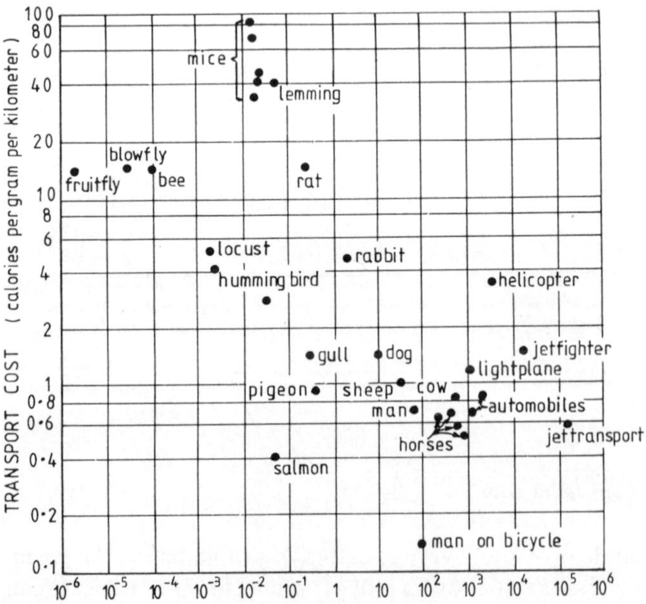

Fig. 8.6 *Relative energy cost of transport (from S. S. Wilson, 'Bicycle technology', Scientific American, March 1973)*

We conclude that muscle power is still an important source of mechanical power in many areas and is likely to remain so for a long time to come. It is often the most suitable method of providing power to enable the introduction of new devices to a region. Efforts should be made to use this source of power more efficiently and in the case of human muscle to look to alternatives in order to remove the ceaseless drudgery to which so many people are condemned for their whole lives.

8.4 Combustion

The direct combustion of biomass is the simplest and earliest method of converting the chemically stored energy into heat. The open wood fire was one of the first significant steps by which man separated himself from the rest of the animal kingdom. It provided warmth, a means of converting food into more readily digestible form and safety from prowling animals. The simple open wood fire is still widely used in the less developed areas but is inefficient in use of fuel. Traditional cooking stoves and ovens have been developed all over the world and have the advantage of requiring less fuel than the open fire, also, by means of a chimney or a flue, they can be used indoors, see Fig. 8.7. More efficient combustion can be achieved by using an enclosed system in which the air supply may be adjusted. Benjamin Franklin is often credited with the invention (1740) of the fuel efficient stove, although this was in fact a reinvention of a system which has been widely used in the Orient for centuries. Franklin developed the 'Pennsylvania Fire Plan' to try to overcome the then serious firewood shortage by arranging for an improved combustion efficiency. In his own words 'By the help of this saving invention our wood may give us fuel as we consume it!' Today, the Franklin stove, widely used in the USA, shows very little resemblance to the original. Nevertheless, Franklin's contribution was to recognise the need to control the air supply. The first really fundamental work was due to Count Rumford,* who carried out an extensive series of measurements. To reduce the fuel requirement of domestic cooking and heating is probably the most important contribution which could be made to the energy problem of developing countries. It is claimed that significant reductions in fuel of factors of two or even higher can be obtained by modifications to the traditional cooker. It is desirable to retain the traditional appearance of the cooker in order to get rapid local acceptance.

With very few excepions it is necessary to cook food in order to make it edible, that is, to be both palatable and digestible. Hence cooking is a universal requirement. The major processes employed are boiling, frying, grilling and baking. In the first two processes a fluid, water and fat, respectively, is used as the heat transfer medium between the container and the food. Grilling

* Benjamin Thompson (1753–1814) was by birth an American who founded the Royal Institution in London (*circa* 1799) and became Count Rumford of Bavaria.

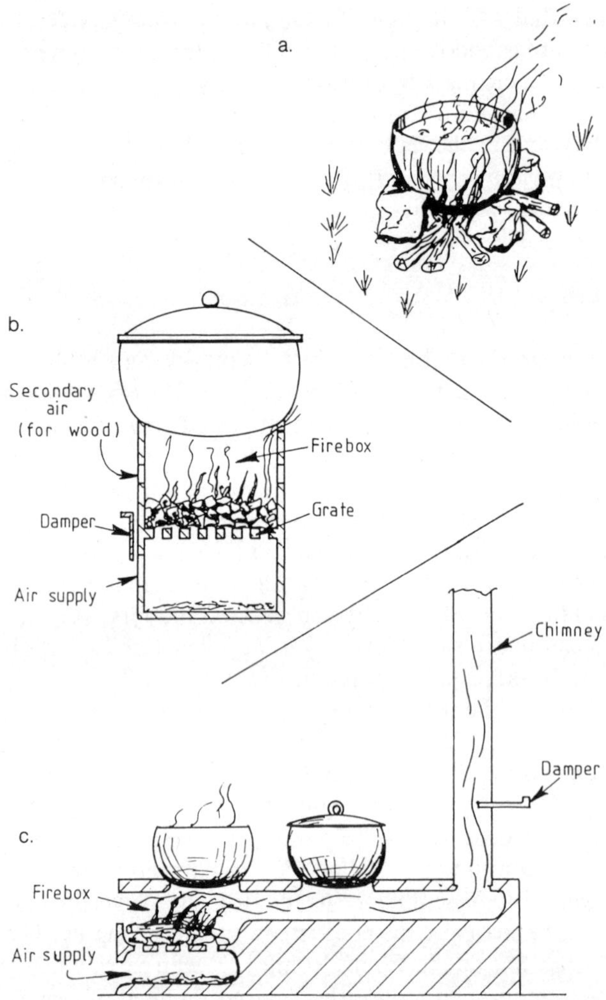

Fig. 8.7 *Traditional cooking stoves*

employs a mixture of direct radiation and gas convection from the fire, and baking uses radiation and gas convection within a container.

Cooking in a pot is very generally employed in the developing countries and most cooker design has concentrated on the improvement of the traditional wood or charcoal fired stoves. The energy required to cause the actual chemical change between cooked and uncooked food is negligibly small and in these terms all cookers are of very low efficiency. Stove 'efficiency' in the literature is usually understood to mean the percentage of the heat released by combustion which enters the cooking pot as useful energy. Different authors assess this 'useful energy' in different ways and quoted efficiencies should be looked at with caution. The technical problems in stove design can be simply stated:

1. To arrange for the complete combustion of the fuel with air in order to release the maximum heat from the fuel.
2. To control the rate of heat release.
3. To transfer as much of this heat as possible to the food container(s).
4. To minimise heat loss from the stove and container(s).

The simplest traditional arrangement is to mount the cooking vessel on three stones, Fig. 8.7*a*. The rate of burning can be varied by moving wood in or out of the combustion volume. Control is difficult and heat losses owing to wind can be high.

A basic traditional stove design which incorporates the most important features is shown in Fig. 8.7*b*. The fuel is burnt on a grate which allows the air to be uniformly distributed throughout the fuel bed. Air flow is controlled by a slide or damper. The container is heated by radiation from the fire surface and convection from the exhaust gases. For this arrangement most heat is transferred by radiation. Loss of steam can be reduced by the use of a lid on the container. If wood is used as a fuel, secondary air is sometimes admitted by means of holes placed above the bed. Fig. 8.7*c* shows the second basic traditional design in which a chimney is incorporated. This enables the cooker to be situated in the house. In this design the heat transfer is due to convection. Control of air flow can be either by a damper at the air inlet or by means of a damper in the chimney. The literature is rich in different stove designs, although all can be classified as variants of the two basic designs. A good survey is provided in References 12–15.

Although simple in construction and operation, understanding the physical and chemical processes which occur in the stove and the principles governing stove performance optimisation is a complex and difficult problem. The air flow is due to pressure differences set up by bouyancy forces and these depend on the rate of combustion, which itself is coupled to the air flow. The calculation of heat transfer rates is difficult and in the case of convective heat transfer data is not readily available. Fig. 8.8[16] shows an experimental arrangement for the measurement of stove performance incorporating online data processing. Fig. 8.9 gives a heat balance for a Thai charcoal stove obtained from this equipment.

Designs are available for various construction materials, mud/clay, ceramic, sheet metal and cast iron. There is no universal solution to the stove design problem. Different cultures and practices have different needs. Some wish to cook quickly, others over a long period. Space heating may be advantageous to some communities but not others. In particular, as mentioned earlier, references in the literature to 'efficiency' should be treated with considerable suspicion and not used to compare different designs. The applicability of a

a.

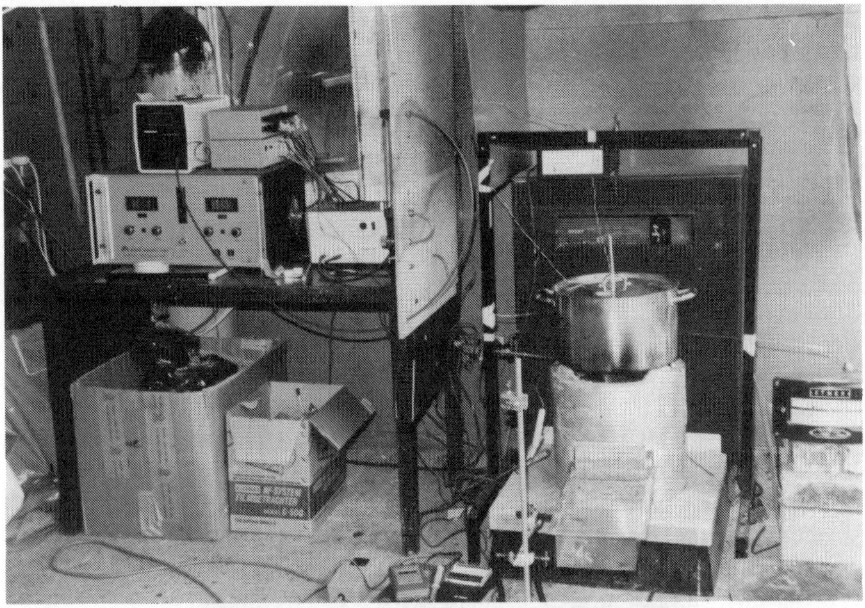

b.

c.

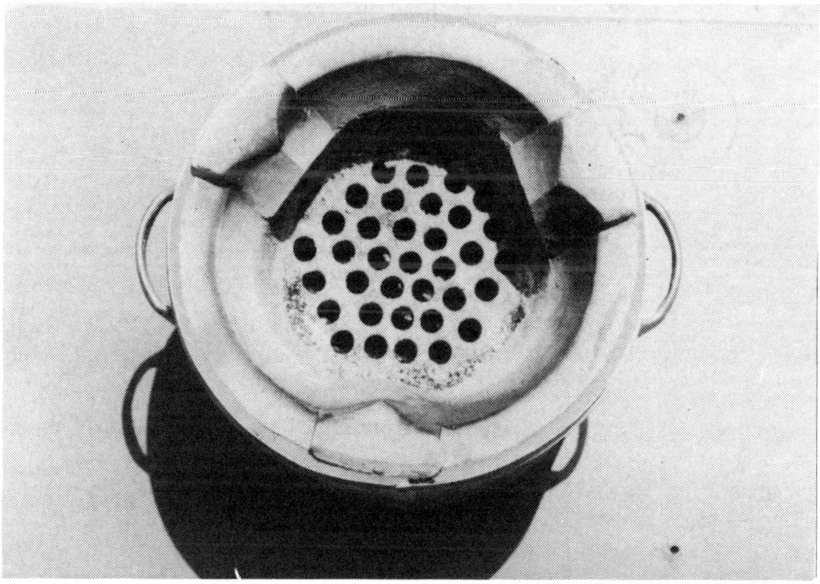

d.

Fig. 8.8 *Experimental arrangement for the measurement of stove performance*
a Stove measurement arrangement, Reading University
b Weighing cooking pot
c Thai stove from above
d Thai stove from the side

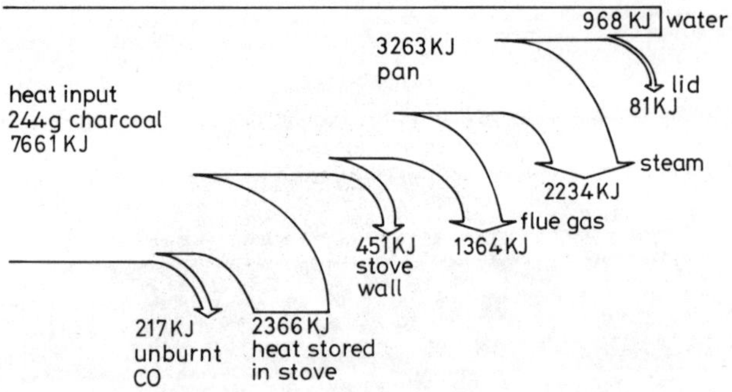

Fig. 8.9 *Heat balance of the Thai stove with the best effectiveness (hot stove)*

particular stove design to a particular community must be assessed by field trials in the area itself. In addition to the technical design, the problem of efficient cooking has major social and economic aspects. The question is 'What is a good stove?' A stove design, to be acceptable, must satisfy the following criteria:

1 Cook to an acceptable standard.
2 Be simple to use.
3 Have low capital cost.
4 Require only simple maintenance.
5 Have low fuel cost.
6 If the stove is to be used in an enclosed space, it should not release smoke or fumes.

Having developed a suitable stove for a particular area there still remains the problem of dissemination, particularly in the rural areas. There are various solutions to this problem, including village extension work and the encouragement of local sources of supply through local craftsmen and small industry. The problems of dissemination present the major difficulty in achieving worthwhile improvements.

8.5 Vegetable oils

Vegetable oils have long been used in oil lamps, and more recently they have been experimented with as a fuel for diesel engines. Vegetable oils can be obtained from a number of plants including legumes (soybeans, ground nuts) palms (oil palm), nuts (coconut) and various others such as sunflower, rape and castor beans (Chapter 2). Fig. 8.10 shows an assortment of oils used in a student project on diesel performance. Vegetable oils are very similar to diesel oil in calorific value. The figures for a number of vegetable oils are given in Tables 8.3, 8.6 and 8.8 and show that on a volume basis they are around 94% of diesel oil and 88% on a mass basis.

Fig. 8.10 *Vegetable cooking oils used as fuels for diesel engines*

The crop is harvested in the usual way and then dried to remove moisture. The oil content of ripe seed is around 40% to 45% by weight. This oil can be removed by mechanical pressing or by solvent extraction; the latter method is only suitable for large scale processing, and for small scale use hand or engine operated screw presses are used. Typically, 60% to 70% of the oil can be recovered by mechanical pressing. The 'cake' which remains after oil extraction is rich in protein and can be used as cattle feed. The oil is filtered and can then be used in diesel engines either mixed with diesel fuel or alone without modification of the engine. The efficiency and power output of the engine using vegetable oils will be much the same as with diesel oil. There are some difficulties arising from the use of vegetable oils in diesel engines. The viscosity of vegetable oils at low (room) temperature is some ten times that of diesel fuel at the same temperature and this affects the injection process and

Table 8.6 *Comparison of some vegetable oils with diesel oil*

Fuel	Calorific value MJ/kg (gross)	Density kg/l	Kinematic viscosity mm^2/s at 30°C	Cetane no.
Corn oil	39·5	0·91	34·9	37·6
Soyabean oil	39·6	0·91	32·9	37·9
Diesel	45·3	0·84	2·7	47

leads to carbonisation. It is adviseable to start cold engines on pure diesel fuel. Gums are present in vegetable oils; their removal or conversion is possible but expensive on a small scale and if not removed may result in carbon deposits which will require more frequent cleaning than when the engine is used with diesel oil. Fig. 8.11 shows a deposit on the injector from a single cylinder 5½ hp engine after 5 hours running. It is also advisable to switch to pure diesel oil at the end of running to avoid the build up of gum deposits. Nevertheless, in spite of these problems vegetable oils used either with diesel oil or alone may well become an important source of liquid fuel for remote communities where transport costs for diesel are high and delivery uncertain.

Fig. 8.11 *Carbon Deposit on injector of single cylinder engine after 5 hrs. running*

8.6 Pyrolysis

Pyrolysis, or destructive distillation, is the word used to describe the physical and chemical breakdown brought about by heating biomass or carbonaceous material in the absence of air. The products of pyrolysis arise as solids, liquids and gases, Table 8.7. The actual proportions will depend on the temperature and rate at which the reactions occur. In the case of biomass, initially (as the temperature is raised) moisture will be given off and some tar released. Above 200°C carbon monoxide and dioxide are formed and at temperatures around 400°C the gases will contain methane and other hydrocarbons, hydrogen, acetic acid, acetone and methanol.* Below 300°C the main products are char,

* Methanol is also known as wood alcohol.

Table 8.7 *Products of pyrolysis*

Solids	Char, ash
Liquids	Tar, Pyroligneous acids
Gases	Water vapour
	carbon monoxide, carbon dioxide
	hydrogen, hydrocarbons, nitrogen

tar and liquids, and above 500°C the products are primarily char and gases. Heat must be added for pyrolysis to occur. This can be provided by burning wood or using some of the products of the reaction itself. Table 8.8 gives the results obtained with oak by Churchouse[19] using a small externally heated kiln at a temperature of 500°C.

Table 8.8 *Pyrolysis of English oak at 500°C (Reference 19)*

Component	% by weight
Pyroligneous acid and tar	60
Charcoal	27·4
Gas	12·6

The earliest example of pyrolysis on a commercial scale was the production of coal gas in the early 19th century. Coal is heated in a sealed retort and the resulting mixture of gases and vapours form a combustible gas of calorific value around 15–20 MJ/m^3. About 7% of the mass remains in the retort in the form of coke which is used in iron smelting and as a solid fuel.

Later on methods were developed for the complete gasification of coal by the admission of a small quantity of air to give partial combustion. Owing to the presence of nitrogen in the air, the resulting gas, called producer gas, has a lower calorific value (around 4·5–5·0 MJ/m^3) than coal gas. The 'gasification' of biomass is discussed in Section 8.7.

8.6.1 Charcoal production

In the traditional method of charcoal production wood is stacked in a pit in the ground and the stack covered by turf to exclude the air. The stack is then fired and the ingress of air carefully controlled. In the process roughly 15% by weight of the wood is converted to charcoal. Assuming charcoal to have twice the heat value of wood (Table 8.3) the conversion efficiency is about 30%.

In spite of this loss of energy the use of charcoal for cooking is widespread.

The reasons for this use are mainly those of convenience and ease of transport. Open wood fires produce smoke which may cause respiratory and eye complaints and also dirties walls and ceilings in the more expensive houses. These disadvantages more than offset the advantages of insect control and preservation of the construction materials of the house. While the smoke problem can be overcome by using chimneys, charcoal cooking is cleaner and more convenient. If the greater efficiency of charcoal over wood stoves, together with the smaller bulk for transport, is taken into account, the loss in energy compared with the original wood is not as serious as might at first be supposed. On the other hand, if the efficiency of charcoal production can be improved this would be of immediate value. A factor of two or more improvement in charcoal yield seems possible. Commercial charcoal production operates at about 400–500°C and yields (on a dry load basis) 10–40% by weight, and simple plant designs are available. Charcoal is made either in a kiln or a retort. Charcoal kilns operate in the same manner as the traditional pit method. The wood is partially burnt by the admission of air, the resulting heat enables the charcoal reaction to convert the remainder of the wood. The wood is prevented from burning completely by restricting the ingress of air. In the retort method wood is sealed in a vessel, the retort, which is then heated externally. Kilns are normally employed for small scale charcoal production.

One of the simplest, but nevertheless effective, designs is the Tongan kiln which is based on the 200 l (40 gallons) oil drum, Fig. 8.12. A slot about 20 cm

Fig. 8.12 *Tongan kiln, Lae, Papua New Guinea*

wide is cut along the cylindrical side of the drum which is then placed on its side with the opening facing the direction of the wind. A small quantity of wood is placed in the drum and ignited. further wood is added until the drum is full and the wood burning vigorously. After about an hour the drum is then rotated so that the opening is at the bottom and sealed by packing with earth or clay. After cooling about 12 kg of charcoal can be extracted.

A larger kiln design by TPI* is shown in Fig. 8.13. Although more expensive than the Tongan kiln it is recommended where a higher production rate is required. The kiln consists of two concentric steel cylinders with a conical top having four air vents which can be sealed by means of lids. There are also eight air channels with attachments for chimneys on which the kiln is supported. The TPI kiln will produce 300–400 kg of charcoal per firing cycle of 1–3 days (depending on the dryness of the wood and the cooling time).

Fig. 8.13 *TPI transportable charcoal kiln (Courtesy T.P.I.)*

* TPI – Tropical Products Institute, Culham, Berkshire, UK.

In traditional charcoal production the gas/liquid products are burnt to provide heating and the only usable product is the charcoal itself. It is possible to devise relatively simple, small scale plant for the collection of some of these other products. These are discussed below.

8.6.2 *Small pyrolysis plant*

Pyle describes a design developed in Ghana.[20] The basic unit is a 2 ft diameter 10 ft high cylinder made from sheet metal and insulated by char. The daily input of 6 tonnes (dry) feed is said to produce 1300 kg char, 1000 kg oil, with an energy recovery of 70%. The char can be used as a domestic fuel, the gas for drying the feedstock and the oil is to be used as a fuel. Plant of this type should be capable of an output of 30% charcoal, 20% gas and 20% oil (all by weight) which compares very favourably with the 15% charcoal only production of the traditional method. Fig. 8.14 shows a small pyrolysis plant using[21,22] coconut husks and sited at Lae, Papua New Guinea.

Fig. 8.14 *Small pyrolysis plant, Lae, Papua New Guinea (Courtesy D. Fussey)*

8.7 Gasification

In pyrolysis the feedstock is decomposed by heating in the absence of air, in the gasification process[23,24] a small amount of air is admitted and in the case of wood and other biomass the resulting product is a mixture whose principal components are carbon monoxide, hydrogen, methane and the inert gases nitrogen and carbon dioxide. Fig. 8.15*a* shows a small scale gasifier suitable for use with a feedstock such as wood, coconut husk, etc. this is an updraught type, in which air is heated in the combustion zone and passes vertically upwards through the bed and the feedstock is fed in from the top and the ash removed at the bottom. On entering the gasifier moisture is removed from the feedstock in the drying zone. In the reduction zone pyrolysis occurs and the material is broken down to release tars and other volatiles. The principal reactions in the reduction zone are

$$C + CO_2 = 2CO$$

and

$$C + H_2O = CO + H_2$$

and are endothermic.

a.

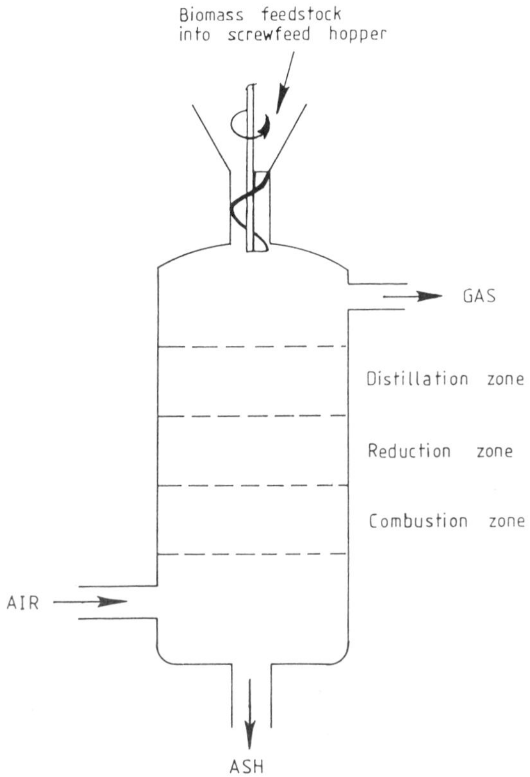

Fig. 8.15a *Updraught gasifier*

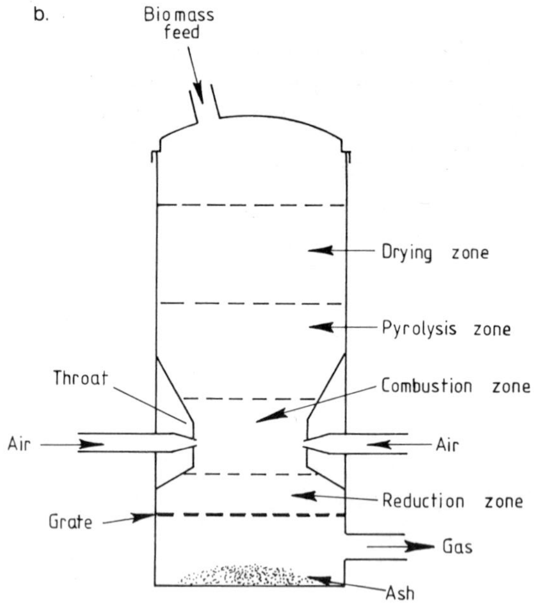

Fig. 8.15b *Downdraught gasifier*

By the time that the combustion stage is reached the feedstock is largely carbon which burns with the incoming air according to the exothermic reaction

$$C + O_2 = CO_2$$

If charcoal is used as feedstock then the pyrolysis is not significant. If, however, wood or other biomass is used, then the product gas contains tars, and acid vapours, which may require to be removed before the gas can be burnt in an engine. For this reason it is usual to arranged for the airflow to move down the gasifier, Fig. 8.15*b*, so that the pyrolysis products are burnt in the combustion zones. In the downdraught gasifier the zones, from top to bottom, are drying, pyrolysis, combustion and reduction. The gas has a typical energy content of 5 MJ/m^3 and can be burnt in a spark ignition engine. Fig. 8.16 shows a small commercial gasifier of this type. Fig. 8.17 shows a tractor in which the gas is stored in a gas bag on the vehicle.

Fig. 8.16 *Seltec biomass gasifier using charcoal to operate a 40 h.p. S.I. engine (Courtesy A. G. Harris)*

Fig. 8.17 *Gas operated tractor, China (Courtesy P. Fraenkel)*

8.8 Anaerobic digestion – biogas production

8.8.1 introduction

The production of gas by rotting vegetation in stagnant ponds is a common phenomenon. In the early 1800s William Henry analysed this 'marsh gas', as it is called, and found that it was largely made up of methane. Methane is produced by the bacterial attack, or digestion, of cellulose. The methanogenic bacteria cannot live in the presence of oxygen and a reaction which requires this condition is known as 'anaerobic'. At the bottom of a stagnant pond oxygen is used up by various chemical reactions and is not replaced, thus giving rise to the anaerobic state suitable for methane production. Methane is also produced in this way in the digestive track of animals, particularly ruminants. A cow is said to produce up to 500 l of gas a day by this process. The bacteria which cause the breakdown of vegetation in the compost heap require oxygen and such reactions are known as 'aerobic'. Wherever organic material is found in the absence of oxygen, methane production is a possibility. One well known example is the presence of 'fire damp' in coal mines, which has been the cause of many serious mine explosions in the past. Commercially, the most important example is the production of natural gas, which is about 85–95% methane. The actual methane producing reaction which has resulted in natural gas has taken place over thousands or perhaps millions of years. Although the possibility of methane production from dung, and vegetation, has been known for many years, there has been little work done on any large scale. During and after the Second World War there was some interest in Europe, particularly in Germany, where a number of small plants were set up to provide energy for powering tractors and for other agricultural uses. However, the increased availability of the cheap and more convenient oil fuels resulted in this work being discontinued. Several of the larger municipal sewage authorities in the UK and elsewhere collect methane and use it to drive gas engines which generate power for on-site use.

The Indian Gobar* Gas Development represents the first major effort to encourage the widespread use of small domestic scale gas generators in rural villages.[25] The difficulties which the Indians faced, and which were the reasons for setting up the Gobar Gas programme, were twofold. Firewood was becoming increasingly scarce and as a result village prople were using more dried cow dung for cooking. Secondly, fertiliser costs were rising and making it even more desirable that the cow dung should be returned to the fields as fertiliser. The use of cow dung in digesters seemed to offer a solution to both problems. The gas produced, biogas, is about 60% methane and 40% carbon dioxide, and can be burnt to provide heat for cooking and also be used to give lighting. Very little nitrogen is lost in the digestion process and in fact some of the nitrogen, which in the dung would not be available for fertiliser, is made suitable for assimilation by plants by the digestion process. The effluent also is

* Gobar is a local name for cow dung.

in a form which is generally more suited for spreading than is the original dung.* A small scale digester suitable to provide the cooking and lighting needs of a typical family would require the daily dung produced from about five cattle, see Fig. 8.18.

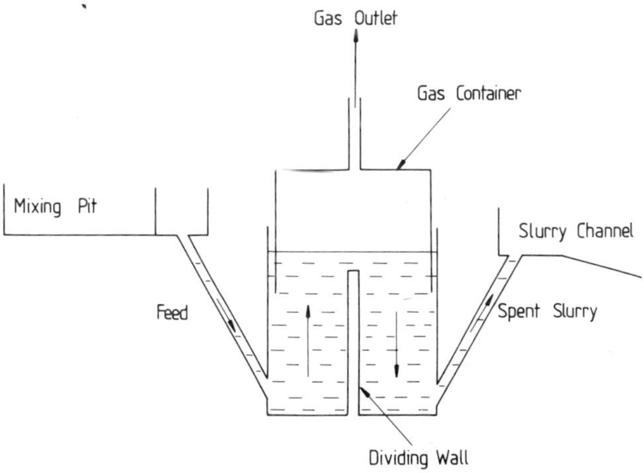

Fig. 8.18 *Indian Gobar gas plant*
The gas container is metal (steel with anti-corrosion paint). The digester body is lined with bricks.

The Indian programme has been vigorously promoted; the number of digesters is around 120,000, and the programme is continuing. The Chinese have a massive programme on family type digesters, Fig. 8.19, and as many as seven million are said to be in use.[28,29,30]

Large programmes are also in hand in Korea, the Philippines[26] and Taiwan and smaller scale work is being carried out in most developing countries;[27] there is also some interest in the developed countries. It is not yet clear whether the domestic scale digester is the most suitable size to adopt. If a larger community size is chosen then it is possible to arrange for a higher level of skill in its operation. Against this must be set the problems of arranging for the gas to be distributed and used fairly and for a fair distribution of the effluent.

8.8.2 The digestion process

During the digestion process polysaccharides (cellulose and starches) are converted mainly to methane and carbon dioxide. The overall chemical reaction is given below:†

$$C_6H_{10}O_5 + H_2O \rightarrow 3CH_4 + 3CO_2$$

* It is necessary to expose the effluent to the air for two weeks or so in order to remove toxins, principally H_2S.

† The reaction equation suggests that equal volumes of methane and carbon dioxide should be produced. In practice some of the carbon dioxide reacts to form carbonates which remain in solution, hence the rather greater proportion of methane in the biogas.

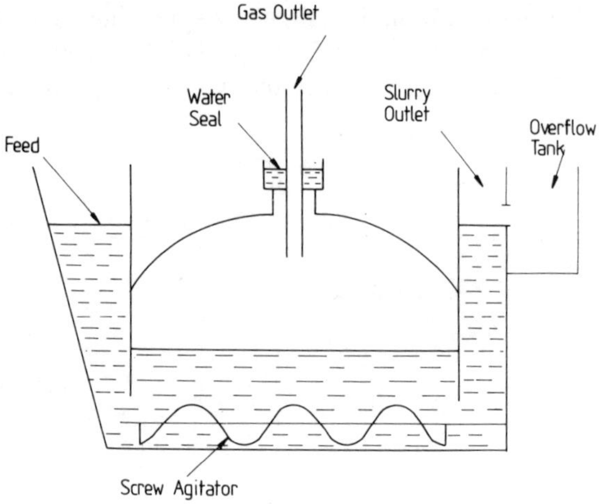

Fig. 8.19 *Chinese biogas digester*
The digester is constructed from masonry, and the dome is fitted with a water seal at the gas outlet.

This is an oversimplification of what actually occurs, since the feedstock contains many different substances and the proportions are to some extent variable. Table 8.9[32] gives a typical analysis for cow dung and pig dung. In addition the process takes place in at least three stages and the detailed chemical reactions are complex and not fully understood. However, the equation describes broadly what occurs and is a useful generalisation.

The digester acts in a way which is similar to that of the stomach of an animal. It provides an environment which is suitable for a bacterial population to live in and to reproduce. This environment must have the correct physical

Table 8.9 *Biochemical analysis of cow and pig faeces (Reference 32)*

Constituents	cow	pig
Ash	17	16
Lignin	11·6	9
Crude fat	—	7
Crude protein	2·4	19
Fibre	41·4	—
Neutral detergent fibre (essentially cellulose, lignin and hemicellulose)	—	49
Acid detergent fibre	—	3
Cellulose	21·4	—
Lipid	3·2	

and chemical conditions, such as temperature and acidity, and in addition the bacteria must be provided with a supply of food. This food supply, dung or vegetation, should contain both carbon and nitrogen in a chemical form in which the bacteria can break it down. The ratio of carbon to nitrogen for optimum performance is around 30 to 1. If the ratio is greater than this then the bacterial population will be unable to build up sufficiently and the reaction rate will be slowed down. If the ratio is very much lower then there will be a build up of ammonia which is toxic to the bacterial population Digesters normally operate at C/N ratios, rather higher than the optimum. Table 8.10 lists the carbon to nitrogen ratio of some common feedstocks. In estimating the likely

Table 8.10 *Carbon to nitrogen ratio (various sources)*

Cow	18*
Pig	20
Human	6–10
Poultry	15
Hay Grass cuttings	12
Potato peel	25
Oat straw	48
Wheat straw	150

* Non-lignin carbon dry weight

yield from a particular loading it is important to understand the composition of the feedstock and in particular to identify the component which is biodegradable. Wet cow dung, for example, contains about 80% water. of the remaining 20% about 80% is organic. Not all the organic matter is capable of digestion and lignin in particular is difficult for the bacteria to break down. Table 8.11 lists several feedstocks and gives rough proportions for water content, total solids and percentage of organic matter. In the literature there are frequent references to the daily output of dung per animal. This will vary widely with

Table 8.11 *Wet dung content: typical relative proportions of water, organic and non-organic solids (Reference 31)*

	% water	% total solids	Organic	Non-organic
			% of total solids	
Cow	80	20	80	20
Pig	80	20	85	15
Human	80	27	92	8
Poultry	65	35	65	35

size of animal, the way in which it is fed, and, most important, the method of collection. Output of free ranging animals will usually be restricted to the dung produced when they are in stalls at night, whereas those animals which are permanently confined will yield all the daily dung output. Since animals reject nitrogen in the form of urine, where the dung includes urine the nitrogen content will be higher than for faeces alone. Table 8.12 gives somes rough figures for daily dung yield. The dung should be mixed with water before adding to the digester. Opinions differ as to the proportions to be used: Indian practice is to add equal proportions of water and wet dung, others add rather less water. The aim should be to get a creamy, well-mixed consistency with a solids content of between 8% and 12%.

Table 8.12 *Dung yield* (various sources)*

	Weight of animal kg	Weight of wet dung per day kg	Weight of dry dung per day kg	Weight of volatile solids kg
Cow	450	40	5	4
Pig	45	3·2	0·5	0·42
Poultry	1·8–2·2	0·12	0·03	0·02
Buffalo (Indian)		14		
Cow (Indian)		10		

* Faeces and urine

The digestion process takes place in three main stages:

1 Hydrolysis or liquefaction stage
2 Acid formation
3 Methane formation

Initially bacteria break down the large molecules of starch and cellulose by hydrolysis, this is followed by a conversion into fatty acids of which acetic acid is the principal component produced. These processes occur fairly rapidly, in a day or so, and the bacteria concerned are not particularly sensitive to environmental conditions. It is an acid producing process and in a newly set up digester the acidity level will increase and the pH* of the digester fall to a value of around 6, see Fig. 8.20. The methanogenic bacteria which carry out the next stage of the process to give methane are much more sensitive to environment; they require an almost neutral condition of 7·5 to 8 and their population builds up slowly. This phase of the process can take from 20 to 30 days depending of temperature and other factors. When a digester becomes established at this pH

* The acidity or alkalinity of a solution is usually expressed in terms of pH. A neutral solution has a pH of 7, whereas acids have a value below 7 and alkaline solutions a value greater than 7. Fig. 8.20 indicates typical pH values.

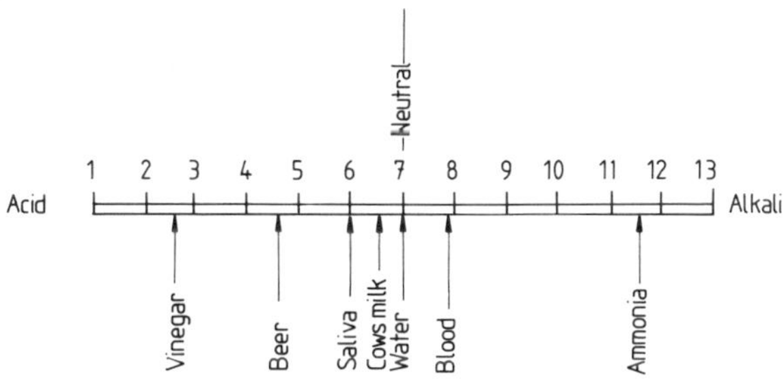

Fig. 8.20 *The pH scale*

it is said to be buffered. This means that as further feedstock is added the acid forming bacteria can work but the acid so formed is neutralised to maintain the condition suitable for the methanogenic bacteria. Initially the gas evolved is mainly carbon dioxide but as the digester becomes established the methane content rises until, for cow dung, the proportions are around 60% methane and 40% carbon dioxide. It is difficult for a newly started digester to achieve the well-buffered state, with its complex mix of bacterial population, and it is recommended that a starter liquid of around 20% of the volume of the digester is added from an operating digester. Generally the composition of biogas is 35–70% methane and the remaining gas is mainly carbon dioxide. Small quantities of other gases such as hydrogen and hydrogen sulphide are also formed. The optimum temperature of the digester is around 35°C but the bacteria will function from 18°C to 40°C. The rate of methane production will double for a rise of 10–15°C in temperature. Below 18°C the output of gas becomes negligible. It really is most important to keep up the mean temperature of the digester. It is sometimes overlooked that this temperature is determined by the mean temperature over the 24 hours and not the mean day temperature. In many of the developing countries night temperatures are low and resulting average temperature is too low for acceptable performance unless measures are taken to add heat in some way. This is also the case in more temperate climates, and the addition of relatively cold feedstock represents a significant heat loss for the digester. External heating can be applied using a biogas burner to heat water which is then circulated, through pipes, in the digester. Solar flat plate heaters can also be used but a circulating pump will be required.

8.8.3 Practical digesters

Fig. 8.21 shows an Indian-type digester situated in Papua New Guinea and having an output suitable for the cooking needs of a family, and Fig. 8.22 shows a Chinese family-sized plant. In both designs the operation is continuous and there is a daily input of feedstock and an equivalent daily removal of effluent. Fig. 8.23 shows a large scale industrial digester situated on a pig farm

a.

b.

Fig. 8.21 *Indian-type digester, Lae, Papua New Guinea*

a Digester
b Gasburner

Fig. 8.22 *Chinese type family-sized digester, Sri Lanka (Courtesy Shive Rajapakse)*

Fig. 8.23 *Large scale industrial digester, Maya Pig Farm, Phillipines*

in the Philippines. The Indian design consists of a vertical cylinder having a diameter of from 1.2 to 6 m and a depth from 3·5 to 6 m; the smaller dimensions would provide the domestic needs of a single family of five people. The cylinder is divided into two halves by a vertical wall. The gas is collected in a steel cylindrical cap which floats on the slurry as shown in the figure. The gas pressure is in the range of 7 to 14 cm of water.* The pressure can be increased by placing stones on the top of the steel cap. Feedstock in the form of slurry is added to the bottom of the digester by means of a pipe shown in the left-hand side of the figure. When slurry is added an equal volume of spent liquid is displaced through the pipe on the right-hand side. This spent material is dried and used as fertiliser. The volume of biogas produced daily is roughly equal to the volume of the digester (for a temperature approximately 30°C) and for small digester of the type indicated a daily volume of around 2–3 m^3 would be expected. The feedstock required is about 45 kg wet dung or the output of 5 cows. The gas is used for cooking and lighting Fig. 8.24. It is important to use burners designed for use with biogas, because if town gas burners are used the jet sizes are incorrect and performance is poor. The gas can be used in spark ignition engines. If it is required to fuel compression ignition (diesel) engines then it is necessary to provide 15% or so of the fuel in the form of diesel fuel in order to obtain ignition. Typical gas requirements are given in Table 8.13.

* Atmospheric pressure will support a column of water 10 m in height.

a.

b.

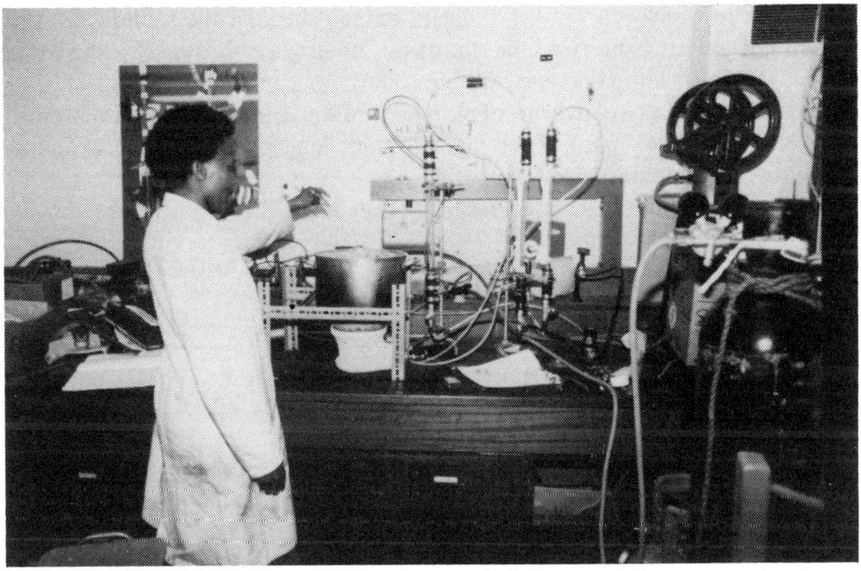

Fig. 8.24 *Biogas stoves*
a Cooking, Sri Lanka
b Development of biogas burner, University of Reading

Table 8.13 *Biogas use*

Cooking (2 in and 4 in burners)	225–450 l/hr	350 l/day/person
Lighting	115–140 l/hr for 100 candle power	
Mechanical power	425 l/hp hr	

The Chinese design[29–30] shown in Fig. 8.19 differs from the Indian design in one important respect; the digester is covered by a masonry dome. Particular care is necessary in the construction of such a dome in order that it should be gas tight. The dome is fixed and the gas collected under it displaces slurry into a reservoir above, and the side of the digester pit. The pressure is variable and lies in the range 20 to 140 cm of water.

In addition to providing a convenient source of domestic fuel and in giving a valuable fertiliser the digester also makes a useful contribution to community health. Since human excreta is a suitable feed there is encouragement to make use of it and thus to some extent remove this potential source of infection from the community. The digestion process reduces, though does not eliminate completely, the disease producing organisms in the dung and other feedstock. It is not as effective in this regard as composting, since the maximum temperature which is achieved is lower, and in particular the roundworm egg, which has a thick protective sheath, passes unaffected through the digester.

It is too early yet to come to definite conclusions on the future impact of the biogas generator as a convertor of biomass and waste products. Undoubtedly it will have a role to play. There is still some uncertainty on the economic viability of the domestic scale plant.

Digester characteristics are summarised in Table 8.14.

Table 8.14 *Summary of digester characteristics*

Feedstock	Yield m^3/kg	Methane %	Calorific Value MJ/m^3
Dry Cowdung	0·3	60	23
Dry Pig/Chicken Dung	0·4–0·5	65	26–28
Conversion Efficiency 30%–70%			
Daily Yield per unit volume of digester	$m^3/m^3 \sim 1$		

8.9 Fermentation to alcohol

8.9.1 Introduction

The production of alcohol from the fermentation of sugar by yeast has been known since the dawn of civilisation. References to wine and its effects on the individual abound in the early literature. The grape constitutes a particularly compact and convenient package of ingredients for the wine maker, since it contains a considerable fraction of sugar together with nitrogen and other essential elements. Yeast is present in the form of the bloom on the grape surface and the grape juice contributes sufficient acidity to enable the fermentation to proceed vigorously. In fact it would be difficult to construct a better kit for the wine maker. During the fermentation process the enzymes in the yeast convert the sugar into alcohol, mainly ethyl alcohol, and provided that sufficient sugar is available then the alcohol concentration will continue to rise until it reaches a value of around 9% by volume. Above this figure the yeast dies and the process ceases. If a higher concentration of alcohol is required then it is necessary to separate out the alcohol from the water by distillation. If the alcohol water mixture is heated, then, since alcohol has a higher vapour pressure than water, it will evaporate more readily and can be recondensed in a second cold vessel, the condenser. Spirits are made in this way and the distillation device is known as a still.

While alcohol production by fermentation for wines and spirits has a long history it is only relatively recently that attention has been given to its large scale production from sugar and other biomass for use as a fuel. The most ambitious programme of this type has been mounted in Brazil where in 1982/83 the annual output was $5{\cdot}7 \times 10^6$ m^3 of which $3{\cdot}5 \times 10^6$ m^3 was anhydrous and $2{\cdot}2 \times 10^6$ m^3 hydrous. By the year 2000 it is hoped that this figure will have risen to provide 75% of the total liquid fuel used in the country. Sugar cane is used as the source of sugar, and a major energy requirement in the process is heat for the distillation of the alcohol. In the case of sugar cane the residues which remain after the cane has been crushed to remove the sugar, known as bagasse, can be burnt to provide this heat and thus avoid the need for a conventional fuel.

8.9.2 Land requirement

In Section 8.2.3 it was shown that 1 kg of glucose will yield rather more than half a litre of ethyl alcohol. Typical yields of sugar cane per hectare in Brazil are around 50 tonnes per hectare. On conversion to alcohol this yields around 70 l/tonne or 3500 l/ha. In order to provide the 3×10^9 l/year necessary to provide a 20% addition to petrol a total land area of $8{\cdot}6 \times 10^5$ ha will be necessary. This is not a particular problem in a country such as Brazil but can be a serious disadvantage in an area where land is scarce and alternative uses must be considered.

8.9.3 Alternative Feedstocks

Sugar can readily be fermented to produce alcohol. If starch is used as the feed it is necessary first to hydrolyse the material in order to convert it to sugar. This can readily be achieved by means of an enzymatic reaction.*

Cassava (also known as manioc or tapioca) is a bushy herb-like plant with roots which resemble sweet potatoes and which have a high starch content (25%). The yield of alcohol is higher per tonne than for sugar cane (170 l/tonne compared to 70 l/tonne) but since the yield per hectare of cassava is less than for sugar (15 tonnes/ha compared to 50 tonnes/ha), the overall yield of alcohol per hectare is less, around 2500 l. Sorghum will give typically 32·5 tonnes/ha yr, which when converted to ethanol yields 3775 l/ha yr.

8.9.4 Energy ratio

The annual yield of ethanol from sugar cane is around 3500 l/ha or an energy equivalent of 84 GJ/ha. In order to produce this alcohol, energy inputs are required in the form of fertiliser, herbicides, pesticides and mechanical power for land preparation and irrigation. Further energy is required for harvesting, transport and the processing of the sugar cane to alcohol and its separation by distillation. Table 8.15 shows a summary of these figures for a Brazilian plant.[34] the second column gives these figures expressed as a fraction of the energy of the ethanol produced and the third column lists the energy inputs per

Table 8.15 *Ethanol from sugar cane (Brazilian data)*

Process	Energy kJ/kJ in ethanol		Energy GJ/ha/yr	
1 Growing				
Fuel	0·07		5·14	
Fertiliser/chemical	0·04		2·94	
Sub-total		0·11		8·08
2 Extraction/fermentation/ distillation				
Bagasse	1·26		92·61	
Chemical	0·01		0·73	
Sub-total		1·27		93·34
3 Total production energy		1·38		101·42
4 Energy in ethanol		1·00		73·8

* The hydrolysis reaction involves the addition of a water molecule to the starch molecule:

$$C_6H_{10}O_5 + H_2O \rightarrow C_6H_{12}O_6$$

In principle cellulose can also be hydrolysed but at present the methods used have proved uneconomic.

hectare. From Table 8.15 it is seen that growing the crop requires 8·08 GJ/ha yr and the subsequent processing to form alcohol a further 93·34 GJ/ha yr to give a total of 101·42 GJ/ha yr. Now 101·42 GJ compared to the solar input of around 80 000 GJ/ha yr is negligible. If, however, we neglect the solar input and consider only the energy used in the cultivation, harvesting and processing we see that 1·38 times as much energy is required in the process compared to the energy value of the ethanol produced (the energy cost of the factory and agricultural machinery has not been taken into account). On the other hand it might be said that since 1·26 of this figure is obtained from the sugar cane in the form of bagasse the true ratio is only 0·12 or an energy gain of 8.

Cassava has a less favourable energy balance than sugar cane, since, unlike sugar cane, the residues cannot be used to provide the major part of the process energy. However, if wood is grown and used in the processing the energy ratio in terms of fossil fuel input is similar to the figure for sugar.

The criterion used depends very much on the manner in which the energy is to be used. If the site is remote then the energy content in residues may be of no use; in other situations it might have considerable market value. Table 8.16 compares some characteristics of biogas and alcohol plants. The figures given here are intended only to illustrate the argument. Published data on the energetics of alcohol production differ markedly.

Table 8.16 *Comparison of fermentation methods*

	Biogas	Alcohol
Feedstock	Cellulose	Sugar Starch can be used after a simple hydrolysis, cellulose requires a more difficult hydrolysis process
Process time	20 to 30 days	Few days
Product	60% methane plus carbon dioxide	Ethyl alcohol in water
Separation	Appears as gas, none unless CO_2 requires to be removed by scrubbing	Distillation necessary to separate out alcohol
Transport and storage	Expensive, use *in situ* as produced	Simple and cheap

8.9.5 Use of alcohol

Alcohol has uses in industry both as a solvent and in pharmaceuticals and in some countries as a feedstock, for the plastics industry, it also can be used as a fuel for internal combustion engines.

After the fermentation process ethanol is separated from water by distillation. When an alcohol/water mixture is distilled, the vapour is condensed and the resulting distillate is found to have a higher concentration of alcohol than the original mixture.[35] This process can be repeated a number of times resulting in successively higher alcohol concentration. However, when the concentration reaches 95% alcohol 5% water, a constant boiling mixture is obtained and no further concentration can be obtained. This is called the azeotropic mixture. It is only possible to produce 95% ethanol by simple distillation and the resultant ethanol/water mixture is called 'hydrous' alcohol. The remaining water can be removed and a 99·86% pure 'anhydrous' alcohol obtained by means of a further distillation in the presence of an 'entrainer' such as benzene. Anhydrous alcohol can be mixed with petrol to give 'gasohol'. Mixtures of up to 20% ethanol are commonly used and can be burnt in a conventional spark ignition engine without any modification. Anhydrous alcohol is more expensive to produce than hydrous alcohol. The latter cannot be mixed with gasolene since the water causes separation of the alcohol and the gasolene. 100% hydrous alcohol can be used in special, high compression, spark ignition engines, but in addition to arranging the higher compression ratio it is also necessary to take precautions against corrosion.

Ethanol is unsuitable for use in compression ignition (diesel) engines.

References

1 'Firewood crops. Shrubs and tree species for energy production (Nat. Ac. Sciences, Washington DC, 1980)
2 REAY, D.: 'The history of man powered flight' (Pergamon, Oxford, 1977) p. 148
3 'Energy and protein requirements', Technical Report Series No. 522, report of a Joint FAO/WHO Ad. Hoc Expert Committee, Geneva, 1973
4 SPIERS, H. M. (ed) : 'Technical data of fuel', 5th edn British National Committee, World Power Conference, London, 1955
5 RAMASWAMY, N. S.: 'Animal Power in India. The management of animal energy resources and the modernisation of the bullock cart, system' Indian Institute of Management, Bangalore, 1979)
6 BARWELL, I. and AYRE, M.: 'The harnessing of draught animals' (ITDG, London, 1982)
7 PAPANEK, V.: 'Design for the real world' (Bantam Books, New York, 1972)
8 MCCULLOGH, J.: 'Pedal power in work, leisure and transportation' (Rodale, 1977)
9 WEIGHTMAN, D.: 'The use of pedal power in developling countries' (ITDG, London, 1976)
10 WEIR, A.: 'The dynapod. A pedal power unit' (Vita, Maryland, 1979)
11 WILSON, S. S.: 'Pedal power lectures in socially appropriate technology' (Technische Hochschool, Eindhoven, Netherlands, 1975)
12 De LEPELAIRE, G., PRASAD, K. K., VERHART, P. and VISSER, P.: 'A wood stove compendium' (Eindhoven University, Eindhoven, 1981)
13 'Wood conserving cook stoves. A design guide' (VITA/ITDG, London, 1980)
14 PRASAD, K. K. and VERHART, P. (eds): 'Wood heat for cooking' (Indian Academy of Sciences, Bangalore, 1983)
15 MICUTA, W.: 'Modern stoves for all' (Bellerive Foundation, Geneva, 1981)

16 DUNN, P. D., SAMOOTSAKORN, P. and JOYCE, N.: 'The traditional Thai cooker', 2nd EEC Conference on Biomass, Berlin, Sept 1982
17 'Proceedings of the International Conference on plant and vegetabler oils as fuels', ASAE, St. Joseph, Michegan, USA, Aug 1982
18 BRUWER, J. J., BOSHOFF, D. van D., HUGO, F. T. C., DU PLESSIS, L. M., FULS, J., HAWKINS, C., WALT VAN DE, A. N. and ENGELBRECHT, A.: 'Sunflower seed oil as an extender for diesel fuel in agricultural tractors', Symposium of South African Institute of Agricultural Engineers, June, 1980
19 LEVY, J. P.: 'Distillation, pyrolysis and alcohol production from wood' Proc. Int. ISES Conference, Solar Energy in Agriculture, Reading 1976
20 PYLE, D. L.: 'Biomass as a fuel source in developing countries', Conference on Small Scale Energy for Developing Countries, organised by ITDG and University of Reading, Reading, June 1979
21 TATOM, J. W., HARAHAP, F., CHIANG, T. J. and APANDI, R. M.: Report on AID Project A-1914, 1977
22 FUSSEY, D. E.: Private communication
23 FOLEY, G. and BARNARD, G.: 'Biomass gasification for developing countries', Technical Report No. 1 Earth Sciences, IIED London 1983
24 BREAG, G. R. and CHITTENDEN, A. E.: 'Producer gas. Its potential and application in developing countries' (Tropical Products Institute, London, 1979)
25 SATHIANATHAN, M. A.: 'Biogas, achievements and challenges' (Association of Voluntary Agencies for Rural Development, New Delhi, 1975)
26 MARAMBA, F. D.:'Biogas and waste recycling. The Philippine experience' (Maya Farms, Metro Manilla, 1978)
27 BARNETT, A., PYLE, L. and SALRAMIAN, S. K.: 'Biogas technology in the Third World. A multidisciplinary review' (IDRC, Canada, 1978)
28 'Biogas in China' (National Office for Biogas Development and Extension, Peoples Republic of China, Beijing, 1973)
29 BUREN, A. V.: 'A Chinese biogas manual. Popularising technology in the countryside' (ITDG, London, 1981)
30 MCGARRY, M. G. and STAMFORTH, J.: 'Compost, fertiliser and biogas production from human and farm wastes in the Peoples Republic of China' (IDRC, Canada, 1978)
31 FRY, J. L.: 'Practical building of methane power plants for rural energy independence' (D. A. Knox, Andover UK, 1974)
32 HOBSON, P. N. and SHAW, B. G.: 'The role of strict anerobes in the digestion of organic materials. Microbial aspects of pollution' in SYKES, G. and SKINNER, F. A. (eds), Society for App. Bact. Sym. No. 1, pp. 103–121 (Academic Press, London, 1971)
33 Dr POMPILIO, Furlando, Filho: Private communication
34 CARVALHO, Av. de., YANG, V. and MILFONT, W. N.: 'Energetics of ethanol production from sugar cane and cassava, International Symposium on Alcohol Fuel Technology, Methanol and Ethanol, Solfsburg, Germany, Nov 1977
35 ATKINS, P. W.: 'Physical chemistry' (OUP, 1978)

Chapter 9

Wind energy

9.1 Introduction

Man has made use of the power in the wind to propel his ships from earliest times. Windmills also have a long history, they are reported to have been in use in China and Japan from 2000 BC and they are also known to have been used in Babylon for water pumping around 1700 BC. These mills appear to have been of the vertical axis type, that is their sails rotated about a vertical axis rather than the horizontal axis of the conventional Dutch or English type. In the vertical axis machines the wind blows the sails round for half a rotation, they must then return to their original position by moving against the wind. One method of achieving this is to arrange that the sails are hinged so that they do not present such a large area to the wind on the return half cycle. Fig. 9.1 shows such a design which is in use in China today and may well be of the same design as the very early mills in this region. The Greek mathematician and engineer Hero invented the horizonal axis machine shown in Fig. 9.2. This was intended to operate a small piston/cylinder air pump to provide air to drive a small pipe organ. Hero's windmill design would seem to have been regarded as an interesting toy and neither the Greeks nor the Romans appear to have made use of wind powered machines although water powered machinery was used extensively for corn grinding and other purposes. The horizontal axis windmill was developed in Europe in the twelfth century and by the thirteenth century was very widely used for corn grinding and water pumping.[2] In a horizontal axis machine it is necessary to arrange that the sails can be turned to face the wind direction to take account of wind direction changes. This was achieved by rotating the complete mill which was mounted on a vertical post, hence the name post mill. In another design, the Tower mill, only the cap rotated, Fig. 9.3.

It is interesting to note that, until the invention of the steam engine, wind and water were man's most powerful sources of mechanical energy. The windmill played a dominant role in opening up the mid-west of the USA for agriculture by providing a source of power for irrigation and water supply. In

a.

b.

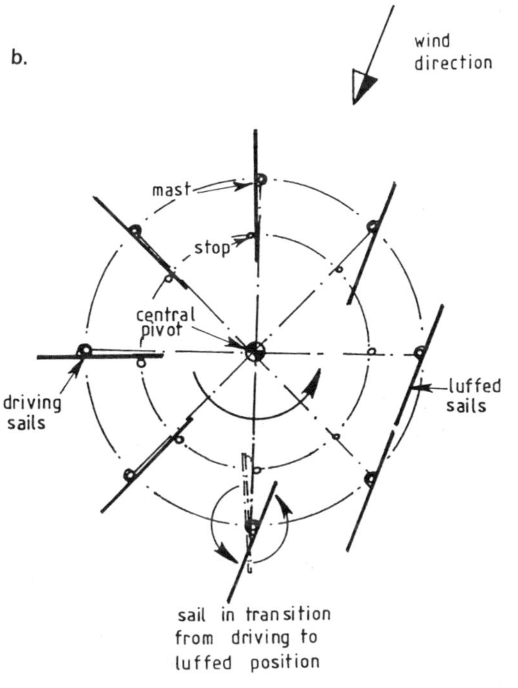

Fig. 9.1 *Chinese vertical axis windmill pumping brine. (From Farmers of Forty Centuries. F. H. King, Johnathan Cape Ltd.)*

b Principle of Operation

Fig. 9.2 *Small horizontal windmill blowing an organ (from 'The pneumatics of Hero of Alexandria', Bennett Woodcraft)*

the twentieth century the development of the internal combustion engine together with the availability of cheap petroleum fuels has tended to displace the windmill in agricultural communities. A second major effect has been the spread of rural electrification.

Fig. 9.3 *Tower mill*

However, even today small windmills are still extensively used for water pumping in Australia, South Africa and the USA Fig. 9.4. There are also a number of small electricity generating windmills which provide power for isolated farmhouses and other rural communities, Fig. 9.5. During the first half of this century studies were made on the feasibility of large scale electricity generation from the wind. Several generators ranging in output from 50 kW to 1250 kW were constructed and brought into operation. The largest of these wind generators was the Smith–Putman machine which was designed by Putman and built at Grand Pa's Knob in Vermont, USA,[3] Fig. 9.6.

Fig. 9.4 *Southern Cross Water Pumping Mill, Australia (Courtesy Southern Cross)*

Fig. 9.5 *Dunlite 2 kW Wind generator. National Centre for Alternative Technology, Wales (Courtesy P. Fraenkel)*

Construction of the machine was commenced in the winter of 1940–41 and was completed in August 1941. The machine was two-bladed with a rotor diameter of 175 ft (53.3 m) and each blade was 11 ft (3.6 m) wide by 70 ft (21 m) long and weighed 8 tons. The rotor was mounted downwind of the 175 ft (53.3 m) tower and the total turbine weight was 250 tons. The turbine was rated at 1250 kW in a wind speed of 13.5 m/s and power control was achieved by changing the pitch of the blades. The machine was operated satisfactorily for some time and actually produced a peak power of 1600 kW on one occasion. A bearing defect caused a shut down in February 1943 and war time supply difficulties delayed completion of the repair until March 1945. Shortly after recommissioning a blade failed and the project was discontinued. It is unfortunate that this

Fig. 9.6 *1250 kW aerogenerator at Grandpa's Knob, Rutland, Vermont (From Power from the Wind, P. C. Putman. 1948. D. Van Nostrand and Co. Inc.)*

splendid pioneering effort should have taken place at a time when war time priorities prevented its proper assessment. The combination of early mechanical failures and uneconomic costs led to a loss of interest in these large machines. Following the energy crisis of the early seventies this situation has changed dramatically and there is now an extensive and world wide development interest into large scale wind energy. These programmes are concerned with the development of machines which can be used for the generation of electricity and are intended for connection to national grid systems. The connection to the large grid system enables the fluctuation in output of an individual windmill generator to be compensated for by other generating plant in the system. A new abbreviation has appeared in the literature, the WECS or Wind Energy Conversion System. The US Wind Energy programme is the largest of these programmes but other countries are active. It is not easy to appreciate the size of the later wind convertor designs, for example, the US MOD 1 has a rotor diameter of 61 m; this is approximately the same distance as the distance between the wind tips of a Boeing 747 Jumbo jet aircraft, Fig. 9.7, and produces 2 MW of electricity in a wind of velocity 11 m/s. The rotation speed is only 35 rev/min, but, because of the very large diameter, the rotor tip speed is around 100 m/s.

Fig. 9.7 *US MOD 1 wind generator (Courtesy L. Devine)*

In addition to the interest in the horizontal axis machines for large wind convertors there are also studies on vertical axis machines of the Darrieus type with either curved blades, Fig. 9.8, or straight blades, Fig. 9.9.

The literature is rich in wind powered machine inventions from the past and more are continuing to appear. Some of these are described at the end of this chapter.

Fig. 9.8 *Vertical axis wind convertor of Darrieus type with curved blades (Courtesy L. Devine)*

Fig. 9.9 *Offshore array of multimegawatt straight blade wind turbines (Courtesy P. Musgrove)*

In addition to its interest for large scale power production, wind is attractive as a source of small power, particularly for use in the rural areas of the developing countries. In order that wind power can be effectively exploited in these areas it is necessary to provide designs which are capable of local manufacture and hence can be maintained by locally available skills.

One possibility is to use a combined system consisting of a wind generator, diesel generator and storage battery. Studies show that savings in diesel fuel, when compared with a diesel generator, of around 50% diesel fuel is possible, for an average electricity cost/unit similar to that for the stand-alone diesel system.

The major characteristics of wind power are given in Table 9.1. The

Table 9.1 *Wind power characteristics*

Advantages	Disadvantages
Free	Diffuse – low power density
Widely available	Variable in amount – requiring storage
Inexhaustible	Random in direction (generally but not at all sites
Non-polluting	Unreliable
	Site specific

principal problem which arises in using wind power is its variability and, in general, some form of storage is necessary, for example by the provision of a water tank for a wind pump or an electric battery for small electricity generator. In the case of large scale electricity the problem is overcome by grid connection, but only at the cost of additional standby plant. Possibly the only application in which the wind variability is less important is in the intermittent requirements of corn grinding, but even here it is at some cost, convenience and the availability of labour at the right time. This chapter outlines the basic principles underlying the design of wind machines.

9.2 Wind energy and power

Wind energy is the kinetic energy of the moving air mass. A mass m having a velocity v will possess a kinetic energy equal to $1/2\ mv^2$. If we consider unit area at right angles to the direction of the wind then each second a mass equal to ρv will cross this area, where ρ is the density of the air, Fig. 9.10. The

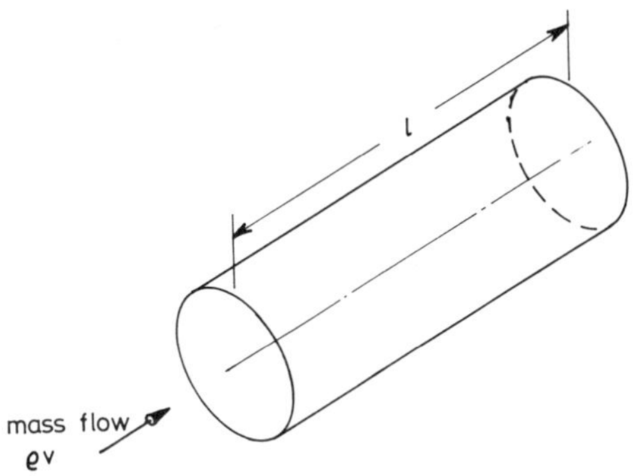

Fig. 9.10 *Mass flow per unit area per second*

corresponding energy flow per unit area or power per unit area P/A is therefore given by:

$$P/A = 1/2\ \rho\ v^3 \tag{9.1}$$

For example, a 15 mph (6.7 m/s) wind will give a power flux of 180 W/m^2; this is a figure comparable to the solar flux falling on a horizontal surface. Fig. 9.11 shows how P varies with wind velocity. This v^3 dependence of power on velocity causes wind energy to be highly sensitive to variation in wind velocity. A doubling of the wind velocity will cause the power output to increase by a factor of eight, whereas a halving will cause an eight-fold reduction. If the rated power is 1 kW such a swing in power would range from 125 W to 8 kW.

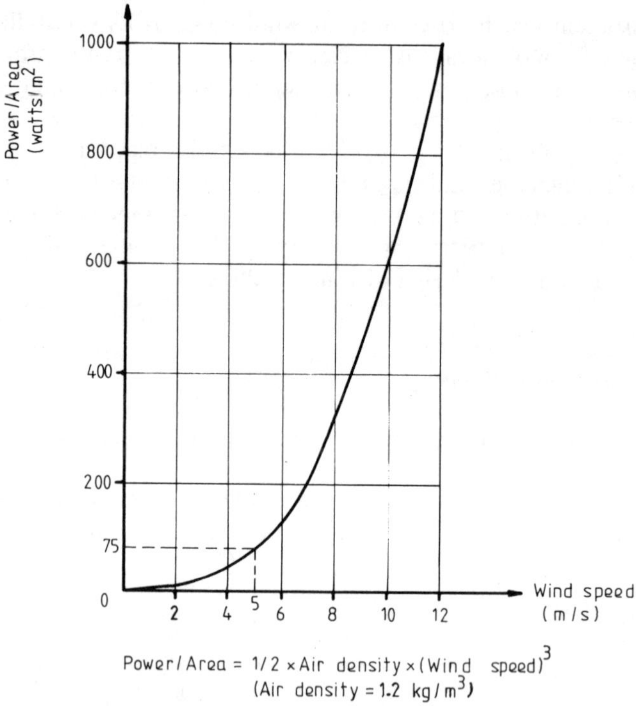

Fig. 9.11 *Variation of specific power with wind speed*

9.3 Wind data

9.3.1 General considerations

In assessing an area for its wind power potential it is necessary to have information on the amount of wind in the area and its seasonal and diurnal variation both in magnitude and direction. This information is usually available from meteorological records. The actual data is provided in the form of isovent maps based on hourly mean wind speeds collected by a network of meteorological stations. For example, the average annual wind velocity in the UK ranges from less than 4.5 m/s in central England to 7.8 m/s in the Western Islands. This data is used, together with eqn. 9.1, to estimate the available wind energy. Because of the cube law dependence of power on velocity the substitution of the average annual wind velocity $\bar{v}$ in eqn. 9.1 will lead to an underestimate of the power density and a correction factor is necessary. The precise value of this factor will depend on the site, it has a value approximately equal to two. Hence average annual power per unit area:

$$P_{av} = \rho \bar{v}^3 \tag{9.2}$$

Since the instantaneous wind power varies as the cube of the velocity it might be questioned whether estimating wind energy on the basis of hourly means is likely to lead to significant error. Studies have shown that the structure of the wind is such that the use of the hourly average leads to an underestimation of useful wind energy by about 5% to 10%.

Wind data are taken using instruments at varying heights above the ground, and the results are corrected to refer to a standard height of 10 m. The instantaneous wind velocity is sometimes expressed on the Beaufort Scale, see Table 9.2.

Isovent maps are valuable in building up a picture of the regime in an area and in understanding the likely seasonal and other variations. However, they do not assist in the selection of actual sites. Because of the velocity cubed relationship it is extremely important to select windy sites, for example, if by suitable site selection it is possible to obtain an average velocity 10% higher than the average for the area the power available will be almost 30% greater.

Table 9.2 *Beaufort scale*

Beaufort number	Description	Wind speed MPH	Wind speed m/s	
0	Calm	<1	<0·4	Still, smoke rises vertically
1	Light air	1–3	0·5–1·5	Smoke drifts
2	Light breeze	4–7	2–3	Wind felt on face, leaves rustle
3	Gentle breeze	8–12	3·5–5	Leaves and small twigs move, streamer extends
4	Moderate breeze	13–15	5·5–8	Raises dust and papers, moves twigs and small branches
5	Fresh breeze	19–24	8·1–10·9	Small trees in leaf begin to sway
6	Strong wind	25–31	11·4–13·9	Large branches move, overhead wires sing
7	Moderate gale	32–38	14·1–16·9	Whole trees move, resists walkers
8	Fresh gale	39–46	17·4–20·4	Breaks off twigs, impedes walkers
9	Strong gale	47–54	20·5–23·8	Blows off roof tiles, chimney pots
10	Whole gale	55–63	24·4–28	Trees uprooted, widespread damage
11	Storm	64–72	28·4–32·5	Widespread damage (rare)
12	Hurricane	73–82	32·6–35·9	

Electricity generation allows more flexibility in site selection since the output power can relatively easily be transmitted to a load some distance away. Water pumping, if mechanical drive is used, will restrict the choice of site to a much greater extent, and, for example, may rule out the use of an otherwise suitable hill. We will now look in more detail at the problem of site selection.

9.3.2 Site selection

9.3.2.1 Effect of height: It is found that wind velocity increases with height above the ground according to the relationship:

$$\frac{v_1}{v_2} = \left[\frac{Z_1}{Z_2}\right]^x$$

where v_2 is the wind velocity at some reference height Z_2 and v_1 is the wind velocity at height Z_1. The constant x will depend on the nature of the surface. A typical value for x for flat grassy surfaces is 0.17.

Wind velocities are normally measured at a height of 10 m, hence for a height of 20 m the velocity will be increased by a factor $(2)^{0.17}$ or 1.12. Since power varies as the cube of the velocity the power will be increased by a factor $(2)^{0.51}$ or 1.42, an increase of 40%.

9.3.2.2 Hill sites: Hills can sometimes be used to give an enhancement of wind velocity over that on the surrounding plain. The best hills are those having smooth sides and a fairly steep conical shape, Fig. 9.12. The effect of height above the top of the hill is complex and the previously quoted relationship cannot be used; it is necessary to make actual measurements on the site.

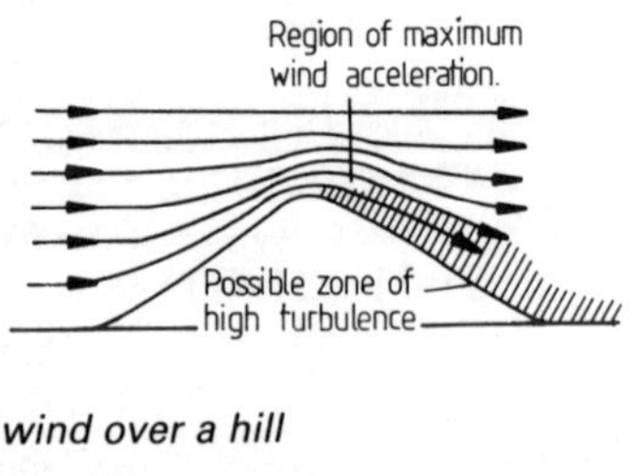

Fig. 9.12 *Acceleration of wind over a hill*

9.3.2.3 Land–Sea breezes – diurnal breezes: At the sea shore the general wind pattern will have superimposed on it an air flow from the sea to the land at night, and in the reverse direction during the day. The reason for this is shown in Fig. 9.13*a* and *b*. Briefly what happens is that at night the land mass cools relative to the sea and the warmer air over the sea rises to be replaced by air from the land. During the day the sun causes the land surface to rise in temperature more quickly than that of the sea. These land/sea breezes can be

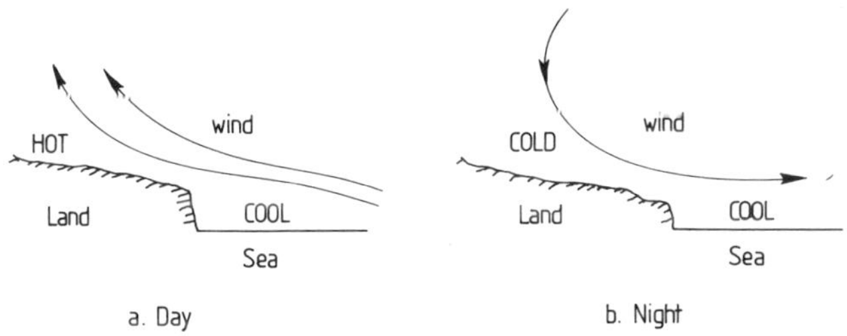

Fig. 9.13 *Diurnal wind*
a Day
b Night

quite strong and can dominate the wind pattern. Fig. 9.14 shows a wind rose for such a situation in Thailand. In this case it is unnecessary to arrange for a horizontal axis mill to change direction and very simple designs can be used, Fig. 9.15. A rather similar situation to the sea breeze can arise in valleys and in this case the mill would be sited on the hillside.

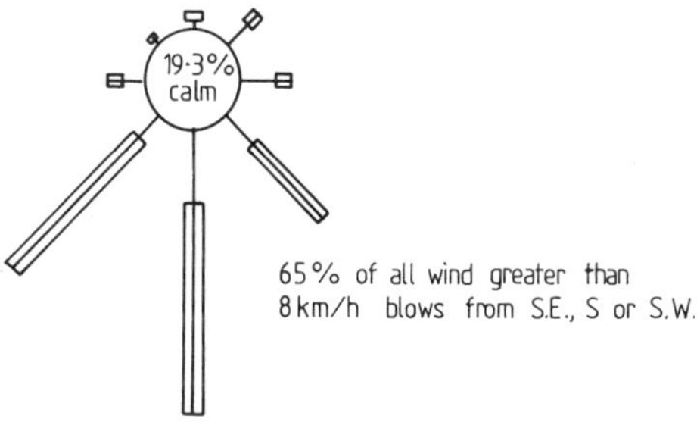

Fig. 9.14 *Wind rose, Bangkok, March, April, May*

9.3.3 Wind statistics

If the annual duration or total time that the wind has a particular velocity range is noted over a year it is found to vary as shown in Fig. 9.16. This refers to Medani in the Sudan and each column represents the number of hours that the wind has a velocity in the range of ±0.25 m/s about the mean velocity. The average velocity can be obtained from Fig. 9.16 and has a value of 4 m/s. Histograms in the form of Fig. 9.16 can be replotted as a smooth curve which gives the probability distribution, Fig. 9.17. Care should be taken in the interpretation of the vertical axis. If this is expressed as numbers of hours then the range of velocity must be stated. The total area under the curve will then be 8760 hours which is the total hours in a year. If the vertical axis is given as a probability of the wind velocity lying within a certain range about a mean value then the area under the curve should equal unity.

Fig. 9.15 *Water pumping wind mills, Bangkok, Thailand.*
a Wooden bladed mills
b Canvas bladed mills

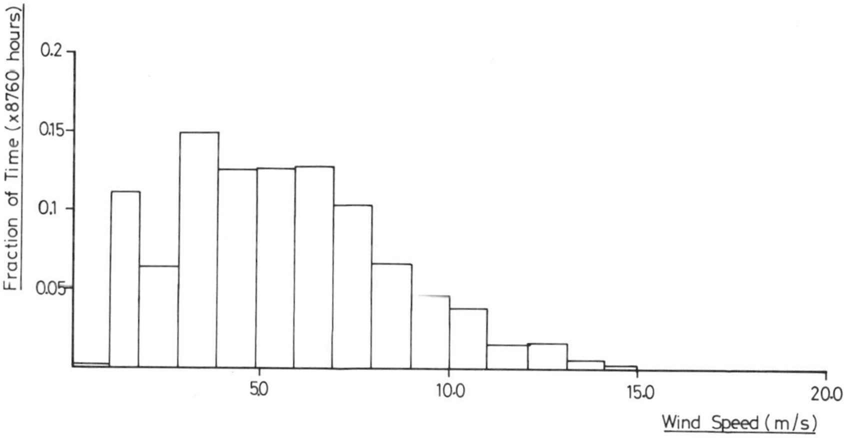

Fig. 9.16 *Velocity histogram for Medani, Sudan*
$\bar{v} = 5{\cdot}0$ m/s

From the data in Fig. 9.17 it is possible to derive a curve which relates the percentage of the year when the wind speed will be greater than or equal to a particular value. This is called a velocity duration curve, Fig. 9.18. Such a plot can be used to derive the total wind power that can be expected in a year, Fig. 9.19.

For some purposes the seasonal variation in wind power is important, in which case it is necessary to break down the annual figures into monthly totals.

Curves such as Fig. 9.19 present the windmill designer with a dilemma. In order to extract the maximum possible energy from the wind using a particular mill it is necessary to design the system so that it can operate at maximum wind velocity. This, however, will only occur for a small fraction of the total operating time, and for the rest of the operation the windmill will be producing

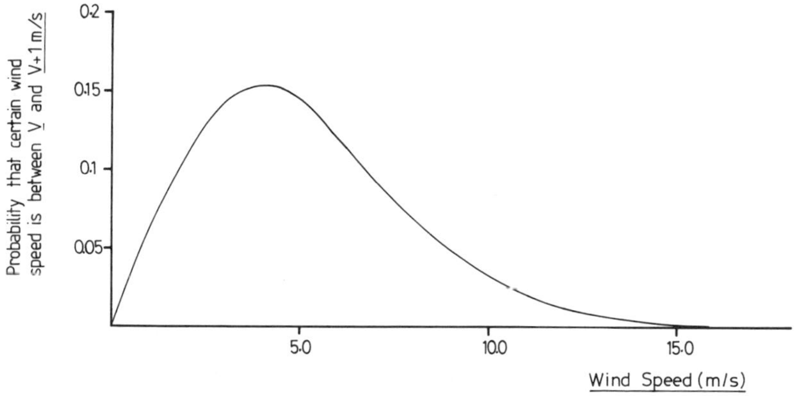

Fig. 9.17 *Wind probability distribution for Medani (Sudan)*
$k = 2$, $c = 5{\cdot}7$ m/s
Weibul parameters

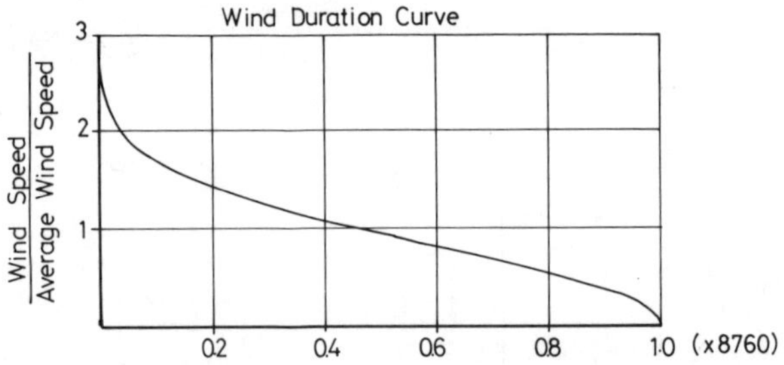

Fig. 9.18 *Velocity duration curve*

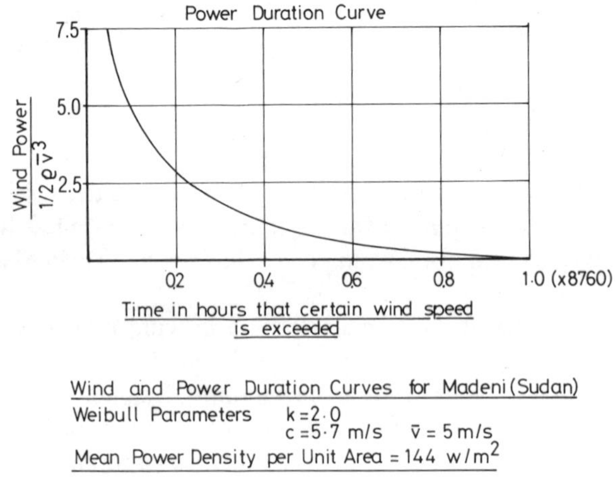

Fig. 9.19 *Power duration curve*

power at a lower value than the peak rating of the generating equipment. This means that the ratio mean power/rated power will be low and hence the cost of unit energy high. The choice of peak rating value is an important design consideration. It is discussed in Section 9.5.4.

9.4 Wind machine design

9.4.1 Maximum power extraction: the Betz limit

It is clearly not possible to extract all the energy from the air which passes through the windmill disc. If this were so then the air would cease to move and pile up behind the windmill. In practice the airflow through the disc is as shown in Fig. 9.20. It was shown by Betz[4] that under conditions of maximum power extraction the air velocity at the windmill disc will have fallen to 2/3 of the value of the upstream wind velocity v_∞ and further decreases to a final value of

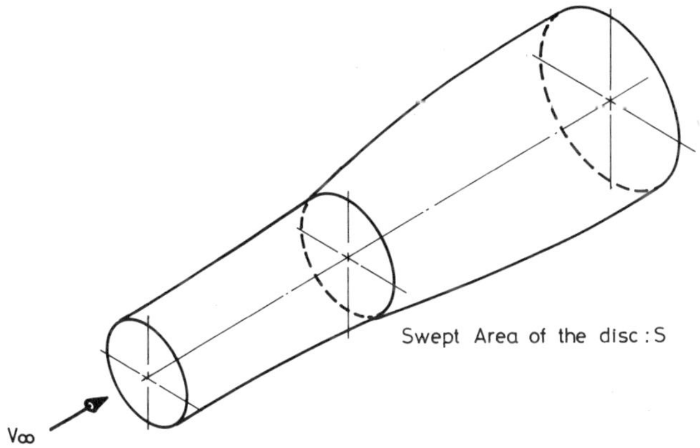

	Upstream	Disc	Downstream
Velocity	V_∞	$\frac{2}{3}V_\infty$	$\frac{1}{3}V_\infty$
Area of streamtube	$\frac{2}{3}S$	S	$2S$
Power	$\frac{1}{3}\rho V_\infty^3 S$	$\frac{4}{27}\rho V_\infty^3 S$	$\frac{1}{27}\rho V_\infty^3 S$

Fig. 9.20 *Condition for maximum power extraction, Betz limit*

1/3 the initial wind velocity well downstream of the windmill. This means that we can only extract a fraction 24/27 of the initial wind energy. It is more convenient to express the performance of a windmill as the power output divided by the power in the wind passing through an area equal to the swept area of the mill. From Fig. 9.20 this is seen to be a fraction 16/27. This fraction works out to be 59% and is called the Betz limit. Sometimes windmill performance is given as a fraction of the Betz limit.

Practical windmills will not achieve the Betz limit, thus the wind velocity at the disc will be somewhat greater than the value of 2/3 v_∞. The value of the wind velocity at the disc is expressed in terms of an inflow factor a such that the wind velocity is given by $v_\infty(1 - a)$. The inflow factor will lie between 0 when the mill is not operating and 1/3 when the mill is operating at maximum efficiency. In windmills the blades rotate, giving a further limitation to efficiency. In order that the rotor should rotate it is necessary for the wind to exert a torque about the rotor axis; by the principle of angular momentum the rotor will exert an equal and opposite torque on the air stream causing it to rotate in the opposite direction to that of the rotor. This rotating wake is particularly important at low rotor speeds. The rotating wake will contain kinetic energy which will need to be subtracted from the energy available to the rotor. The effect of the rotating wake on maximum efficiency is shown in Fig. 9.21.[5] The efficiency is plotted in terms of a dimensionless speed λ which is defined as tip speed/wind speed. The effect of wake rotation is seen to be less

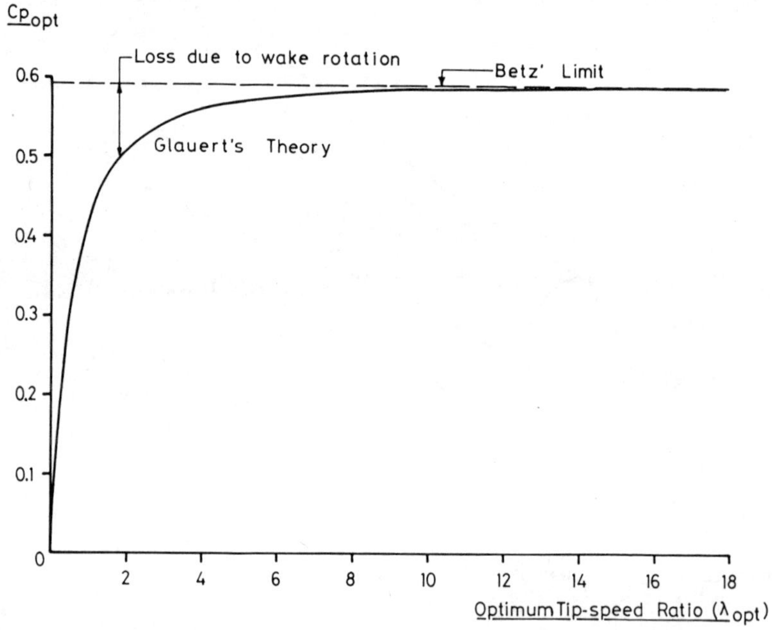

Fig. 9.21 *Effect of rotating wake on maximum efficiency*

important at higher values of λ where efficiency approaches to the Betz limit, but is a severe limitation at low speeds. In a similar manner to that for the axial reduction of wind speed the increase in angular velocity of the wake is expressed in terms of the rotation speed of the rotor Ω and a radial inflow factor a_1 such that the average wake angular speed at the rotor is $a_1\Omega$. It should be noted that because of the axial and radial inflow effects, the magnitude and direction of the wind as seen by the rotor blade is not the same as the undisturbed wind velocity v_∞.

9.4.2 Rotor design

So far we have considered the maximum fraction of the wind energy it is theoretically possible to extract using a wind rotor. The next question is how should it be constructed to achieve energy extraction? There are obviously many practical solutions, and, as in most engineering situations, there is no single 'best' arrangement and the choice will depend very much on the application; for example, the design selected for a small scale pumping system will be quite different from that chosen for large scale electricity generation. In addition to technical factors cost will be a major consideration in making a selection. We will now look at the general principles underlying windmill design.

9.4.3 Drag and lift

It is common experience that an object placed in the wind will experience a force which can be quite large – trees blow down from time to time. This force depends on a number of factors: the wind velocity v; the density of the air ρ; the area of the object at right angles to the wind direction and its shape. In Fig. 9.22 it is seen that if the flow is stopped then a force/unit area (a pressure) of

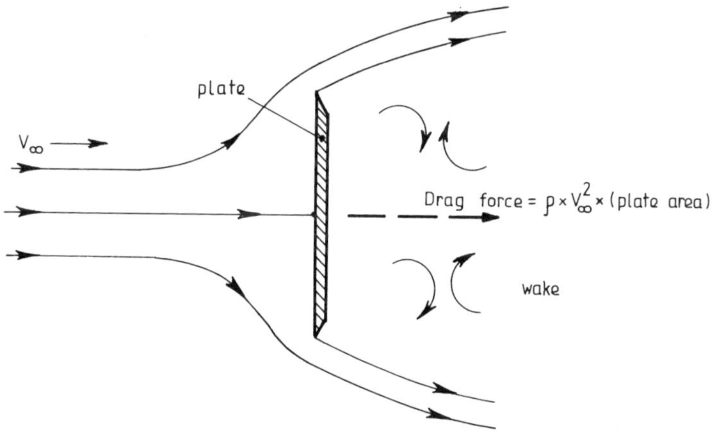

Fig. 9.22 *Drag Force*

ρv^2 will be exerted on the body. This is called the drag force. If the air is not stopped completely but continues to flow round the body then the drag will be somewhat less. Drag force is usually expressed in terms of a dimensionless quantity, the drag coefficient C_D, as follows:

$$C_D = \frac{\text{drag force}}{1/2\, \rho v^2 \times \text{area}}$$

Drag coefficients for various shapes are given in the literature and those for a few simple shapes are given in Table 9.3. For a given shape the drag coefficient is not entirely independent of air velocity, and the range for which a given value applies is defined by a further dimensionless number, the Reynolds number, Re:

$$\text{Re} = \frac{\rho v d}{\mu}$$

where ρ is the fluid density, v is the fluid velocity, d is a characteristic dimension, for example, the diameter of a spherical body, and μ is the viscosity of the fluid.

Reynolds number is an important quantity in fluids problems and occurs frequently. Fig. 9.23 shows the striking effect that streamlining can have on the drag coefficient. The cup anemometer, which is used for wind measurement,

Table 9.3

Drag Coefficients for some standard Geometries.

Body Shape	C_D	Reynolds Number
Circular Cylinder	1·2	10^3 to 10^5
Square Rod	2·0	$> 10^4$
Semicircular Shell →)	2·3	$> 10^4$
Semicircular Shell → (	1·2	$> 10^4$
Elliptical Cylinder 2:1	0·6	10^4 to 10^5
Hemispherical Shell → D	1·40	$> 10^4$
Hemispherical Shell → (	0·39	$> 10^4$

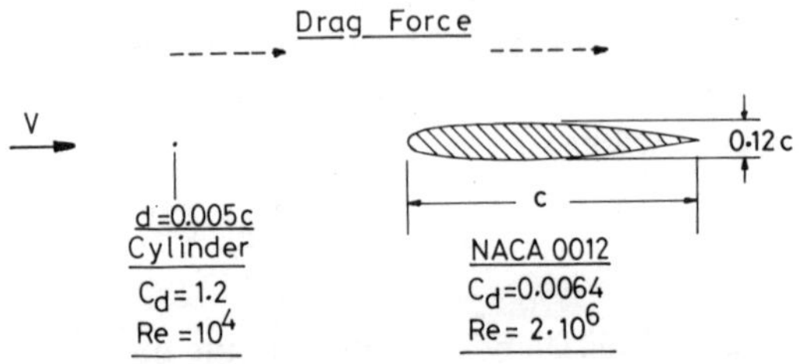

Fig. 9.23 *Comparison in size of a circular cylinder and a symmetrical aerofoil producing the same drag force. (Same incident wind velocity v)*

shown in Fig. 9.24, depends largely on drag for its operation. There is a second effect, lift, on which most modern windmills are based, Fig. 9.25. When a plate is placed at a small angle to the direction of an air flow then in addition to the drag force in the direction of an air flow the plate will also experience a force at right angles to the flow. This is called the lift force. Lift force is due to a pressure difference between the upper and lower surfaces of the plate. Lift arises from an increased pressure on the lower surface and a reduced pressure on the upper surface; the latter makes the greater contribution to the lift force.

Fig. 9.24 *Cup anemometer (Courtesy Casella London Ltd)*

If the angle α, called angle of attack, is increased too much then the flow will break away from the upper surface and the lift force will be considerably reduced. This condition is known as stall. In a similar manner to drag, lift can be expressed in terms of a dimensionless number, the lift coefficient C_L, where

$$C_L = \frac{\text{lift force}}{1/2\, \rho\, v^2 \times \text{area}}$$

C_L will depend on the Reynolds number and the angle attack α.

Lift/drag ratio can be considerably increased by streamlining the plate in the form of a modern aerofoil, Fig. 9.25*b* and *c*.

Graphs of C_L, C_D, C_L/C_D as functions of the angle of attack α for a particular aerofoil are plotted in Fig. 9.26. The area for both C_L and C_D is the area/unit length of aerofoil, that is the chord multiplied by unit length along the aerofoil.

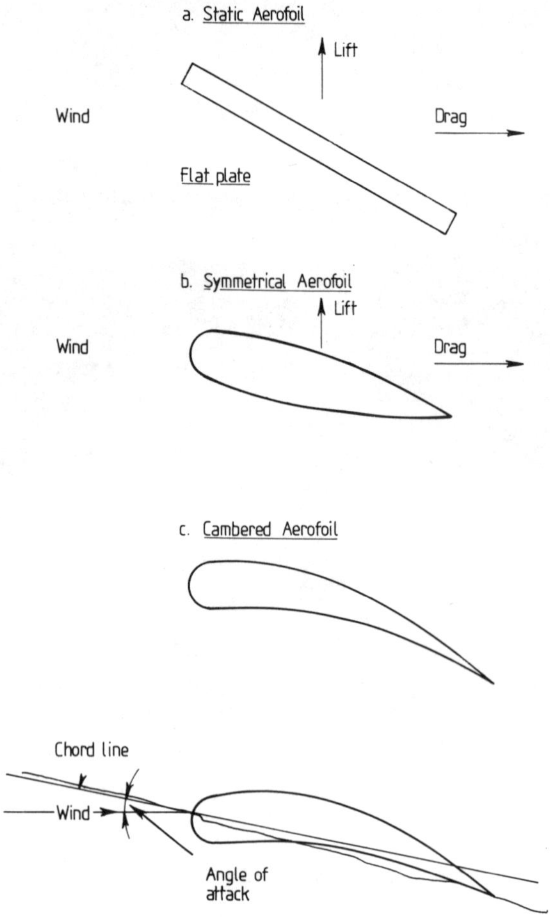

Fig. 9.25 *Lift and Drag*

9.4.4. Forces on the blade

When an aerofoil, such as an aeroplane wing, is moved through the air it will experience lift and drag forces as shown in Fig. 9.27. Exactly the same forces would be experienced by the aerofoil if it were to be held stationary in a moving air stream, Fig. 9.28*a*. If in addition there exists a wind velocity v at right angles to the direction of motion of the aerofoil, Fig. 9.28*b*, then the aerofoil will see a resultant wind velocity w made up from these two components u and v. in the case of the windmill blade a similar situation will arise, since, when a windmill blade moves, each sector of the blade will move relative to the air with a velocity given by Ωr where Ω is the angular velocity of the rotor and r the distance of the sector from the rotor axis. Also, if a wind exists at right angles to the plane of rotation of the blade one might expect the

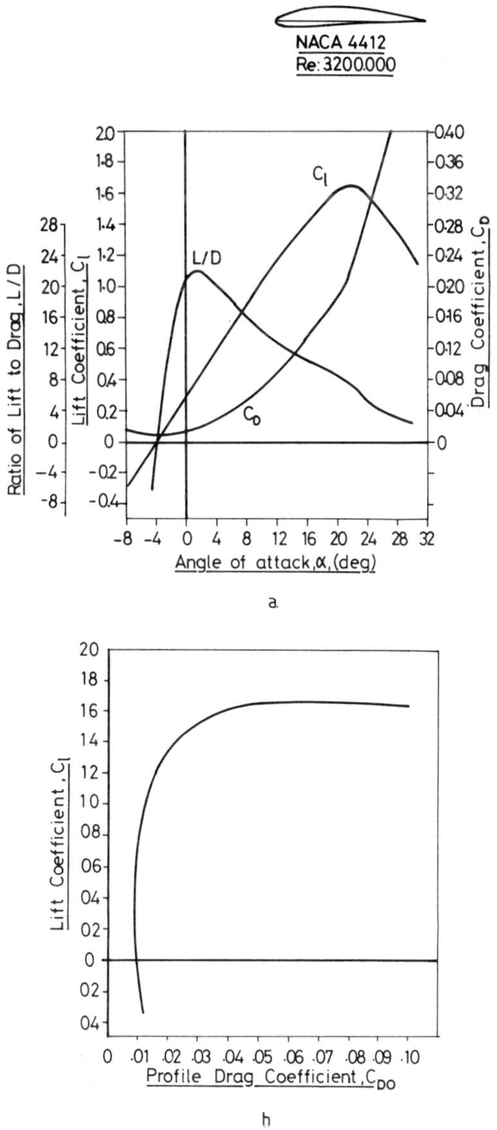

Fig. 9.26 *Aerofoil lift and drag characteristics (Ref. 6)*
a Lift/drag ratio *vs* angle of attack
b Lift coefficient c_L *vs* Drag coefficient c_D

resultant velocity seen by the blade to be made up from these components. In practice the two velocity components are modified since the windmill is extracting energy from the wind. The apparent wind velocity component will be reduced by the inflow factor a to a value $v_\infty (1 - a)$, and because the wake is rotating in the opposite direction to the rotor, the component Ωr will be

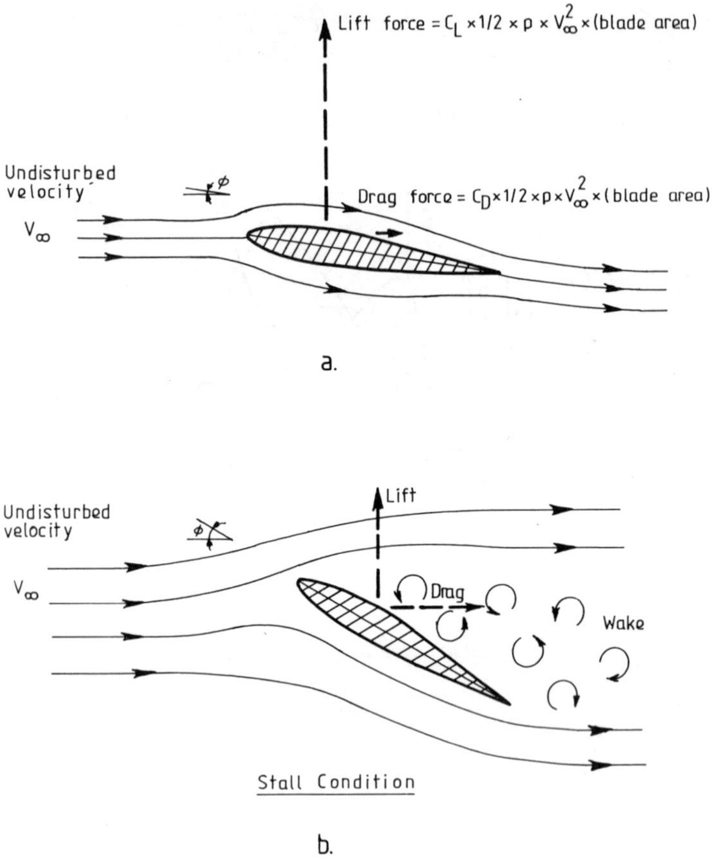

Fig. 9.27 *Lift and drag forces on an aerofoil*
a Normal condition
b Stall condition

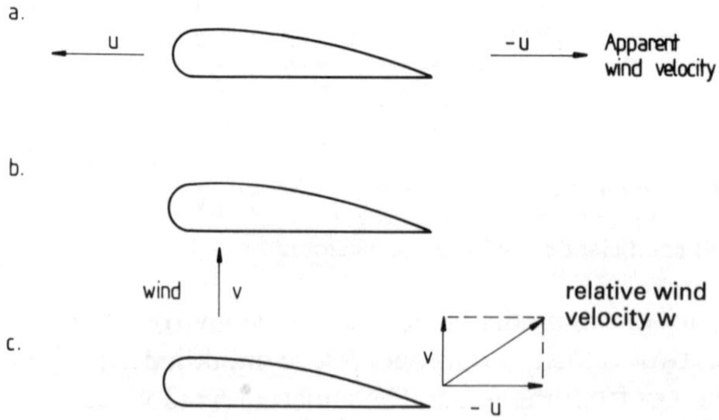

Fig. 9.28 *Apparent wind velocity due to aerofoil motion*

increased by the rotational in-flow factor to $\Omega r\ (1 + a')$. Hence the wind velocity seen by the blades w will have the two components $v_\infty\ (1 - a)$ and $\Omega r\ (1 + a')$, Fig. 9.29. In a practical windmill the blade will be arranged to make an angle β with the direction of rotation. The forces acting on the blades are

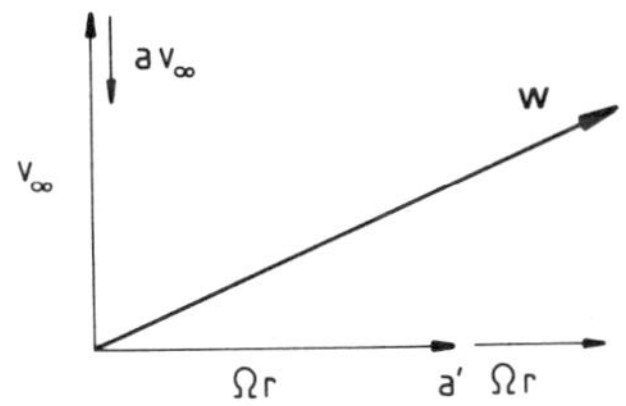

Fig. 9.29 *Apparent wind velocity*

shown in Fig. 9.30*a*. So far, for simplicity, we have referred only to the horizontal axis rotor. The same considerations apply to the vertical axis machines and Fig. 9.30*b* indicates the forces acting on the vertical axis blades.

If the apparent wind velocity w were to remain constant, then as more blades of a given area and shape are added to the rotor more energy will be extracted from the wind. However, as we have seen, this is not the case, and as more energy is removed from the wind the direction and magnitude of the wind as seen by the rotor will change markedly. These changes are expressed in the form of the two inflow factors.

In order to evaluate the performance of a rotor it is necessary to satisfy simultaneously the equations which define the air flow through the rotor, together with the equations which define the forces on the blades due to that air flow. An aerofoil of known characteristics is selected and a value of the tip speed to wind speed ratio λ chosen.

$$\lambda = \frac{\Omega R}{v_\infty}$$

where λ is the tip speed ratio, Ω is the angular velocity, R is the rotor radius and v_∞ is the undisturbed wind speed

The optimum blade angle β and the total chord length (that is, the solidity or fraction of the disc occupied by the blades) can then be obtained for each value of radius r together with the overall power and torque produced by the rotor, Figs. 9.31 and 9.32. The manner in which power and torque will vary if the rotor is operated at other values of λ can then be evaluated. The actual numerical solution of the equations is somewhat tedious and is usually carried

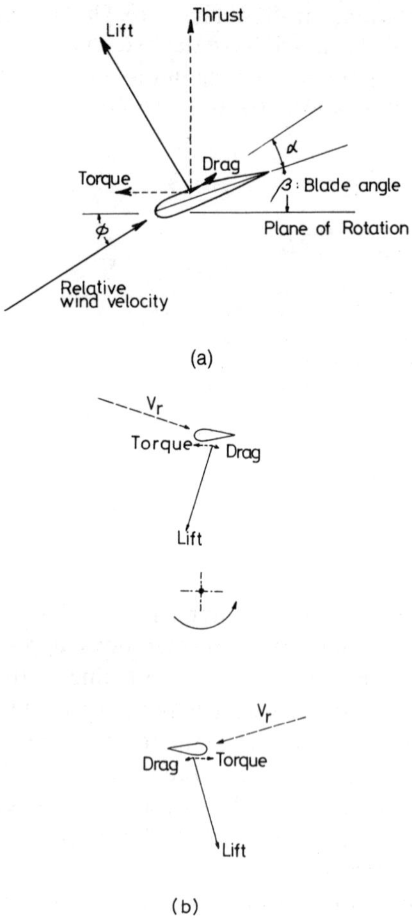

Fig. 9.30 *Forces on the blade*
a Horizontal axis machine
b Vertical axis machine

out on a digital computer. The results can be expressed in terms of a dimensionless power coefficient C_P which is defined:

$$C_P = \frac{\text{power extracted}}{1/2\,\rho\, v^3 S}$$

where S is the area of the disc, or swept area.

A typical characteristic is shown in Fig. 9.33.

9.4.5 Choice of λ and solidity

For a given machine there will be a value of λ at which C_P will be a maximum,

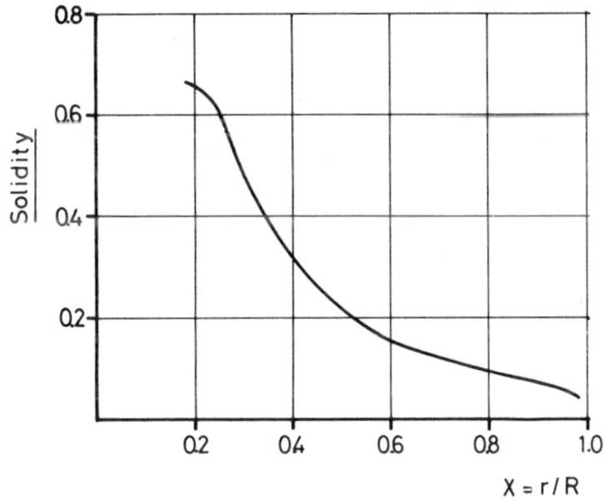

Fig. 9.31 *Solidity versus radius for Horizontal axis wind turbine, 4 bladed, NACA 0012, $\lambda = 4$*

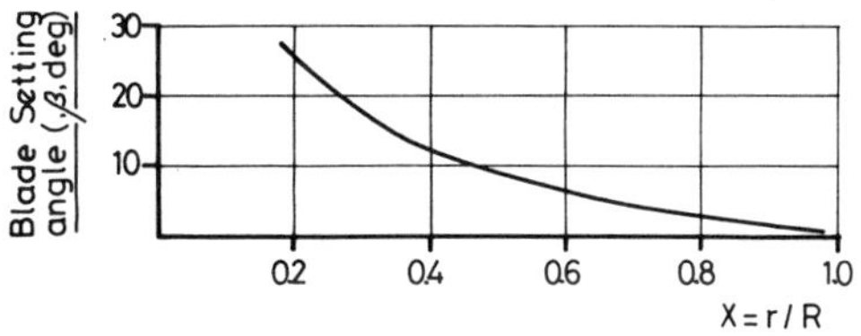

Fig. 9.32 *Setting angle along the blade of a Horizontal axis wind turbine, 4-bladed, NACA 0012, $\lambda = 4$*

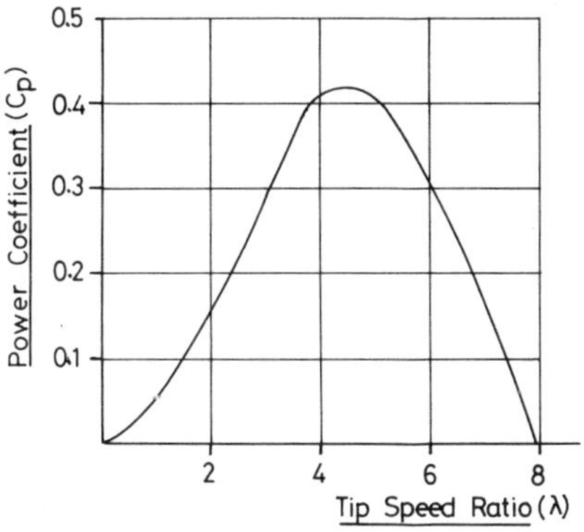

Fig. 9.33 *Variation of the power coefficient with the tip speed ratio*

and hence will produce maximum power, Fig. 9.33. For small machines C_{Pmax} is around 0.3, rising to 0.4 for large machines, Fig. 9.34.

A further useful parameter is the torque coefficient C_T which is defined:

$$C_T = \frac{\text{torque}}{1/2\,\rho v^2 SR}$$

and R is the radius of the swept area.

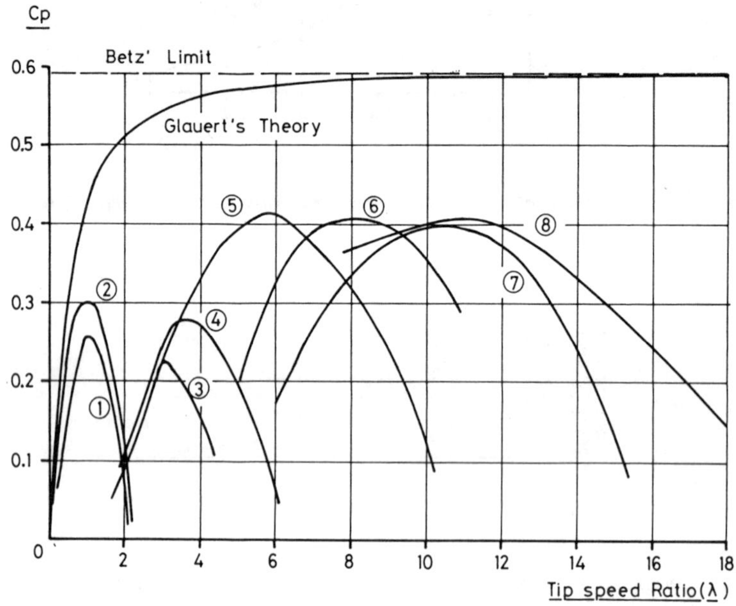

1. ITDG Rotor 6 m Dia. 24 blades. Ref. 16.
2. American Multibladed rotor 3 m Dia. 18 blades. Ref. 17.
3. Musgrove VAWT 6 m Dia. 2 blades. Ref. 18.
4. Pionier-I VAWT 15 m Dia. 2 blades (Darrieus Type). Ref. 19.
5. Sandia National Laboratories 17 m VAWT 2 blades. Ref. 20.
6. 25 m HAT, ECN, Patten. Ref. 21.
7. 5 m Dia. 2 bladed. Ref. 22.
8. NASA Mod-OA 37.8 m HAWT 2 blades. Ref. 23.

Fig. 9.34 *Typical performance of wind power machines*

C_T is related to C_P by the equation

$$C_P = \lambda\, C_T$$

C_T will rise to a maximum value, this occurs for a lower value of λ than that for C_{Pmax}, Fig. 9.35.

So far we have seen how to calculate the output of a windmill employing blades of given aerodynamic characteristics for a chosen value of λ. The solution specified the total solidity but did not indicate how this was to be achieved, that is, it did not say how many blades should be employed.

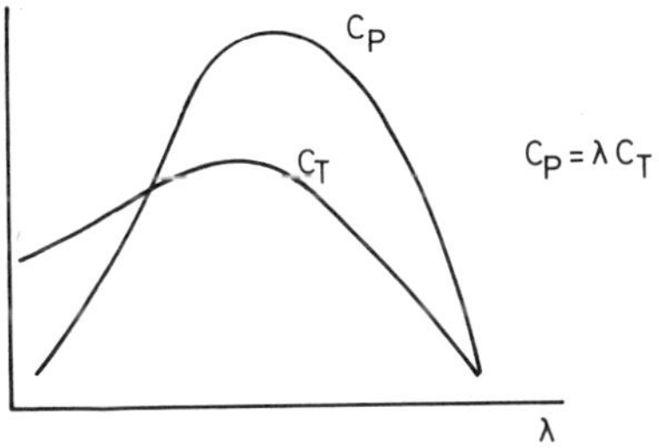

Fig. 9.35 *Power and torque coefficients for different tip speed ratios*

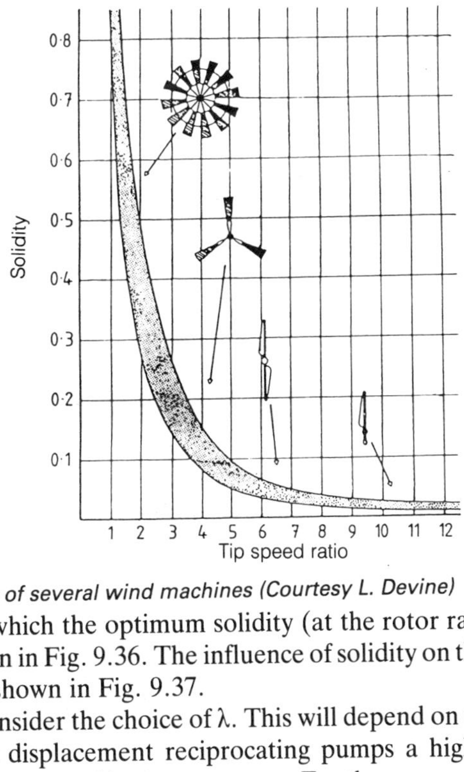

Fig. 9.36 *Solidity of several wind machines (Courtesy L. Devine)*

The way in which the optimum solidity (at the rotor radius) varies with tip speed λ is shown in Fig. 9.36. The influence of solidity on the power and torque coefficients is shown in Fig. 9.37.

First let us consider the choice of λ. This will depend on the application. For use with direct displacement reciprocating pumps a high starting torque is required and a low speed is also necessary. For these reasons a small value of λ would be selected, around 2 to 4. Such a rotor should have a large number of blades, partly for mechanical convenience but also because it is found that at low values of λ the losses from the blade tips are large. This effect is reduced as the number of blades is increased. If the application is for electricity generation then the starting torque requirement is small but high speed is desirable for the electrical generator. Hence a high value of λ is suggested, say 8 to 10. For such high values of λ it turns out that the solidity is low and also tip losses unimportant. For these reasons a two or three blade rotor would be chosen.

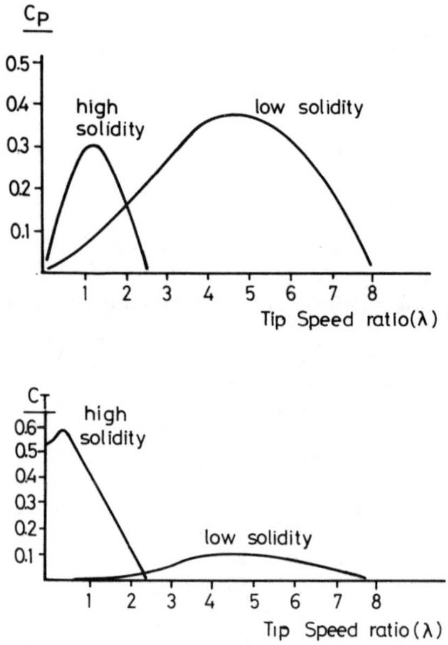

Fig. 9.37 *Influence of solidity on power and torque coefficients*

9.5 Operation

9.5.1 Start up, normal and high speed running

At start up the rotor will be stationary, thus the apparent wind velocity w, will be in almost the same direction as the undisturbed wind velocity v_∞. There will still be a small component in the rotor plane due to the radial inflow factor a'. Nevertheless, the angle ϕ between the direction of the blade rotation and the wind v_r will be almost $\pi/2$. Hence most of the blade will be in the stalled condition, Fig. 9.38. As the rotor speed increases, then ϕ will decrease, the torque will reach a maximum value and somewhat later the power output will reach a maximum. If the speed is allowed to increase further then the power will decrease. This would occur if the windmill was not loaded. The power falls off at high speeds because the angle of attack and hence lift coefficient falls. The angle of attack α can go sufficiently negative over part of the blade to cause the lift to act in such a direction as to retard the rotor.

9.5.2 Performance characteristics

In practice we will wish to know the performance of a given windmill. This can be obtained from the C_P versus λ characteristics, such as Fig. 9.35. For a given wind velocity v_∞ the power output will be given by:

$$P = C_P \, 1/2 \, \rho \, v_\infty{}^3 \, S$$

where S is the swept area.

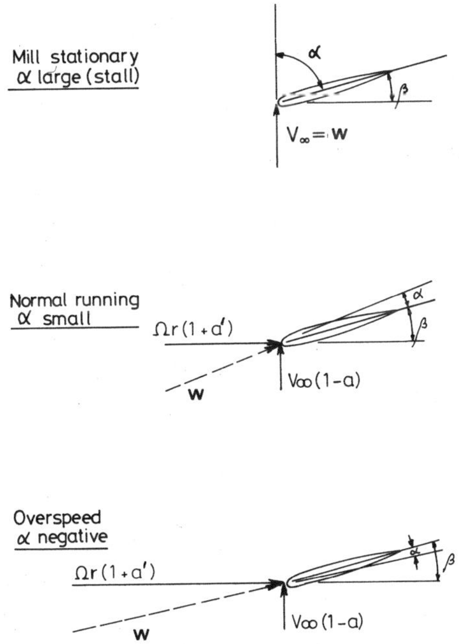

Fig. 9.38 *Start up, normal running and overspeed conditions*

The corresponding speed n is obtained from the relation:

$$\lambda = \frac{\Omega R}{v_\infty} = \frac{2\pi n R}{v_\infty}$$

or

$$n = \frac{\lambda\, v_\infty}{2\pi R}$$

Similarly, torque T is given by

$$T = C_T\, 1/2\, v^2{}_\infty SR$$

Fig. 9.39*a* shows how power output and torque vary with rotation speed for a fixed wind velocity v_∞. A different set of curves will be required for other values of v_∞, Fig. 9.39*b*.

Referring to Fig. 9.40, P_{max} indicates the point of maximum power output and corresponds to a rotation speed n_o. This is unstable and a point such as P_1 should be selected. That P_1 at speed n_1 is stable can be seen as follows. If the speed is slightly less than n_1 then there will be additional power available to accelerate the rotor to speed n_1. Similarly, if the speed is too high then the power available will be less than that required by the load. In a similar manner it can be shown that P_2 is unstable.

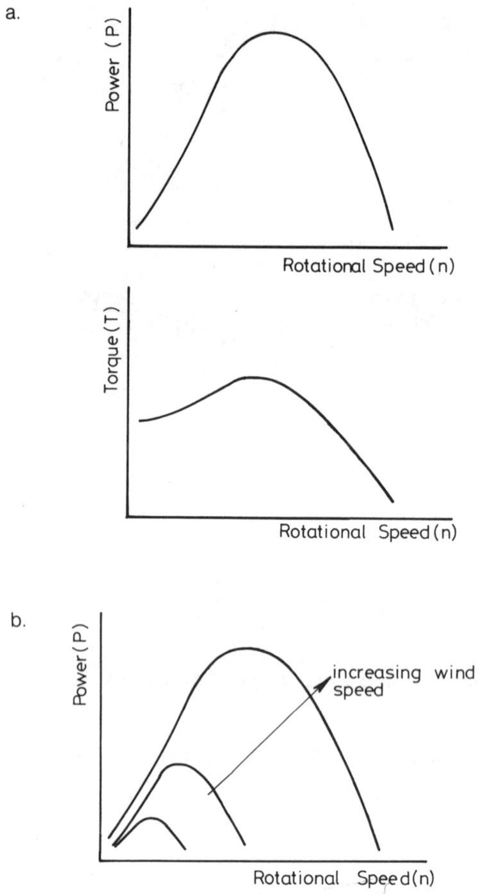

Fig. 9.39 *a* Power and torque characteristics of a wind machine versus rotational speed at a fixed wind velocity
b Power characteristics of a wind machine versus rotational speed with the wind speed increasing

Fig. 9.41 indicates the qualitative difference between a low speed (low λ) machine designed for example for water pumping with that of a high speed (high λ) machine suitable for electricity generation.

9.5.3 Load matching

Fig. 9.40 indicates the power which could be obtained from a properly matched windmill, that is one in which the load has been so chosen that it will absorb the maximum power available at the corresponding rotation speed of the windmill. In fitting pumps and electrical generators to windmills this does not necessarily happen, and one sometimes finds an unhappy coupling of incompatibles forming a combination which gives only a small fraction of the

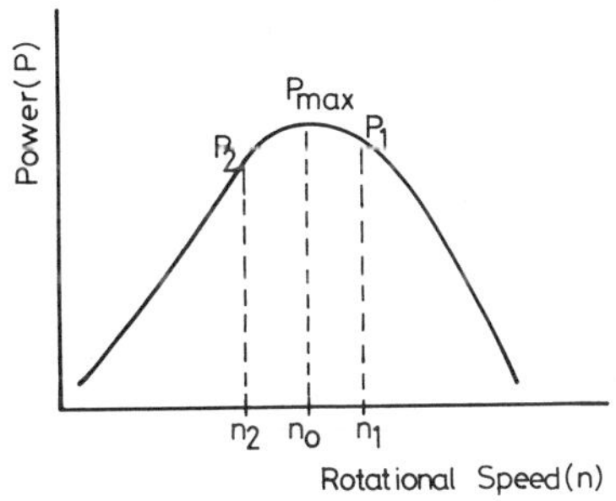

Fig. 9.40 *Selection of working point*

power output that the windmill could provide. The problem arises because the power output characteristics of the windmill change with changes in wind velocity, Fig. 9.39*b*. The optimum operating point will move as shown in Fig. 9.42. That is, the load should have this power/speed characteristic; this may not in fact be the case. For example, Fig. 9.43 shows that the power required for a reciprocating pump is linearly proportional to its speed. Superimposing this characteristic on the power output curves for a windmill we see that the operating point *P* will frequently be at a value of power considerably less than the maximum the mill could generate for the same value of wind velocity. This problem of achieving better matching and hence more total energy from windmills has caused designers to develop pumps and generators with characteristics which more closely match those of the windmills.

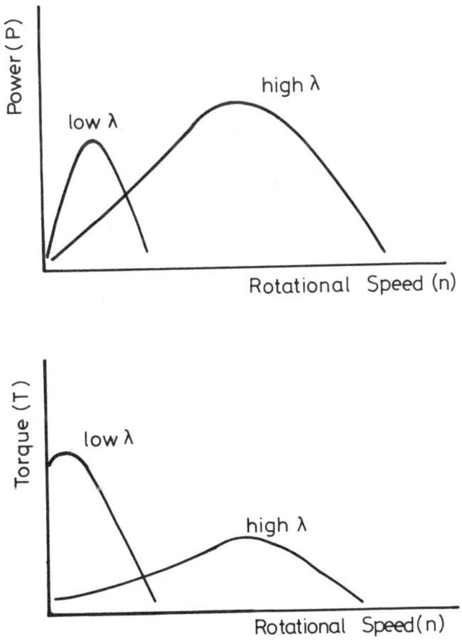

Fig. 9.41 *Variations of power and torque with rotational speed at constant wind speed*

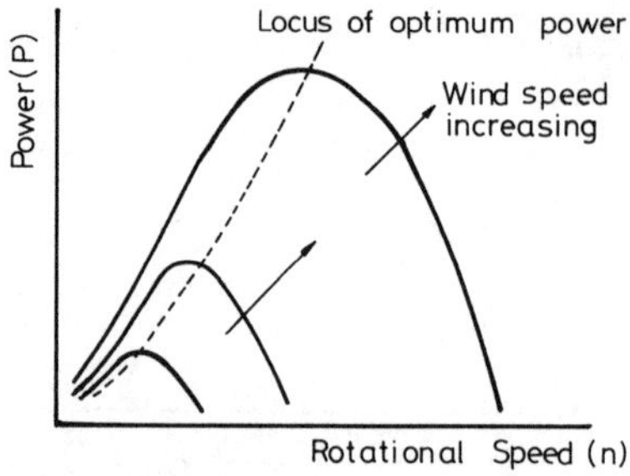

Fig. 9.42 *Optimum power point for different wind speeds*

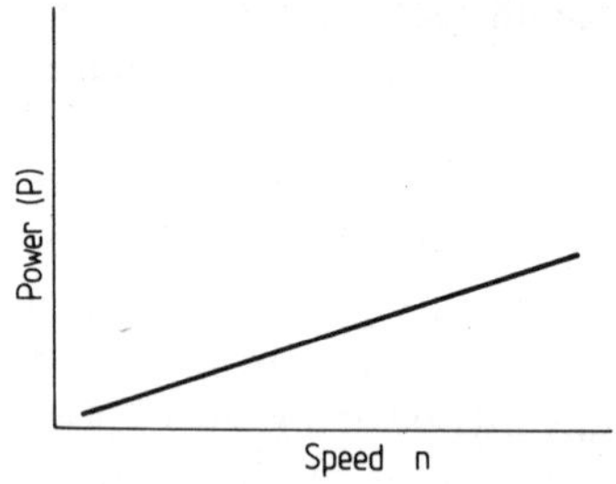

Fig. 9.43 *Characteristic of a reciprocating pump versus speed*

9.5.4 Speed range selection

When designing a wind pumping system it is usual to select a wind speed below which the torque is insufficient to operate the reciprocating pump. This is called the cut-in wind speed. Owing to the v^3 dependence on power, the loss of energy at these low wind speeds is small. The upper cut-out, or furling speed v_F is determined by limits to the strength of the rotor set by centrifugal and other forces.

The situation is somewhat different when the external load may have a cost comparable or even greater than that of the windmill, Fig. 9.44. This applies in the case of electrical generation. In this case it is usual to limit the total power to some value considerably less than that which could be generated at the furling speed. The value of the speed for the rated power is called the rated speed v_r, Fig. 9.44*b*. As pointed out in an earlier section, the higher wind speeds occur less frequently and there is little point in installing additional capacity in the form of expensive plant which will only be used relatively infrequently.

9.6 Small wind machines: *mechanical design and materials choice*

Refer to Figs. 9.4, 9.5, 9.15, 9.45 and 9.46.

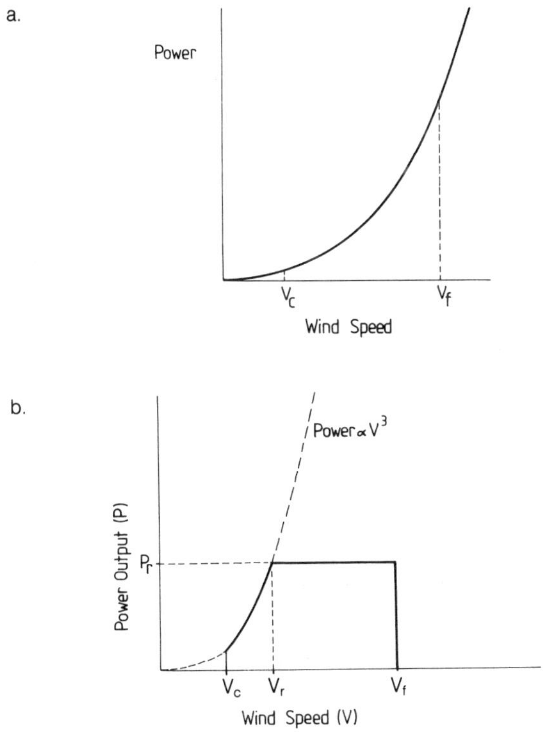

Fig. 9.44 *a* Selection of cut-in and furling speeds for wind pumps
b Selection of cut-in, rated and furling speeds for electricity generating windmills

9.6.1 Mechanical forces

9.6.1.1 Rotor: The three main forces experienced by the rotor are:

1 Axial force due to the wind, Fig. 9.47*a*.

2 Circumferential force due to air resistance tangential force from the air.

3 Centrifugal forces due to blade rotation.

In addition there will be forces tending to twist the blade, and gyroscopic forces will arise due to gusting of the wind causing a change of direction.

A major source of trouble with rotors which are not properly balanced is vibration which can seriously damage a machine.

9.6.1.2 Tower: The tower is subjected to the same axial wind force as the rotor. This causes a bending movement, Fig. 9.47*b*. It will also be required to carry the weight of the head and will be subject to wind forces acting on its own structure.

Fig. 9.45 *Cretan type wind pump. S. Ethiopia (Courtesy P. Fraenkel)*

9.6.1.3 Power Transmission

In simple pumping mills the power is often taken by means of a reciprocating vertical shaft which is operated from a crank at the top of the mill and drives a reciprocating pump at the bottom of the shaft, Fig. 9.48. The buckling load on this rod can be high and may require slideways to be spaced down the shaft. For deep wells it will be necessary to use single acting pumps having a working stroke in which the pumping rod is in tension. The extension of such rods can be quite high for deep wells and must be allowed for.

Fig. 9.46 *Prototype wind pump developed by ITDG (Courtesy P. Fraenkel)*

If the windmill shaft speed is too high for a reciprocating pump then a gearbox is required. This tends to be heavy and also requires to be topped up with lubricating oil from time to time. Belt drives are sometimes used, see Fig. 9.15. More flexibility in siting is achieved if electricity is generated. Windmills

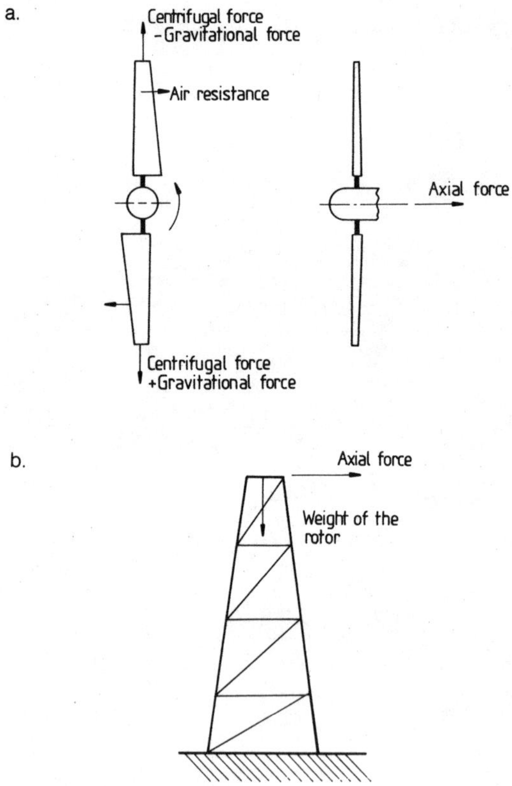

Fig. 9.47 *Forces on wind machines*
a Forces on rotor
b Forces on tower

using hydraulic power transmission and also pneumatic power take off have been used occasionally.

9.6.2 Materials
Designs should always be chosen with local materials and manufacture capabilities in mind.

9.6.2.1 Rotor: Rotors are usually made from:
Steel plate (Fig. 9.4);
Wood – the Thai wooden-bladed machine is interesting in that the blades operate at a very high tipspeed/windspeed value of $\lambda \simeq 10$. The blades are made from hard wood and have dimensions 8 m long by 20 cm wide by 5 cm thick. They are crudely carved to give a twisted aerofoil shape (Fig. 9.15*a*)
Wood and canvas sails (Figs. 9.3, 9.15b, 9.45)

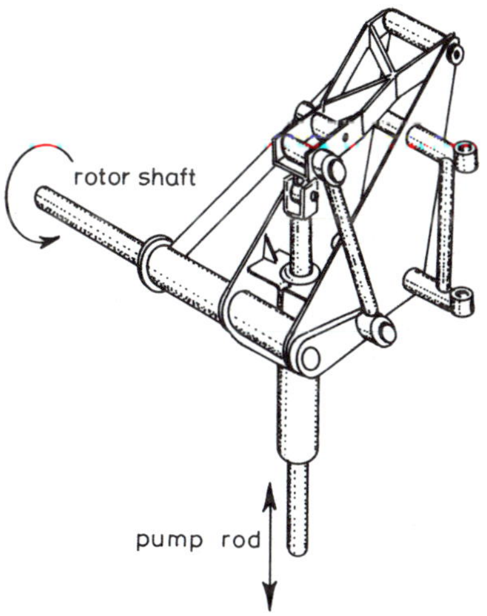

Fig. 9.48 *Schematic of the ITDG windpump transmission system for converting rotary to reciprocating motion (Reference 9)*

In some areas Dural is available and can be considered. The larger and more advanced designs have used glass fibre reinforced epoxy resin.

9.6.2.2 *Bearings:* Wooden, oil impregnated, bearings have been found suitable for small low speed pumping mills; plain bearings, ball bearings and self-lubricating bearings have all been used but difficulties are reported with achieving adequate lubrication in some cases.

9.6.2.3 *Tower:* Towers are generally between 5 and 20 m high, see Fig. 9.49. The main materials include:

angle iron, bolted or welded
tubular steel, bolted
cross-braced wood Fig. 9.49a
guyed wooden poles Fig. 9.49b
concrete

9.7 Orientation, control and power take off

9.7.1 Orientation

In the case of vertical axis machines this is not required, although some means of furling in excessively high winds will be necessary. In the case of horizontal axis machines, orientation into the wind is necessary except for sites where the wind direction does not change, for example in situations where sea breeze is the dominant wind.

Various methods are employed for orientation:

1 Tail vane
2 Secondary rotor
3 Downstream mounting
4 Servo motor

Most small machines employ a tail vane such as the one used in Fig. 9.5. Some machines use a small horizontal axis rotor set at right angles to the main shaft. This rotor operates a worm and pinion drive which turns the main rotor into the wind. In some of the larger machines the main rotor is mounted downstream of the tower and thus automatically takes up a position normal to the wind direction, see Fig. 9.7. In the larger machines orientation is often achieved by means of an electric motor drive with electronic control.

9.7.2 Control and regulation

Control to avoid damage from very high winds is required for all wind machines. Several solutions to this problem have been developed. The process is sometimes called furling and this is literally what was done on the older mills such as the tower mill shown in Fig. 9.3, where the canvas sails were rolled up or furled. This method is still used on mills such as the cretan mill, Fig. 9.45.

For small pumping machines, such as shown in Figs. 9.4 and 9.49b, it is usual to turn the main rotor out of the wind. Fig. 9.50 shows a common method.

a.

b.

Fig. 9.49 *Wind pumps with wooden towers*
a Small wind pump, Thailand
b Gaviotas wind pump developed in Colombia

Electricity generating machines normally arrange to change the pitch of the blades. In the smaller machines this is often achieved by the centrifugal force acting on weights mounted on the rotor. The Winco machine, Fig. 9.5, employs air brakes operated by centrifugal force. On large machines the control is by electronically operated electric motors.

In addition to the requirement to avoid damage by overspeed in high winds there may be a requirement to operate the machine at constant rotation speed. This need can arise in electricity generation, and speed regulation is normally achieved by pitch control. An interesting alternative for the straight-bladed vertical axis machines has been proposed by Musgrove.[11] This involves changing the geometry by allowing the blades to move as shown in Fig. 9.9. and Fig. 9.52e.

9.8 Alternative designs

Over the years there have been many proposals for new wind machine geometries. Figs. 9.51 and 9.52 show a number of these ideas.

Although the vertical axis machine was the first to appear it is the horizontal axis machine having a rotor of the type shown in Fig. 9.51 (*b,c,d*) which is by

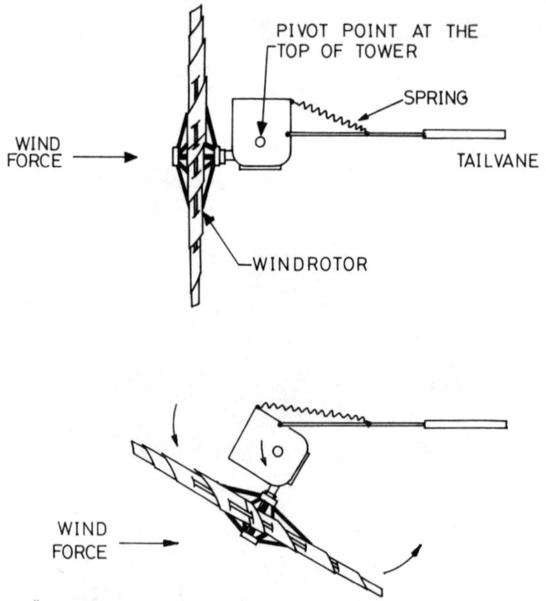

Fig. 9.50 *Arrangement for turning the rotor out of the wind at high speeds*

far the most numerous today. Small machines of this type are used for water pumping in numbers which must be at least millions. Electricity generating machines having a power of up to a few kilowatts and two or three blades are also quite widely used. Most of the high power prototype machines are also two or three blade horizontal axis machines. The MOD range in the USA are amongst the largest of these to be built, Fig. 9.7, and there are similar programmes in Europe. The vertical axis machine has received less attention until recently, apart from its very wide use as a wind indicator in the form of the anemometer, Fig. 9.52. Fig. 9.8 shows a prototype curved blade vertical axis machine and Fig. 9.9 an array of multimegawatt machines having straight blades. This latter design, proposed by Musgrove, has a variable geometry which gives a power output which is less dependent on wind velocity than the fixed blade geometry design and is considered to be a strong contender for large power output applications.

Of the rest, some remain ideas,[12] others, such as the multirotor, Fig. 9.51 *h*, have been taken to an advanced design stage, and some, such as the Enfield–Andreau, Fig. 9.50 *f*, and the contra rotating blades, Fig. 9.51 *g*, have been built in prototype form. In the Enfield–Andreau machine the blades are hollow and open at the tips, the reduction in pressure due to rotation (venturi effect) causes air to be drawn through the hollow mast in which an air turbine is mounted. The output of this turbine provides the useful output of the machine. One machine was constructed with an output of 200 kW.

The contra-rotating blade design has the advantage over single rotor machines of high effective speed, since the stator is connected to one set of

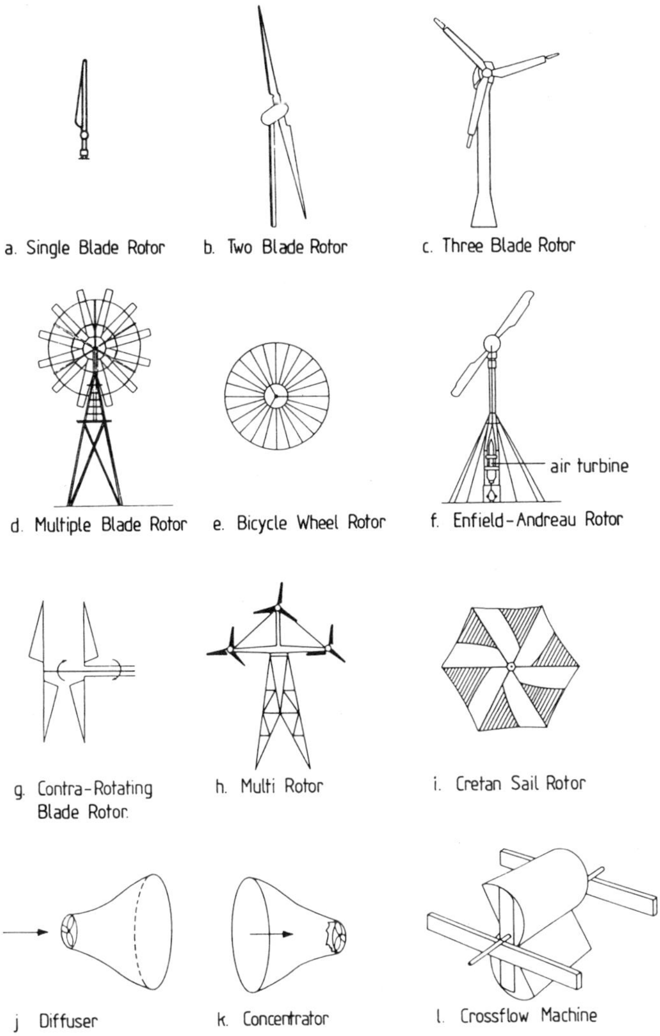

Fig. 9.51 *Horizontal axis wind machines*

blades and the rotor of the generator to the contra-rotating blades. One machine was constructed with a rating of 70 kW.[12] The cross-flow designs, Fig. 9.51 *l*, are of low performance; Fig. 9.51 *j* and *k* sketch suggested designs which employ means to guide the airflow to or from the rotor.

If a cylinder which is rotating about its axis is placed at right angles to an air flow then it will experience a force as shown in Fig. 9.53. This effect, the Magnus effect, gives a lift force similar to that experienced by an aerofoil. The Magnus effect will also occur with a rotating sphere and is responsible for 'slice' in the flight of a golf ball. The Magnus effect was applied by Anton Flettner in the 1920s to the propulsion of ships and also in windmill design,[13,14]

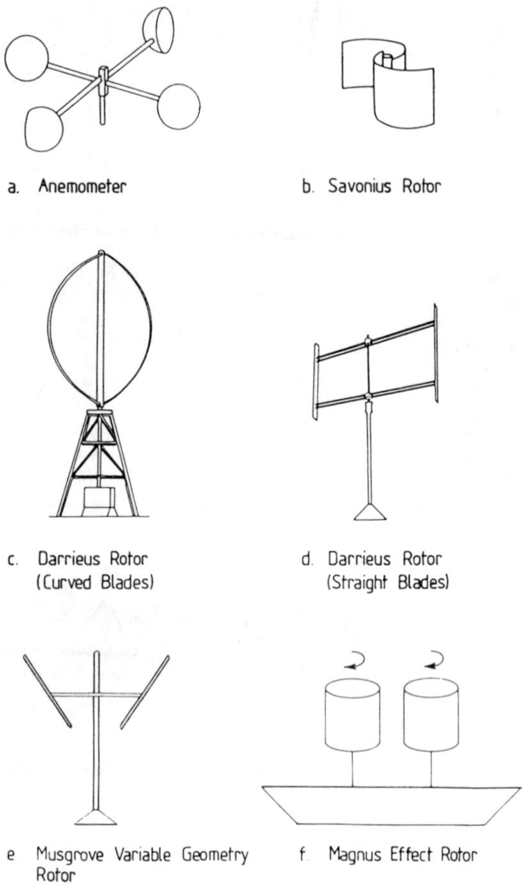

Fig. 9.52 *Vertical axis wind machines*

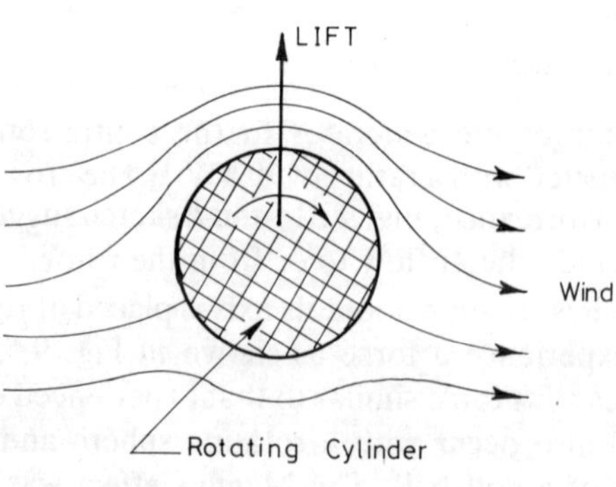

Fig. 9.53 *Magnus effect*

Fig. 9.54 *Flettner's rotor ship (Courtesy Haynes Publishing Group)*

Fig. 9.55 *Flettner's rotor mill (Courtesy Haynes Publishing Group)*

see Figs. 9.54 and 9.55. A ship, the *Buckau*, was constructed with two rotors rotating at 750 rev/min. It successfully completed trips from Danzig to Scotland and to New York, carrying timber. The windmill had a diameter of 66 ft. Although both devices worked they were not considered to be viable commercially and work on them ceased. More recently[15] interest in the application to both wind power and ship propulsion has revived.

References

1 WOODCROFT, Bennet.: The Pneumatics of Hero of Alexandria. Pub. Taylor Welton and Maberly. London. 1851
2 GIMPEL, J.: The Medieval Machine. Pub. Gollancz. London. 1977.
3 PUTMAN, P. C.: Power from the Wind. Pub. Van Nostrand N.Y. 1948
4 BETZ, A.: Windmills in the Light of Modern Research. N.A.C.A. Technical Memo. 474. 1928
5 GRIFFITH, R. T.: The Effect of Aerofoil characteristics of Windmill Performance. Aeron Jour. pp 322–326 July. 1977
6 JACOBS, E. N. Ward, K. E. Pinkerton, R. N.: The Characteristics of 78 Related Airfoil Sections from Tests in the Variable Density Wind Tunnel. Langley Memorial Aeronautical Laboratory. N.A.C.A. Technical Report. No. 460 1933
7 FRAENKEL, P.: Food From Windmills. Pub. I. T. Publications, London. 1975
8 VAN DE VEN, N.: Construction Manual for a Cretan Windmill. Pub. S.W.D. Amersfoort. 1977
9 FRAENKEL, P.: An International Development Programme to Produce a Wind Powered Water Pumping system suitable for small scale, economic, manufacture. 2nd Int. Symp. Wind Energy System. B.H.R.A. Cranfield. U.K. 1978
10 SAVONIUS, S. J.: The S-Rotor and its Applications. Mechanical Engineering. Vol. 53. No. 5 May 1931.
11 MUSGROVE, P. J.: The Variable Geometry Vertical Axis Windmill. Proc. 1st. Symp. Wind Energy systems. B.H.R.A. Cranfield. U.K. 1976.
12 GOURÍERES, D. LE.: Wind Power Plants. Theory and Design. Pub. Pergammon. Oxford. 1982
13 FLETTNER, A.: The Story of the Rotor. Pub. Crosby, Lockwood and Son. 1926.
14 TOKATY, G. A.: A History and Philosophy of Fluid Mechanics. Pub. Foulis. Henley-on-Thames. 1971
15 PIKE, Dag., Rotor Power Ousts the Aerofoil. *New Scientist*, 31–33. Aug. 30th. 1984
16 WEGEREEF, E.: Scale model of the ITDG 6 metre wind tunnel tests for different blade setting angles. WM – 076. Twente University of Technology, The Netherlands February 1984
17 A performance comparison of various wind-turbine types for powering water pump. Proc. 17th IECEC, Paper 829345, pp 2082–2088, 1982
18 Ed. Peter Musgrove. The performance of the Rutherford 6 metre Diameter Vertical Axis Wind Turbine, Wind Energy Conversion, pp 110–129. Camb. Univ. Press 1984
19 MACHIELSE, L. A. H.: Prestatiemetingeu aan de Pionier-I ISMUAU, Energieonderzoek Centrum Nederland, Nationale Windenergie Conferentie, Noordwijkerhout, The Netherlands, 1983
20 Sandia National Laboratories, HAT., ECN, Petten
21 DEKKER, J. W. M., DE GROOT, C. M. RIEFFE, H. Ch.: Energieouderzoek Centrum Nederland, Nationale Windenergie Conferentie, Noordwijkerhout, The Netherlands. 1983
22 POWLES, S. J. R., ANDERSON, M. B., WILSON, D. M. A., PLATTS, M. J.: Articulated wooden blades for a HAWT Design, Construction and Testing, Proceedings of the Fourth BWEA, Wind Energy Conference, Cranfield, U.K. 1982
23 MAYDEW, R. C., KLIMAS, P. C. J.: Aerodynamic Performance of Vertical and Horizontal Axis Wind Turbines. Energy 5, pp 189–90, 1981

Chapter 10

Hydro power

10.1 Introduction

Historically, the water wheel ranks with the windmill in importance as a source of power for simple industrial operations such as flour milling and later for sawmills and forges. The earliest reported use of the water wheel, or water mill as it was called since it was initially used solely for milling purposes, was by the Greeks*.[1,2]

The earliest design was a vertical axis machine consisting of a wheel made up of paddles mounted on a vertical shaft and lowered into a fast stream. Obviously there must be a greater current falling on one side than the other or it would not turn and it was usual to arrange to deflect some of the current on to the wheel by means of a mill race. The design spread to Northern Europe where it became known as the Norse wheel, Fig. 10.1. An important

Fig. 10.1 *Simple Norse wheel*

development in water wheel design was made possible by the invention of the crown and pinion gear which enabled the mechanical shaft output from the wheel to be changed through a right angle. This permitted the use of a

* Landels, Reference 2, questions the early evidence that this mill was known to the Greeks.

horizontal axis water wheel to be used with a vertical axis grindstone. The development of horizontal axis water wheels of this type was due to the Romans and is described in the writings of Vitruvius, see Fig. 10.2. The horizontal axis machine was a considerable improvement over the Norse type and enabled much larger machines to be constructed. Nevertheless, because of their simplicity the Norse machines still persisted and were to be found in use in isolated communities in islands such as the Shetlands and Faeroes until the beginning of this century. Horizontal axis machines were used extensively in the Middle Ages[3] and simple designs are still employed in many rural areas in the developing countries.

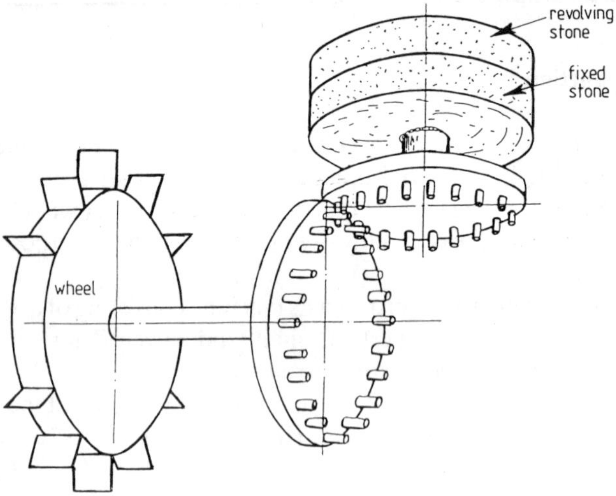

Fig. 10.2 *Horizontal axis wheel with gearing*

The earliest serious study of water wheel performance was by Smeaton[1,3] in 1759. He reported a measured maximum efficiency* of 22% for the undershot wheel, Fig. 10.3, and a figure of 63% for the overshot wheel, Fig. 10.4. Smeaton used radial paddles in his experiments. Later work by the Frenchman Poncelet in the early ninteenth century, using curved blades, resulted in an improvement of the efficiency of the undershot wheel to a value of 65% or about the same as that measured by Smeaton for an overshot wheel.

* Efficiency is defined as the ratio of the work output divided by the sum of the kinetic and potential energy of the water passing through the mill. This is not necessarily the best criterion for the assessment of the value of a small mill. For example, if the water is taken from a larger river, or the mill is placed in such a river, efficiency is of no importance. We are not going to make any impact on the flow energy of a large river, such as the Euphrates, by one mill, and efficiency in itself becomes meaningless. What is much more relevant is some sort of 'goodness' criterion such as the cost per unit of work from the mill. Nevertheless, efficiency may have a secondary significance, since, if the device is inefficient, then for a given output more water must flow through it and hence the construction will need to be larger and stronger than a more efficient machine and hence the cost of energy produced will be higher.

Fig. 10.3 *Undershot wheel*

Poncelet continued his work on hydraulic machines and designed a radial inward flow turbine which was further developed by Francis in the USA to become one of the most popular modern turbines. At about the same time another Frenchman, Fourneyron, was also working on radial flow turbines. His design made use of an outward flow from fixed internal vanes into an annular rotor and resulted in a somewhat complex construction; nevertheless, Fourneyon's turbines operated successfully and he installed over one hundred machines in the mid-1800s.

Fig. 10.4 *Overshot wheel*

Further development in water turbines took place over the latter half of the nineteenth century, a notable advance being the invention of the Pelton wheel in 1880 by an American of that name who was responsible for the operation of hydraulic machinery in the Californian gold fields. The Pelton turbine replaced simple radial vane type machines similar to the Norse wheel shown in Fig. 10.1. A further major invention was the Kaplan turbine by Professor Kaplan in Sweden in 1917. Large scale hydro power is used exclusively for the generation of electricity. The three turbine types, Pelton, Francis, and Kaplan, make up the greater number of modern large scale turbines. They all, have high efficiency (around 90% or higher) and the largest turbines each produce several hundred megawatts of power. Pelton turbines are most suited for high head applications, cross-flow and Francis turbines for medium head, and Kaplan turbines for low head.

The choice of turbine for a particular application will depend primarily on head. For a particular site, and hence head, the flow is fixed by the power requirement and will also affect the selection of the type of turbine. Fig. 105 indicates the approximate range of three major turbine designs, Pelton, cross-flow and Kaplan, for micro hydro applications.

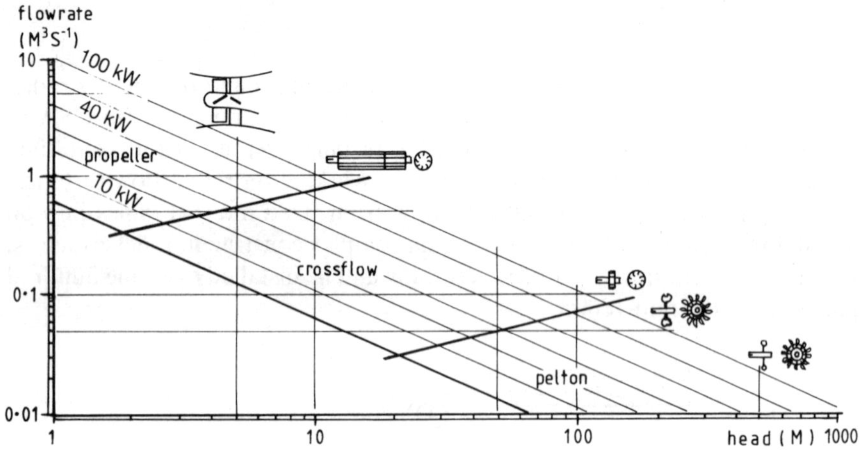

Fig. 10.5 *Selection chart for water turbines under 100 kW (based on Reference 5)*

A great attraction of water power is that it is free, it is also renewable and does not cause pollution. The French call hydro power '*houille blanche*' or white oil. There are no fuel costs, no handling or storage charges and the plant itself is usually simple to operate, easy to maintain, and of long life. A major disadvantage of water power is that it is site specific, that is it is only available in certain sites and these may not be near to centres of population. Capital cost tends to be high owing to the large amount of civil work (dams etc.) required. The development of efficient, relatively cheap, high voltage transmission has enabled large power electrical generators to be driven by water turbines and the power transmitted to the areas where it is needed. Today one-quarter of the total world electricity is generated by hydro power. Much of the hydro potential of the developed areas of the world has been used. There is still a great deal of untapped large scale hydro power available in the developing countries. More important for the developing countries, with their dispersed populations, is the large number of sites suited for mini or micro-hydro power ranging from a megawatt to as little as a few kilowatts. Small hydro plant can be used both for electricity generation and for producing mechanical shaft power.

10.2 Energy due to gravitational head

Water situated on a hill will have potential energy with respect to a lower level. A mass m at height h will have energy:

$$E = m\, g\, h$$

where g is the acceleration due to gravity (9.81 m/s^2).

If the water falls to the lower level this potential energy will be converted to kinetic energy, 1/2 mv^2 where v is the velocity.

Assuming no loss:

$$m\, g\, h = 1/2\, m\, v^2$$

or

$$v = \sqrt{(2\, g\, h)}$$

It was pointed out when considering storage, that gravitational energy is not a good way of storing energy and large volume flows are needed to give useful power.

If the flow is Q then the power P is given by

$$P = \rho\, g\, h\, Q \qquad (10.1)$$

where ρ is the density of water (1000 kg/m^3). Hence

$$P \triangleq 10\, h\, Q \text{ kilowatt}$$

if h is in metres and Q in m^3/s.

In many parts of the world river currents provide a predictable and reliable source of renewable energy. A river current with a current of 1 m/s has an energy flux of 500 W/m^2. The equivalent head h for such a current flow v is given by:

$$h = \frac{v^2}{2g} \qquad (10.2)$$

or substituting into eqn. 10.1

$$P = \rho\, Q \frac{v^2}{2}$$

Table 10.1 shows the way in which velocity and power vary with head.

The formulae show that power output is determined by the product of head and volume flow, hence where the head is large, only a small volume flow is

Table 10.1

Head m	Velocity m/s	Power for 1 m^3/s flow, kW
1	4·44	9·81
3	7·8	29·43
10	14·0	98·1
30	24·3	294·3
100	44·4	981
300	78	2943
1000	140	9810

required, whereas at the opposite end of the scale, where the head is low, a correspondingly large flow is required to generate the same amount of power. Where a choice of site is available the high head site is preferred on economic grounds.

As has been mentioned earlier a single turbine design is not suited to cover the whole range of heads and power level required in practice and the most important turbine designs will be discussed in the next section.

10.3 Water turbines

Water turbines have been developed which are able to convert the energy of the water (potential and kinetic) into mechanical shaft power with a high conversion efficiency. Turbines (and pumps) can be classified in various ways, for example, whether the direction of flow through the rotor is parallel to the axis of the rotor (axial flow) or if it is perpendicular to the axis of the rotor (radial flow). Turbines are also classified as impulse or reaction and this grouping is used here.

In an *impulse* turbine the water pressure, or potential energy of the water, is converted to kinetic energy in a fixed jet from which the water emerges at high velocity but at atmospheric pressure. The high velocity water jet impinges on a vane or bucket placed on the circumference of a wheel causing the wheel to rotate, Fig. 10.6.

In the *reaction* turbine the water is allowed to completely fill the rotor and exert a pressure on the blades. It differs from the impulse turbine in that there is a pressure drop through the rotor itself, and both the kinetic energy of the incoming water and the pressure is used, whereas in the impulse machine only the kinetic energy of the water jet entering the rotor is used. The lawn sprinkler is a simple example of a reaction turbine, see Fig. 10.7.

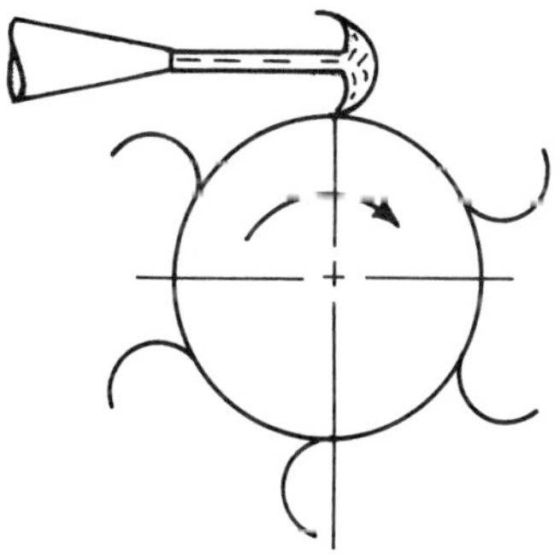

Fig. 10.6 *Pelton wheel*

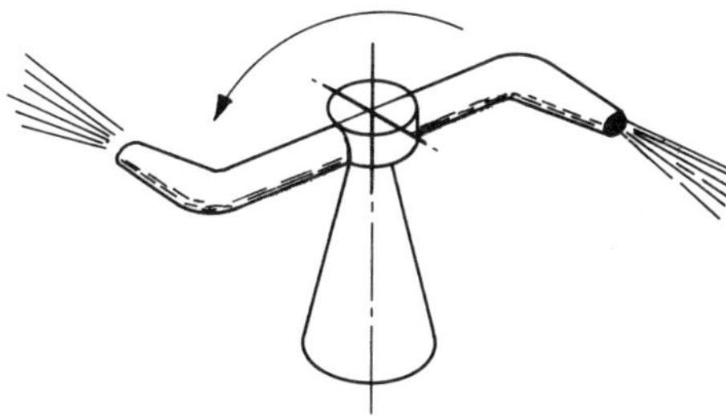

Fig. 10.7 *Lawn sprinkler*

10.4 Impulse turbines

10.4.1 Pelton turbine

The basic arrangement is shown in Fig. 10.6. The turbine consists of a solid disc mounted on a shaft with vanes or buckets arranged uniformly around the periphery. A high velocity jet from a nozzle strikes the buckets roughly tangentially to the disc (or runner) causing it to revolve. The object is to transfer as much as possible of the kinetic energy in the jet to the turbine runner. Ideally, the water should leave the buckets with no kinetic energy at all. Clearly, if the wheel is stationary then although the jet will exert a force, no work will be done, and the efficiency will be zero, Fig. 10.8*b*. Again, if the jet speed v is the same value as the velocity of the bucket, there will be no force produced and so the conversion efficiency will be zero, Fig. 10.8*d*. If, however, the bucket velocity is chosen to be half that of the jet v, then the velocity of the jet, relative to the bucket on arrival, will be $v/2$; on leaving the bucket the jet velocity will have been reversed relative to the bucket to a value $-v/2$ and the jet will leave the bucket with an absolute velocity of nearly zero, Fig. 10.8*c*. For this condition all the jet kinetic energy will have been transferred to the bucket. Practical Pelton wheels have a high conversion efficiency of around

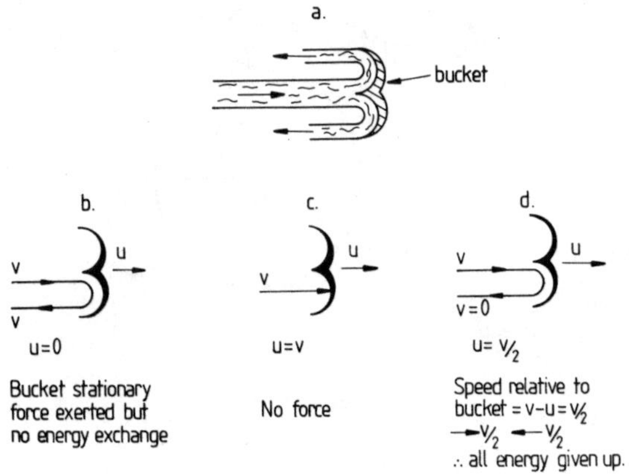

Fig. 10.8 *Impulse Turbine. Effect of relative blade/jet speed*

90% and this figure does not vary appreciably over a large range of flow, Fig. 10.9. It is important that sudden changes in flow should be avoided, since, if the inlet valve at the turbine is suddenly closed, the resulting pressure impulse owing to water hammer can be serious. Water hammer is sometimes observed when a domestic tap is turned off suddenly. It is due to the shock caused by the destruction of the momentum of the moving mass of water in the pipe. In the domestic case the result is simply a noise but in the case of water flowing from a several hundred metre head the resulting pressure rise can be disastrous and fast control of a Pelton turbine is achieved by means of deflection of the jet away from the runner rather than by stopping the flow. A further safety factor which must be taken account of in Pelton turbines is provision for sudden loss of load; if this happens, the unloaded runner will speed up to a bucket velocity nearly equal to the jet velocity, and the resulting increase in centrifugal forces may cause damage to the rotating machinery.

Large Pelton turbines (>20 kW) are used when a high head is available, normally 30 m to 500 m. They may be mounted either horizontally or vertically and the larger machines can have several jets. Like other large water turbines,

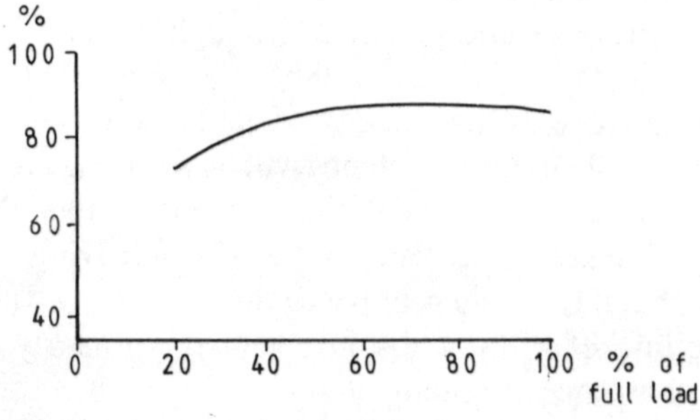

Fig. 10.9 *Efficiency versus load for Pelton turbine*

Pelton wheels are usually coupled to electrical generators and for such generators high rotation speed is an advantage. Fig. 10.10*a* and *b* shows a large Pelton turbine installation. Machines have been constructed in the power range up to 100 MW and with rotation speeds normally between 300 to 750 rev/min. Fig. 10.11 *a* and *b* shows a small Pelton turbine, at this size the design and construction can obviously be greatly simplified. Fig. 10.12 shows a typical layout.

10.4.2 Turgo turbine

This is a variant of the Pelton wheel and was proposed by Crewdson in 1920.[6] In the Turgo wheel the jet enters from the side and a much larger jet area can be used than would be the case for a Pelton wheel of the same diameter, Fig. 10.13. The reason for this is that in the Pelton turbine the jet diameter is restricted by the need to avoid the following bucket, and in practice this means that the jet diameter is between 1/7 and 1/12 of the runner or wheel diameter. By arranging for the jet to enter from the side this restriction is removed and a ratio of jet diameter to wheel diameter as large as 1/4 can be used. The relative velocities between the blade and jet are the same as for the Pelton wheel so that machines of the same diameter and head will be operated at similar rotation speed. However, for the same power it is possible to reduce the diameter of the Turgo runner to less than half that of the equivalent Pelton runner. Hence the Turgo rotation speed will be more than twice that of the Pelton; this is a significant advantage when, as is usual, the turbine is coupled to an electrical generator. Multiple jets may be needed on Pelton turbines whereas if a Turgo arrangement is employed it is possible to use only one jet for the same runner diameter; where power is controlled by variable aperture jets this can result in cost saving. One disadvantage of the Turgo design is the axial thrust on the bearings which must be arranged to operate under these conditions. A further disadvantage, particularly with low power machines, is the cost of replacing the complex Turgo runner compared to the simple and readily replaced buckets.

10.4.3 Cross-flow turbine (Mitchell, Banki or Ossberger)

The cross-flow turbine is an impulse turbine which is particularly suited to small power low head applications. It was originally patented by an Australian, Mitchell, in 1903, and appears to have been independently reinvented in Austria by Professor Banki, who published an account of his work in 1917. The design was taken up by the manufacturing firm Ossberger[7] in Germany and this firm now offers a range of machines from 1 kW to 1000 kW and with heads ranging from 1 to 200 m. It is referred to in the literature by all the four names already mentioned, Fig. 10.14.

The rotor of the cross-flow turbine is made up from a series of long curved blades which are mounted parallel to an axis in the form of a cylinder. A guide vane system feeds a rectangular cross-section jet into the rotor, the water flows

a.

b.

Fig. 10.10 *Large Pelton Installation (Courtesy Robert Gilkes and Gordon Ltd)*
a 240 kW Pelton turbine on 224 m head, 1000 rpm, in foreground and a 1080 kW Turgo turbine on 206 m head, 1000 rpm, in background
b a 55 kW Pelton turbine on 130 m head driving a pump unit at 1500 rpm.

a.

b.

Fig. 10.11 *Small Pelton Turbines*
a 2 Jet 1 kW Pelton turbine
b Pelton turbine coupled to 500 W electrical generator

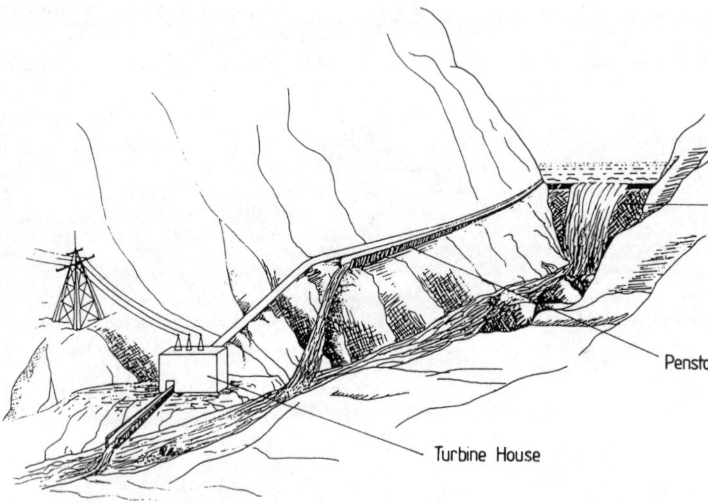

Fig. 10.12 *Typical layout for hydroelectric plant*

through the blades from the outside of the rotor, the jet then traverses the rotor as shown in the figure and falls on the blades on the opposite side of the circumference before emerging from the rotor. Hence the name cross-flow machine. It is seen that the jet interacts with the rotor blades twice on passing through the machine. A side benefit from the cross-flow arrangement is that the jet, on leaving the rotor, has a cleaning action and removes twigs, leaves, etc. which may have entered with the jet. The cross-flow turbine is classified as a radial impulse type machine. This design is very suitable for local manufacture. Quite a large range in power output can be achieved with the

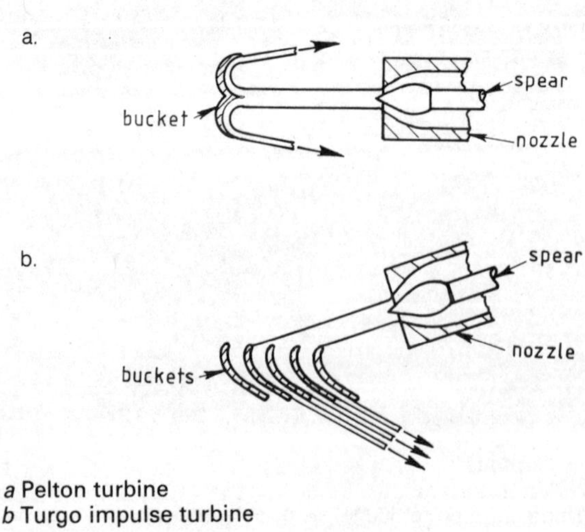

Fig. 10.13 *a* Pelton turbine
b Turgo impulse turbine

a.

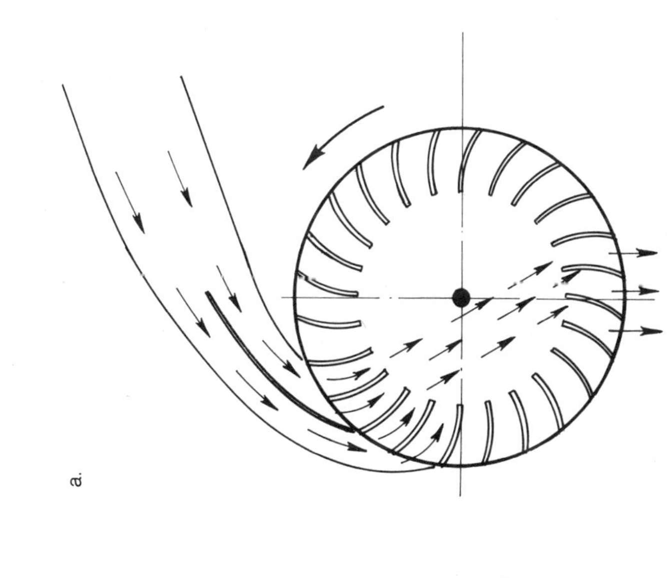

b.

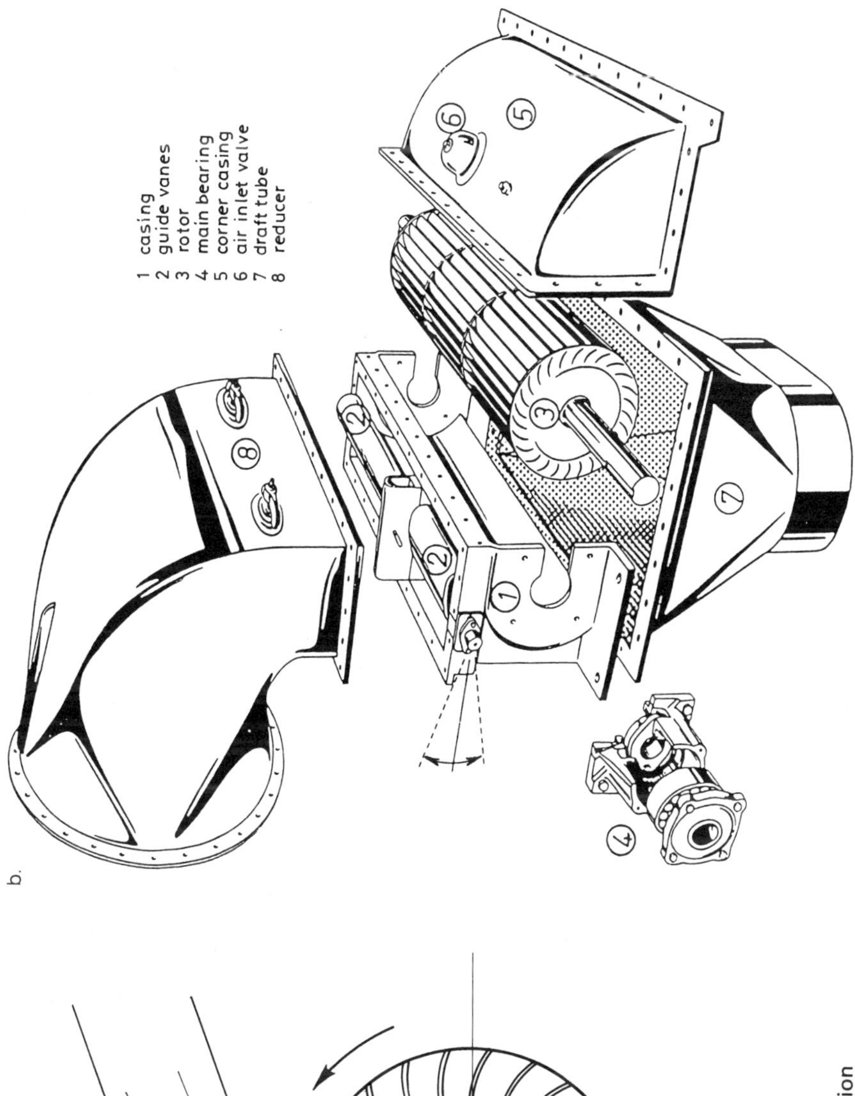

Fig. 10.14 *Cross-flow turbine*
a Principle of operation
b Two-section Ossberger turbine assembly (Courtesy Ossberger)

same diameter rotor by changing the length. The efficiency is high – Ossberger claim 84–88% on full load – and the machine is very insensitive to variation in flow, maintaining a high efficiency down to as little as 10% of full flow, see Fig. 10.15. Fig. 10.16 shows a small locally manufactured cross-flow turbine in Pakistan (Peshawar).

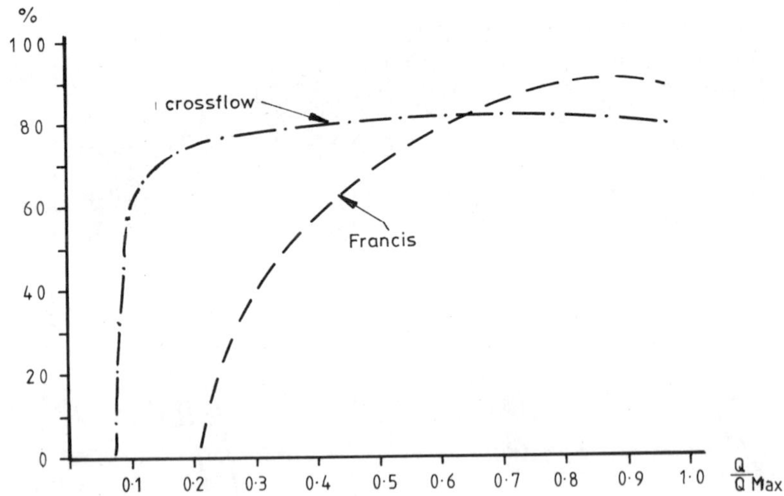

Fig. 10.15 *Efficiency versus load for cross-flow and Francis turbines*

Fig. 10.16 *Locally manufactured cross-flow turbine in Peshawar, Pakistan (Courtesy P. Fraenkel)*

10.5 Reaction turbines

10.5.1 Introduction

In the impulse turbine all the potential energy of the head is converted into kinetic energy before entering the turbine. In the reaction turbine only part of the total potential energy is converted into kinetic energy before entering the turbine so that there will be a pressure drop on passing through the turbine, whereas in a reaction turbine the water completely fills the runner and hence all blades of the runner will be engaged. In the Pelton wheel only one or two blade pairs will be engaged at any one time. Therefore the reaction turbines will be better suited to the large water flows required in low head situations.

10.5.2 Francis turbine

This machine has been referred to earlier, and Fig. 10.17 shows a typical example. Such machines have either vertical or horizontal axis and the water flow is radially inward in direction. Water is fed in uniformly at the outer circumference of the runner by means of a horizontal spiral casing or volute. This snail shell form of the casing is characteristic of reaction turbines and results from the need to feed in the water uniformly around the turbine runner. Guide vanes in the casing direct the water on to the runner blades. The guide

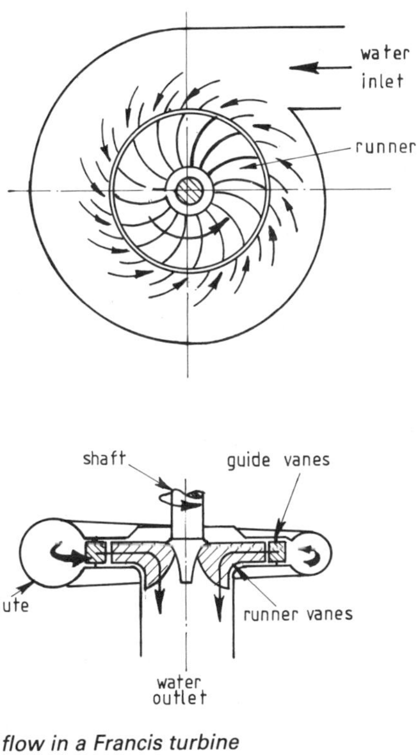

Fig. 10.17 *Schematic of flow in a Francis turbine*

blades may be turned on so as to reduce power output. The runner of a Francis turbine normally has between 5 and 12 blades and on passing through the blades the water leaves along the axis. This water possesses kinetic energy which may be recovered by means of a draught pipe on the turbine exit. This is a pipe of increasing diameter which permits the water velocity to drop. The result is that the pressure at the turbine rotor discharge is reduced and the effective head across the turbine is increased. The part load performance of a Francis turbine is quite good. Francis turbines are usually employed between heads of from 15 m to 300 m and have an efficiency of better than 90% for large machines.

10.5.3 Propeller turbine

Where the head is low and the flow large, an axial flow propeller-type turbine is most suitable. For 10.19 shows the general arrangement. Typical propeller turbines have from 4 to 6 blades, and have a conversion efficiency of better than 90%. Unfortunately, the efficiency of a propeller depends on the flow velocity (in the same way as for the windmill) and the efficiency falls sharply as flow is reduced. Where efficiency is important the Kaplan turbine is used. The Kaplan turbine has movable blades which enable changes in flow velocity to be accommodated. The advantage of the Kaplan turbine over the simple propeller is shown in Fig. 10.18. The Deriaz turbine is a development of the Kaplan turbine and has a somewhat better efficiency versus load characteristic.

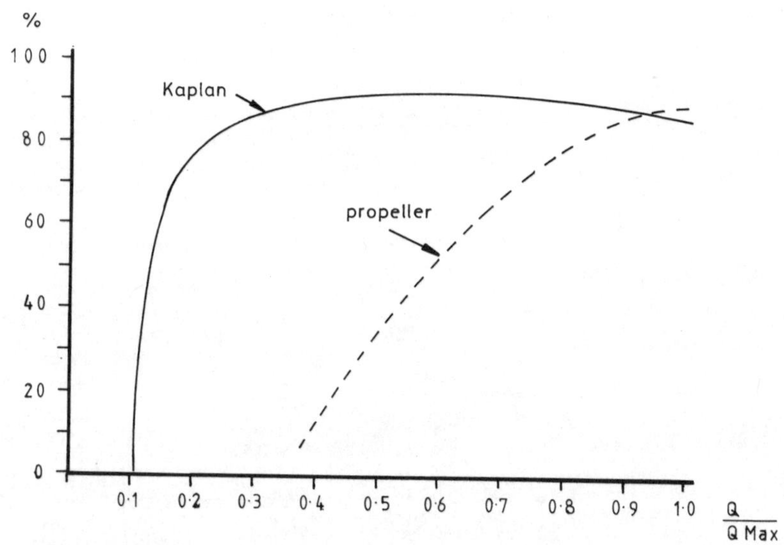

Fig. 10.18 *Efficiency versus load for Kaplan and propeller turbines*

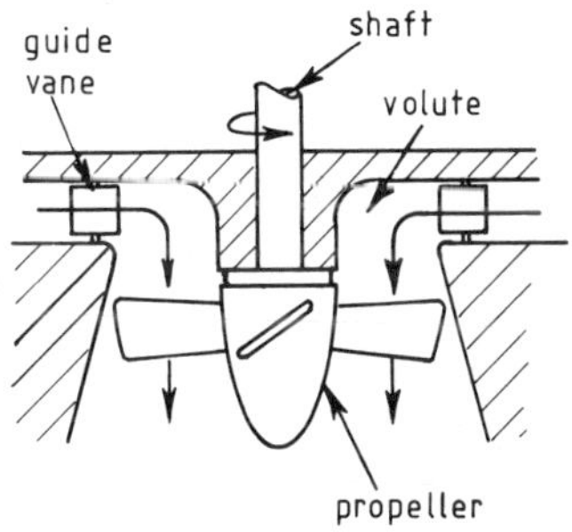

Fig. 10.19 *Kaplan turbine schematic*

10.6 Micro hydro plant

Turbines which produce a power output of less than 100 kW are referred to as micro hydro. While such turbines operate in the same way as turbines in the higher power range, they are subject to economic restrictions which make it necessary to simplify the design, for example, by the use of less complex control arrangements, or by adopting a limited range of standard sizes which will not be optimised for all sites. Most important is the need for a large element of do-it-yourself in site construction and plant installation. Impulse turbines are used more frequently than reaction turbines for micro hydro schemes, partly for reasons of simplicity of construction and partly because of their good part load efficiency, an important consideration where run-of-the-river systems (that is without storage) are used. In some cases the flow may vary by a factor of as much as ten between the dry and wet season.

Fig. 10.5, taken from Reference 5, enables a turbine type to be selected for given conditions of flow and head. The flow rate is plotted vertically and the head horizontally. Lines are drawn for constant power output and the preferred ranges for Pelton, cross-flow and propeller turbines indicated. Naturally there are no sharp boundaries and in practice there is considerable overlap depending upon the desired speed of rotation and power output required. It can be seen that, for most purposes, the Pelton or the cross-flow turbine cover the requirements. As shown on the graph the actual design of the two turbine types will depend on the flow/head. For example, for a given power output as the head is increased, the length of the cross-flow turbine will decrease. Similarly, in the case of the Pelton turbine, the bucket size will be reduced and wheel diameter increased as the head is increased. An interesting suggestion, discussed in Reference 5, is the use of rotodynamic commercial pumps as turbines; they are well made, cheap, and readily available.

Micro turbines can be used to provide mechanical shaft power, or more usually in association with an electrical generator to give an electrical output. They are reliable and require only simple maintenance. If a suitable site is available the micro hydro system should always be seriously considered.

10.7 Speed control

If the load is suddenly removed then the turbine will speed up and this speed increase may result in the increase in forces in the rotor sufficient to cause damage. Hence some form of speed control is necessary. The water flow through the turbine is normally controlled by a variable aperture valve in the case of impulse machines, Fig. 10.13, and in reaction machines the guide blades are arranged to be moved to control flow. Automatic control systems are normally employed on large plant but should be avoided on micro hydro plant in order to keep costs low.

Most turbines are used to generate electricity and it is then necessary to control speed to a more precise extent in order to limit fluctuations in output voltage and frequency from the electrical generator. The mechanical control systems described above can be used for this purpose but electronic control is also adopted for small machines. The electronic controller is arranged to vary the electrical load, in order to control speed. This is achieved by switching electrical resistors in or out to maintain the required frequency.

10.8 Electrical generators

Synchronous generators (alternators) are used with all large water turbines. The frequency of the generated electricity is proportional to the shaft speed which must be controlled to tight limits. For micro hydro machines direct current generators are sometimes used. Induction generators can be used but these require an external frequency source to excite the field winding, hence they are suitable for connection to an AC electrical supply system. Induction generators are not usually selected for use with water turbines.

10.9 Miscellaneous hydraulic engines

10.9.1 Hydraulic ram

If, when water is flowing along a pipe, a tap at the outlet end is turned off suddenly, then the loss of momentum of the column of water in the pipe will cause a sudden pressure rise. This water hammer effect has been referred to earlier as a possible cause of damage to a Pelton turbine. The hydraulic ram makes use of this effect to pump a small quantity of water to a high head using the energy of a large flow falling through a smaller height. The basic arrangement is shown in Fig. 10.20. Water flows from the source to the waste pipe through the valve A, called the impulse valve. The flow causes A to close suddenly. The column of water is stopped and the resulting pressure rise opens valve B and water flows into the high pressure chamber C and up the outlet pipe. The air in chamber C maintains a constant pressure on the output side of

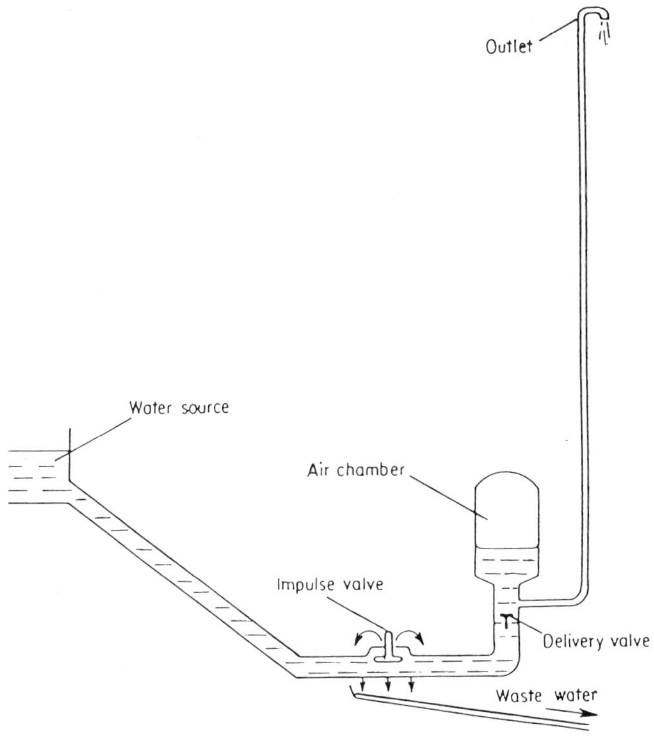

Fig. 10.20 *Hydraulic ram pump*

valve B. An arrangement (not shown) is required to maintain the air in C since small losses occur. The principle of the hydraulic ram was first put forward by Whitehurst in 1775, but his device was not self-acting and required a child to operate the impulse valve. The credit for the invention of the self-acting ram goes to Joseph Michael de Montgolfier (1797) better known for his work with hot air balloons. The hydraulic ram will pump up to a height of 200 m and is extensively used for small pumping requirements. It is cheap, long-lived and requires only simple maintenance.[8]

10.9.2 River current turbines

The relatively high energy flux in rivers[9] such as the Nile was referred to in Section 10.3.

Traditionally energy has been extracted from flowing water (where no static head is available) for water lifting by large undershot waterwheels made from bamboo or other locally available material. A very large waterwheel is necessary to extract a small amount of power and thus this technology would be expensive in areas where suitable materials were scarce.

Since 1978 the Intermediate Technology Development Group's power project has been working on applying windmill technology to extract energy from river currents with a view to providing small quantities of power (up to 1 kW) for water pumping and electricity generation. After model rotor tests on the River Thames at Reading University, a water pumping turbine with a 3 m diameter vertical axis rotor was successfully field tested on the White Nile in Southern Sudan with funding from the Royal Netherlands Government. The prototype rotor drives a centrifugal pump through a two-stage belt transmission and delivers 3.5 l/s through a static head of 5 m in a current speed of 1.2 m/s. The machine was launched in 1980 and has so far run for more than 7000 hours. For the last two dry seasons it has irrigated up to 2 hectares of

a.

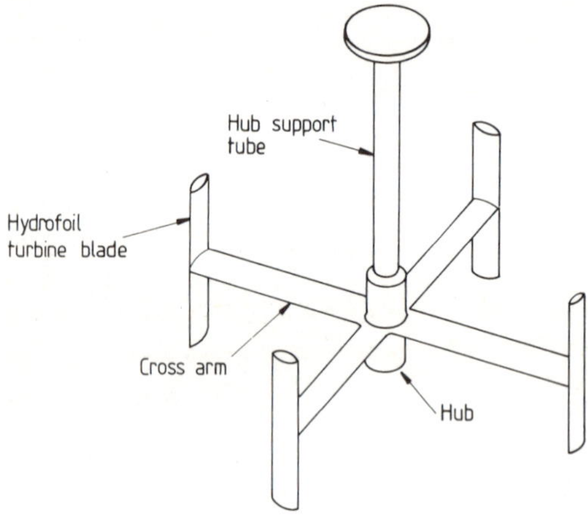

b.

Fig. 10.21 *River Current Turbine*
a Schematic
b Turbine on test at Juba, Sudan (Courtesy P. Garman)

vegetables when run for 8 to 10 hours per day. With a water storage tank large enough to hold 10 hours output the machine could irrigate more than twice this area.

Eight further machines with different rotor, pump and transmission arrangements have been built and tested on the Nile. As a result of this work designs for two different water pumping turbines have been developed. The first of these machines (see Fig. 10.21) is an inclined axis propeller-type turbine driving a centrifugal pump through a two-stage belt transmission. This machine will pump up to 6.8 l/s through a static head of 5 m in a current flow of 1.2 m/s. Static heads of up to 25 m can be achieved if a suitable pump is selected. The overall system efficiency of this machine can exceed 10% if a 75 mm bore delivery pipe is fitted.

The second machine is designed to pump between 1 and 2 l/s through a static head of 5 m; enough for a single smallholder's 1 acre vegetable plot. This machine uses a locally made centrifugal pump and cycle component transmission to keep costs to a minimum. Oil drum floats are used rather than the durable but expensive ferrocement pontoons of the first machine.

The site conditions required for both machines are a water depth of at least 2.75 m and a minimum current speed of 0.85 m/s within 25 m of the river bank.

10.9.3 Stream powered manometric pump

Another pumping device which might be used in rivers and streams is the manometric pump shown in Fig. 10.22. This machine was first proposed by Andrew Wirtz in 1746 and subsequent work has been reported by a number of authors more recently.[10] The pump operates as follows: a helical pipe is wound round a cylinder which is half immersed in water and rotated about its axis. Successive loops of the coil contain water in the lower section and air in the upper, to form a series of coupled manometers. As the cylinder rotates the water columns displace to give a pressure head, the total head being the sum of

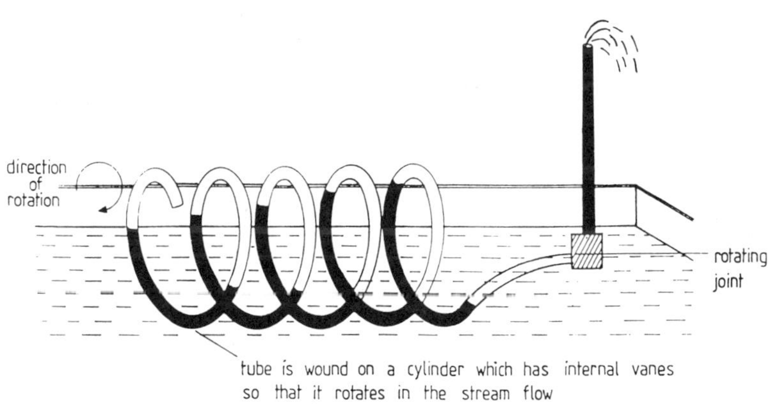

Fig. 10.22 *Stream powered manometric pump*

the individual loop pressure differences. The total pressure difference gives the maximum pumping head. The cylinder can be caused to rotate by the stream using vanes attached to the ends. A small prototype pump has been reported from Juba in Sudan.[11]

References

1 BURSTALL, A. F.: 'A History of mechanical engineering' (Faber, London, 1963)
2 LANDELS, J.G.: 'Engineering in the ancient world', (Chatto and Windus, London, 1978)
3 BOWMAN, G.: 'John Smeaton. Consulting engineers' (CME, 1965)
4 GIMPEL, J.: 'The medieval machine' (Futura, London, 1979)
5 HATHERSALL, R.: 'Micro-hydro – some notes on selection criteria', *Water Power and Dam Construction*, Feb 1984
6 WILSON, P. N.: 'Water turbines' (HMSO, London, 1974)
7 'Small hydro power stations', Sonderdraik. Aus. Energy Developments, 41–43, March 1979
8 WATT, S.A.: 'A manual on the automatic hydraulic ram pump' (ITDG Publications, London, 1975)
9 GARMAN, P.: 'The application of the vertical axis water turbine in developing countries', M. Phil Thesis, Reading University, June 1983)
10 GARMAN, P.: Personal communication
11 STUCKEY, A. T. and WILSON, E. M.: 'The stream powered manometric pump', Proceedings Conference on Appropriate Technology in Civil Engineering, London, April 1980

Appendix 1

Referring to Fig. A 1.1, assume that engine 1 is a non-reversible engine with efficiency η_1 and that engine η_2 is an ideal (perfectly reversible) engine with efficiency η_2. Also assume $\eta_1 > \eta_2$.

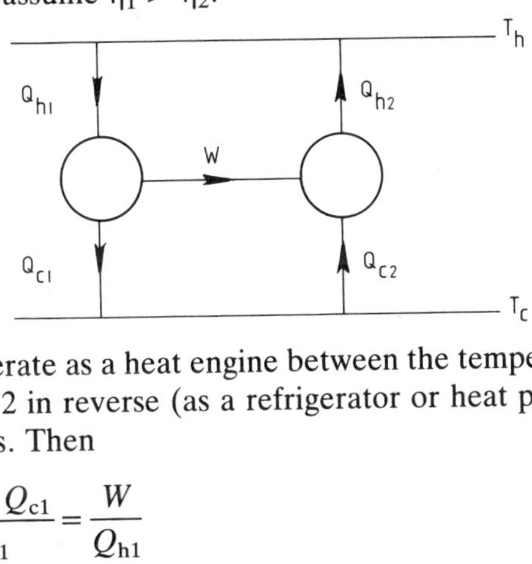

Let engine 1 operate as a heat engine between the temperatures T_h, T_c and let it drive engine 2 in reverse (as a refrigerator or heat pump) between the same temperatures. Then

$$\eta_1 = \frac{Q_{h1} - Q_{c1}}{Q_{h1}} = \frac{W}{Q_{h1}}$$

$$\eta_2 = \frac{Q_{h2} - Q_{c2}}{Q_{h2}} = \frac{W}{Q_{h2}}$$

or

$$\eta_1 Q_{h1} = \eta_2 Q_{h2}$$

but

$$\eta_1 > \eta_2$$

$$Q_{h1} < Q_{h2}$$

and the net effect is to transfer heat from a cold reservoir(T_c) to a hot reservoir (T_h) which is contrary to the second law. Hence, η_1 cannot be greater than η_2. The corollary to this is that all ideal engines must have the same efficiency, the Carnot efficiency.

Appendix 2

Enthalpy H and Gibbs' free energy G

When a fuel is burnt in the normal way the chemical reaction involved is non-isothermal, and if, as is usual, no work is done, the chemical energy released will appear as heat, i.e. as random kinetic energy of the products of combustion. If the products of the reaction are restored to the same pressure and temperature as the reactants then the heat energy released is given by $-\Delta H$, the difference in enthalpy H between products and reactants.

Enthalpy is defined as

$$H = U + PV$$

where U is the internal energy, P is the pressure, and V is the volume.

The change is H in the reaction $-\Delta H$ is the same as the calorific value listed, for example, in Table 8.3. If water is produced in the reaction then if it is condensed the value of $-\Delta H$ is called the gross or higher calorific value and if it is released as vapour then the corresponding $-\Delta H$ will be reduced by the latent heat of the steam and is called the net or lower calorific value.

This random energy can only be converted into work, a directed form of energy, in some form of heat engine in which, of the heat supplied from a high temperature source, only part is converted into work and the remainder rejected to the surroundings. The limiting efficiency for converting heat to work is the Carnot efficiency.

The attraction of the electrochemical cell lies in the fact that, in principle, no random energy stage need be involved and chemical energy, one form of directed energy, is converted direct to work, which is also directed energy.

The maximum energy which can be released from a chemical reaction which takes place at constant temperature and pressure is $-\Delta G$, the difference between the Gibbs' free energy G of the products and reactants. Hence $-\Delta G$ is the maximum electrical energy which can be produced from the electrochemical cell.

This can be shown as follows. By the First Law of Thermodynamics

$$\mathrm{d}W = -\mathrm{d}U + \mathrm{d}Q$$

where dW is the total work done by the reaction, $-dU$ is the loss of internal energy, dQ is the heat released.

$$dW = dW_1 + dW_2$$
$$dW_1 = \text{useful (electrical output) work}$$
$$dW_2 = PdV = \text{work done in expansion}$$

where P is the pressure and dV is the change in volume between products and reactants.

$$dQ = TdS$$

where T is temperature and dS is change in entropy. Hence

$$\begin{aligned} dW_1 &= -(dU + PdV - TdS) \\ &= -\Delta G \end{aligned}$$

where G is defined as $U + PV - TS$ and also

$$\Delta G = \Delta H - T\Delta S$$

For the case considered the change in entropy ΔS depends on the difference of heat capacity at constant temperature and pressure between reactants and products $\Delta(\Sigma\, mC_p)$.

The sign ΔG determines whether a chemical reaction can proceed. For this to be so the Gibbs' free energy of the products G_p must be less than that of the reactants G_r. That is

$$G_r - G_p = -\Delta G$$

This means that energy $+\Delta G$ will be released by the reaction. Normally $-\Delta H$ will also be positive and hence the heat will be released in the reaction by an amount $+\Delta H$.

For the reactions of interest in energy ΔH and ΔG will be large and are usually approximately equal. An exception is hydrogen where the energy released when 1 kg is burnt is 120 MJ/kg ($-\Delta H$) and the energy released, for example in an electrical cell, is 140 MJ/kg ($-\Delta G$).

Bibliography

Chapter 1

HEDLEY, D.: 'World energy. The facts and the future' (Euromonitor Publications Ltd., 1981)
A very comprehensive appraisal of the world energy situation; the book also includes a considerable body of data.

PIMENTAL, D., and PIMENTAL, M.: 'Food energy and society' (Edward Arnold 1979)
Provides an excellent discussion on the interdependence of food and energy and the energy cost of food production systems.

CHAPMAN, P.: 'Fuel's paradise. Energy options for Britain' (Penguin, 1975)
The topics energy analysis, energy costs and energy planning are introduced in a simple, readable, manner.

LEACH, G.: 'Energy and food production' (IPC Science and Technology Press, 1976)

'United Nations Statistical Year Book' (UN, New York, published annually)

'B.P. statistical review of world energy' (British Petroleum, published annually)

Chapter 2

SELLERS, W. D.: 'Physical climatology' (University of Chicago Press, 1965)
A standard textbook which gives a good account of solar radiation and global energy balance.

MONTEITH, J. L.: 'Principles of environmental physics' (Arnold, London, 1973)
An excellent introductory textbook which describes the physical principles underlying micrometeorology.

MCMULLEN, J. T., and MURRAY, R. B.: 'Energy resources' (Arnold, Bath, 1979)
Covers all energy resources.

WALKER, D.: 'Energy, plants and man' (Packard Publishing, Chichester, 1979)
A very short and simple introduction to photosynthesis. Gives a brief but clear account of the required chemistry.

ANDERSON, R. E.: 'Biological paths to energy self-reliance' (Van Nostrand Reinhold, New York)
A comprehensive account of biomass energy. The book is divided into two parts, the first covers all aspects of bioenergy for the non-specialist. The second part covers the scientific aspects such as photosynthesis.

HUBBERT King, M.: 'Energy resources. Resources and man' (National Academy of Sciences, NRC, Freeman, San Francisco, 1969)
Although published 15 years ago this is still relevant and gives an authoritative and valuable account of energy resources, particularly the fossil fuels.

GOLDING, E. W.: 'The generation of electricity by windpower' (Spon, London, 1976)
This book, by an early wind power pioneer, still remains an excellent introduction to the subject.
SIMEONS, C.: 'Hydropower. The use of water as an alternative source of energy' (Pergamon, Oxford, 1980)
SHAW, R.: 'Wave energy. A design challenge' (Ellis Horwood Ltd., Chichester, 1982)
CHARLIER, R. H.: 'Tidal energy' (Van Nostrand Reinhold, New York, 1982)
DUDERSTADT, J. J.: 'Nuclear power' (Marcel Dekker Inc., New York, 1979)
A general introduction to both fission and fusion.
COTTRELL, A.: 'How safe is nuclear energy' (Heinemann Educational Books, London, 1981)
Provides the basic facts on nuclear safety.
BLAIR, I.: 'Taming the atom. Facing the future with nuclear power' (Adam Hilger, Bristol, 1983)
PATTERSON, W.C.: 'Nuclear reactors' (Earth Island Limited, London, 1973)
This book, with the previous book by Dr Blair, together gives a good account of the various factors which arise when considering the desirability, or otherwise, of nuclear power.

Chapter 3

ROGERS, G. F. C., and MAYHEW, Y. R.: 'Engineering thermodynamics, work and heat transfer' 3rd edn (Longman, London, 1980)
The standard undergraduate textbook and an excellent introduction to thermodynamics and its engineering applications.
RAMSEY, J. A.: 'A guide to thermodynamics' (Chapman and Hall, London, 1971)
A very clear, concise but thorough discussion of thermodynamic principles.
ORD-HUME, A. W. J. G.: 'Perpetual motion' (Allen & Unwin, London, 1977)

Chapter 4

ROGERS, G. F. C., and MAYHEW, Y. R.: 'Engineering thermodynamics, work and heat transfer' 3rd edn (Longman, London, 1980)
The standard undergraduate textbook and an excellent introduction to the basic principles and theory.
DAY, J.: 'The search for power' (Hamlyn, 1980)
A comprehensive readable, well illustrated description of engine development.
BRIGGS, A.: 'The power of steam' (Michael Joseph Ltd., London, 1982)
An excellent, non-technical account which is a good complement to the book by Day.
DICKENSON, H. W.: 'A short history of the steam engine' 2nd edn (Frank Cass, London, 1963)
The classic work on early steam development.
DICKENSON, H. W., and JENKINS, R.: 'James Watt and the steam engine' (Clarendon Press, Oxford, 1927)
'The NRDC Ricardo Power Plant', *The Steam Engineer*, Nov. 1954
A short description of a small (4½ hp) steam engine constructed from an I.C. engine cylinder block.
WALKER, G.: 'Stirling engines' (Oxford University Press, 1980)
A comprehensive account of the history, theory and construction of the Stirling engine by a well known authority in the field.
ZARINCHANG, J.: 'The Stirling engine. An annotated bibliography' (ITDG Publications, London, 1972)
A good summary of the early literature.
WEST, C. D.: 'Liquid piston Stirling machines' (Van Nostrand Reinhold Co., New York, 1983)
READER, G. T., and HOOPER, C.: 'Stirling engines' (Spon, London, 1983)

Chapter 5

ROGERS, G. E. C., and MAYHEW, Y. R.: 'Engineering thermodynamics, work and heat transfer' 3rd edn (Longman, London, 1980)
The standard undergraduate textbook on the subject.
RICARDO, H. R., and HEMPSON, J. G. G.: 'The high speed internal combustion engine' (Blackie, London, 1968)
This well known book gives an excellent account of design principle and physical processes in the IC engine.
JUDGE, A. W.: 'Modern petrol engines' 3rd edn (Chapman and Hall, London, 1965)
A clearly written and comprehensive account. (Tizard & Pye).
JUDGE, A. W.: 'High speed diesel engines' 6th edn (Chapman & Hall, London, 1967)
A clearly written and comprehensive account. Includes automotive, stationary and marine engines.
BELL, P. C. (ed.): 'Mechanical prime moves' (Macmillan, Basingstoke, 1971)
This book is made up of factual and authoritative papers and covers spark ignition and compression ignition engines, rotary piston engines and gas turbines.
CUMMINS, C. Lyle: 'Internal fire' (Carnot Press, Oregan, 1976)
A very readable account of the historical development of the internal combustion engine.
SHEPHERD, D. G.: 'Introduction to the gas turbine' (Constable, London, 1962)
COHEN, H., ROGERS, G. F. C., Saravanamutto: 'Gas Turbine Theory' (Longman, London, 1972)

Chapter 6

KALHAMMER, Fritz, H.: 'Energy storage systems', *Scientific American*, 1979, **241**, 6
JENSON, J.: 'Energy storage' (Newnes-Butterworth, London, 1980)
A good general introduction to the subject.
BECKMANN, G., and GILLI, P. V.: 'Thermal energy storage' (Springer-Verlag, N.Y., 1984)
Particularly useful on storage of thermal energy for industrial purposes.
WYMAN, C., CASTLE, J. and KREITH, F.: 'A review of collector and energy storage technology for Intermediate temperature applications', *Solar Energy*, 1980, **24**, 6, pp. 517-540
SWET, C. J.: 'Energy storage for solar applications' *in* KREITH, Jan, E., and KREITH, F. (eds.) 'Solar energy handbook' (McGraw Hill, N.Y., 1981) Chapter 6
BARAK, M.: 'Electrochemical power sources' (Peter Peregrinus, London, 1980)
A comprehensive coverage of all battery types.
SMITH, G.: 'Storage batteries' 2nd edn (Pitman, London 1971)

Chapter 7

BUTTI, K., and PERLIN, J.: 'A golden thread. 2500 years of solar architecture and technology' (Cheshire Books, Palo Alto USA, 1980)
An interesting and well written account of the history of solar conversion.
BRINKWORTH, B. J.: 'Solar energy for man' (Compton Press, 1972)
A good introductory text.
McVEIGH, J. C.: 'Sun power. An introduction to the applications of solar energy' (Pergamon Press, Oxford, 1977)
Also a good introductory text.
DAVIELS, F.: 'Direct use of the sun's energy' (Yale University Press, New Haven, USA, 1964)
PALZ, W.: 'Solar electricity' (UNESCO/Butterworth, London, 1978)

KARLEKAR, B. V., and DESMOND, R. M.: 'Engineering heat transfer' (West Pub. Co., St. Paul, 1977)
An excellent introductory texctbook.
MCADAMS, W. H.: 'Heat transmission' 2nd edn (McGraw-Hill, New York, 1954)
The standard work on heat transfer.
DUFFIE, J. A. and BECKMAN, W. A.: 'Solar thermal processes' (John Wiley, New York, 1974)
KREIDER, J. F., and KREITH, F.: 'Solar heating and cooling' (Scripta Book, New York, 1975)
BAKER, E., FLORO, C. T., GASTELOW, J. P. and McCAFFREY, J.: 'Solar heating and cooling systems design for Australian conditions' *Pergamon Press*, Sydney, 1984)
The book explains many practical problems arising in the choice and design of solar heaters.
MERRIGAN, J. A.: 'Prospects for solar energy conversion by photovoltaics' (MIT Press, Cambridge, Mass., 1976)
An excellent introduction to the topic.

Chapter 8

THRING, M. W.: 'Combustion' 2nd edn (Chapman and Hall, London, 1962)
Although written for furnace designers, this book gives an excellent introduction to combustion chemistry and also to the physical principles underlying the heat transfer and fluid flow processes which are important in combustor design.
SPIERS, H. M. (ed): 'Technical data on fuel' 5th edn (British National Committee, World Power Conference, London, 1955)
An excellent, comprehensive and authoritative collection of data.
MICUTA, W.: 'Modern stoves for all' (Bellerive Foundation, Geneva, 1981)
A clearly written, non-mathematical description of the principles of stove design. This little book deserves to be much better known.
SLESSOR, M., and LEWIS, C.: 'Biological energy resources' (E. F. N. Spon, London, 1979)
A good account of biomass sources and conversion.
HALL, D. O., BARNARD, G. W., and MOSS, P. A.: 'Biomass for energy in the developing countries' (Pergammon Press, Oxford UK, 1982)
Proceedings of the International Conference on plant and vegetable oils and fuels' (ASAE St. Joseph, Michegan, USA, 1982)
Provides a good coverage of current work in the field.
BRAKEL, J. Van: 'The Ignis fatuus of biogas. A critical review of pre-1970 literature' (Delft University Press, 1980)
MEYNELL, P. J.: 'Methane, planning a digester' (Prism Press and Conservation Tools & Technology, London, 1976). A useful, non-technical, introduction to biogas producers.
FRY, J. L.: 'Practical building of methane power plants for rural energy independance' (D. A. Knox, Andover, UK, 1974)
An excellent description of small scale digester design, construction and operation.
MARAMBA, F. D.: 'Biogas and waste recycling. The Philippine experience' (Maya Farms, Metro Manilla, 1978)
A clear, comprehensive account of the principles, design and operation of digesters.
'Alcohol production from biomass in the developing countries' (World Bank, New York, 1980)
A good account of alcohol production using various feedstocks and giving energy ratio data.
'Alcohol fuels. Options for developing countries' (National Academy Press, Washington, 1983)

Chapter 9

British Wind Energy Association. Wind Energy for the eighties. Pub. Peter Peregrinus. 1982
An excellent and comprehensive survey of wind power principles and practice.

GOLDING, E. W.: The Generation of Electricity by Wind Power. Pub. ELF Spon London 1976
Written by a pioneer in wind power. Worth reading, particularly in wind characteristics and site selection.

PUTMAN, P. C.: Power from the Wind. Pub. Van Nostrand. N.Y. 1948. Written by the designer of the wind turbine erected at Grand Pa's Knob.

WARNE, D. R.: Wind Power Equipment. Pub. Spon. 1983

HUNT, V. D.: Wind Power. Pub. Van Nostrand 1981

TAYLOR, R. H.: Alternative Energy Sources For the Centralised Generation of Electricity. Pub. Adam Hilger. 1983
A good description of large scale electricity generation by wind power.

PARK, J.: The Wind Power Book. Pub. Cheshire Books. 1981
Written for the designer of small scale windmills.
Contains much information on design and construction.

MARIER, D.: Wind Power for the Home Owner. Pub. Rodale Press 1981. Describes the siting and installation of small scale electricity generating windmills in the U.S.A. Much of the information is of more general application.

MUSGROVE, P. J.: Ed. Wind Energy Conversion. Proc. 6th BWEA Wind Energy Conf. Pub. C.U.P. Cambridge 1984

Chapter 10

WILSON, P. N.: 'Water turbines' (HMSO, London, 1974)
An excellent, well illustrated, booklet providing a clear introduction to water turbines.

McGUIGAN, D.: 'Small scale water power' (Prisma Press, Dorchester, 1978)

HAMMOND, Rolt: 'Water power engineering' (Heywood & Co., London, 1958)
A good introduction to turbine design principles and hydroelectric power.

MEIER, U.: 'Local experience with microhydro technology' (SKAT, 1981)
Discusses design of small systems and describes several field projects.

ALWARD, R., EISENBART, S., and VALKMAN, J.: 'Microhydro power: reviewing an old concept' (NCAT, Butte. Montana, 1979)
A good practical book describing site assessment together with a description of turbines, generators and ancilliary equipment.

OWENS, W. G.: 'A design manual for water' (VITA, Maryland, 1975)

Index

D

E

F

G

H

I

Q

R

S